Molecular Biology of Plants

edited by

Irwin Rubenstein
Department of Genetics and Cell Biology

Ronald L. Phillips   Charles E. Green

Burle G. Gengenbach
Department of Agronomy and Plant Genetics
University of Minnesota
St. Paul, Minnesota

ACADEMIC PRESS

A Subsidiary of Harcourt Brace Jovanovich, Publishers
New York London Toronto Sydney San Francisco
1979

ACADEMIC PRESS, INC.
111 Fifth Avenue, New York, New York 10003

United Kingdom Edition published by
ACADEMIC PRESS, INC. (LONDON) LTD.
24/28 Oval Road, London NW1 7DX

Library of Congress Cataloging in Publication Data
Main entry under title:

Molecular biology of plants.

Edited proceedings of a symposium held at the University of Minnesota, Mar. 25-27, 1976.
1. Plant genetics—Congresses. 2. Plant physiology—Congresses. 3. Plant viruses—Congresses. 4. Bacteria, Phytopathogenic—Congresses. I. Rubenstein, Irwin.
QH433.M65 581.1'5 79-18510
ISBN 0-12-601950-9

PRINTED IN THE UNITED STATES OF AMERICA

79 80 81 82 9 8 7 6 5 4 3 2 1

Contents

Section III: Plant Viruses and Bacterial Agents

Symposium Speakers and Contributors

Numbers in parentheses indicate the pages on which authors' contributions begin.

Roger Beachy (241), Department of Biochemistry and Biophysics, University of California, Davis, California, 95616

Arnold J. Bendich (1), Departments of Botany and Genetics, University of Washington, Seattle, Washington, 98195

C. A. Bjerknes (73), Biology Department, Brookhaven National Laboratory, Upton, New York, 11973

George E. Bruening (241), Department of Biochemistry and Biophysics, University of California, Davis, California, 95616

Mary-Dell Chilton (291), Department of Microbiology, University of Washington, Seattle, Washington, 98195

J. H. Clinton (73), Biology Department, Brookhaven National Laboratory, Upton, New York, 11973

A. De Picker (315), Laboratorium voor Genetica, Rijks Universiteit Gent, Gent, Belgium

D. De Waele (315), Laboratorium voor Genetica, Rijks Universiteit Gent, Gent, Belgium

Theodore O. Diener (273), Plant Virology Laboratory, Plant Protection Institute, ARS, U.S. Department of Agriculture, Beltsville, Maryland 20705

G. Engler (315), Laboratorium voor Genetica, Rijks Universiteit Gent, Gent, Belgium

Stephen K. Farrand (291), Department of Microbiology, University of Washington, Seattle, Washington 98195

Charles O. Gardner, Jr. (165), Department of Biochemistry and Molecular Biology, University of Florida, Gainesville, Florida 32610

C. Genetello (315), Laboratorium voor Genetica, Rijks Universiteit Gent, Gent, Belgium

Marcella Giesen (197), The Institute for Cancer Research, The Fox Chase Cancer Center, Philadelphia, Pennsylvania 19111

Milton P. Gordon (291), Department of Biochemistry, University of Washington, Seattle, Washington 98195

J. P. Hernalsteens (315), Laboratorium voor Genetica, Rijks Universiteit Gent, Gent, Belgium

Tuan-Hua David Ho[1] (217), Department of Biology, Washington University, St. Louis, Missouri 63130

M. Holsters (315), Laboratorium voor Genetica, Rijks Universiteit Gent, Gent, Belgium

John Ingle (139), Department of Botany, University of Edinburgh, Edinburgh, United Kingdom

Rusty J. Mans (165), Department of Biochemistry and Molecular Biology, University of Florida, Gainesville, Florida 32601

Abraham Marcus (197), Institute for Cancer Research, The Fox Chase Cancer Center, Philadelphia, Pennsylvania 19111

Robert Meeker (93), Department of Molecular Biology and Biochemistry, University of California, Irvine, Irvine, California 92664

Don Merlo (291), Department of Microbiology, University of Washington, Seattle, Washington 98195

E. Messens (315), Laboratorium voor Genetica, Rijks Universiteit Gent, Gent, Belgium

Alice Montoya (291), Department of Microbiology, University of Washington, Seattle, Washington 98195

Eugene W. Nester (291), Department of Microbiology, University of Washington, Seattle, Washington 98195

Ruth Roman (197), The Institute for Cancer Research, The Fox Chase Cancer Center, Philadelphia, Pennsylvania 19111

Jeff Schell (315), Laboratory for Genetics, Ledeganckstraat 35, Gent, Belgium

Daniela Sciaky (291), Department of Microbiology, University of Washington, Seattle, Washington 98195

Samarendra N. Seal (197), The Institute for Cancer Research, The Fox Chase Cancer Center, Philadelphia, Pennsylvania 19111

B. Silva (315), Laboratorium voor Genetica, Rijks Universiteit Gent, Gent, Belgium

[1] Permanent Address: Department of Botany, University of Illinois, Urbana, Illinois 61801

Krishna K. Tewari (93), Department of Molecular Biology and Biochemistry, University of California, Irvine, Irvine, California 92664

S. Van den Elsacker (315), Laboratorium voor Genetica, Rijks Universiteit Gent, Gent, Belgium

N. Van Larebeke (315), Laboratorium voor Genetica, Rijks Universiteit Gent, Gent, Belgium

M. Van Montagu (315), Laboratorium voor Genetica, Rijks Universiteit Gent, Gent, Belgium

Jack Van't Hof (73), Biology Department, Brookhaven National Laboratories, Upton, New York 11973

Virginia Walbot (31), Department of Biology, Washington University, St. Louis, Missouri 63130

Trevor J. Walter (165), Department of Biochemistry and Molecular Biology, University of Florida, Gainesville, Florida 32601

I. Zaenen (315), Laboratorium voor Genetica, Rijks Universiteit, Gent, Gent, Belgium

Milton Zaitlin (241), Department of Biochemistry and Biophysics, University of California, Davis, California 95616

Preface

As the potential for the application of molecular biological techniques to the genetic modification of procaryotes becomes increasingly evident, the quest for the basic information needed to carry out these changes in higher organisms will accelerate. The form, function, and genetic capacity of higher eucaryotes are orders of magnitude more complex than procaryotes, yet many of their basic properties are similar, if not identical. A better understanding and description of eucaryotic molecular biology is obviously necessary before procaryotic techniques can be adapted successfully to the genetic modification of an eucaryotic organism, such as a crop plant. In some cases, such information can be applied directly to the improvement of conventional plant breeding techniques by identifying new selection criteria. Successful modification of plant growth by molecular techniques will be dependent on a better understanding of the endogenous molecular mechanisms for controlling gene expression. This line of reasoning led to the organization of a symposium to examine particular topics in plant molecular biology. The 1976 symposium, held at the University of Minnesota, continued the focus of two previous workshops on the molecular genetic modification of eucaryotes and was attended by approximately 250 persons interested in various aspects and uses of plant molecular biology information.

The enthusiasm evident in the audience's response to the presentations of the invited speakers as well as in the informal exchanges cannot be captured on a printed page. This book does, however, report the formal presentations on plant molecular biology that generated this enthusiasm. It is the editors' hope that the reading audience will find the publication of this information in one book useful.

The topics in this volume are organized around the central dogma of molecular biology. Section I describes the organization and replication of DNA in plant chromosomes, including chloroplast genomes; Section II discusses molecular aspects of transcription and translation, ribosomal RNA gene systems and hormonal control of protein synthesis. Section III examines plant viruses and bacterial agents, in particular the crown gall system, viroids, and the replication of plant RNA viruses. Each of these specific topics contributes to an integrated knowledge of plant molecular biology.

The editors gratefully acknowledge financial support from the National Science

Foundation; the National Institutes of Health; the University of Minnesota College of Biological Sciences; the Departments of Agronomy and Plant Genetics, Biochemistry, Botany, Genetics and Cell Biology, Horticultural Science and Landscape Architecture, Plant Pathology, Biochemistry–Medical School, and Microbiology–Medical School.

SECTION I
Organization and Replication of DNA in Plant Genomes

THE NATURE OF FAMILIES OF REPEATED DNA SEQUENCES IN PLANTS

Arnold J. Bendich

Departments of Botany and Genetics
University of Washington
Seattle, Washington 98195

All eucaryotes probably contain DNA base sequences in their chromosomes which are represented many times and this feature of the genome undoubtedly has biological significance. Two classes of reiterated sequences which are often considered separately are satellite, or simple-sequence DNA and "intermediately repetitive" DNA. Although the function of neither is as yet understood, satellite sequences are thought to be tandemly arranged, to be concentrated in centromeric heterochromatin and possibly to serve a noncoding function in chromosome folding, pairing and movement (Walker, 1971). On the other hand intermediately repetitive sequences are less highly reiterated, and are in part clustered and in part dispersively scattered throughout the genome (Davidson *et al.*, 1975; Efstratiadis *et al.*, 1976). Those repeated sequences which are interspersed with non-repetitious DNA are thought to serve in a gene regulatory capacity (Britten and Davidson, 1969). The terms "repeated" and "reiterated" are somewhat misleading since only rarely do the sequences so designated approach complete base sequence repetition. This is an important point which has not been sufficiently appreciated,

ISBN 012-601950-9

and I will return to this issue later.

From theoretical considerations, Britten and Kohne (1968) proposed that families of repeated DNA sequences were created in rare "saltatory events" (bursts of DNA synthesis) leading to clustered repeats often detectable as satellites in CsCl density gradients. Over a period of time, the repeats accumulate base changes and are translocated throughout the genome. They are observed today as dispersed and partially mismatched intermediately repetitive sequences. This proposal of saltation followed by translocation is an attractive one but the only observation which can be taken as some support for it derives from a low level of homology of mouse satellite DNA with mouse main-band DNA (Flamm *et al.*, 1969). This result was not interpreted at the time as support for the above translocation proposal and later efforts failed to confirm the observation (Cech *et al.*, 1973). The work to be described below concerns both satellite and dispersed repeated DNA sequences in plants and bears directly on the saltation-translocation hypothesis. I have studied the major frequency component of DNA from four vascular plants (barley, daffodil, deer fern and parsley fern) in some detail in order to describe sequence repetition more quantitatively. Sequences complementary to two satellite DNAs from the muskmelon have been detected in its main-band DNA and these observations support the idea that dispersed repeats originated as satellites. There is apparently no other work which deals with the possible satellite origin for dispersed repeats. One reason might be that most of the characterized animal satellite DNAs have simple sequences with a basic repeat of less than 20 nucleotide pairs (NTP) (Bendich and Anderson, 1974), whereas the repeat length of intermediately repeated sequences in many animals is thought to be at least 300 NTP (Davidson *et al.*, 1975).

In the concluding section of this paper, I shall offer an appraisal of the current direction of repeated sequence research and an assessment of whether this direction may lead us to an understanding of functionally repetitive DNA.

COPIES, COMPLEXITY AND FAMILIES

Let us briefly review some of the concepts and terms used in repeated sequence work. For a mathematical description and a glossary of terms see Britten *et al.* (1974). Repeated sequences are recognized by a rate of reassociation that is greater than would be anticipated were each sequence present only once per haploid genome. Figure 1 shows a curve describing the reassociation of DNA from a hypothetical genome containing two components, each reassociating with ideal second order kinetics. This is the

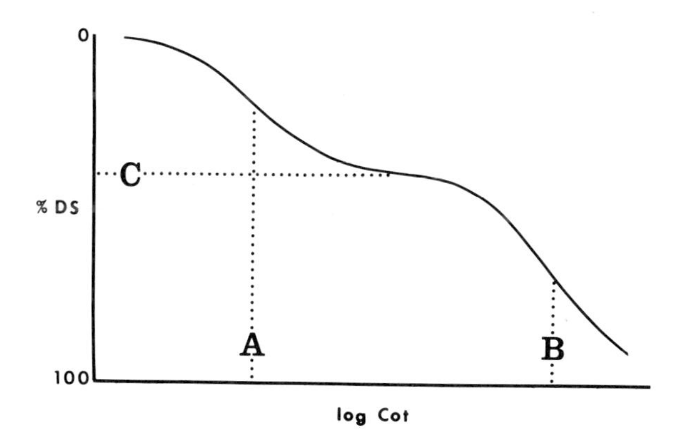

Fig. 1. Reassociation kinetics of a hypothetical DNA containing one repeated and one non-repeated sequence component.

familiar "C_0t curve" of Britten and Kohne (1968). The $C_0t_{\frac{1}{2}}$, (the point at which half the DNA in a component has reassociated) of the slowly reassociating component is equal to B and is that value expected for non-repeated or "single-copy" sequences in this genome. The $C_0t_{\frac{1}{2}}$ for the sequences present many times per genome is equal to A. The number of copies (the repetition frequency) of each sequence in the fast component is then equal to B/A. The fraction of the genome represented by the fast component is equal to C. We now consider the (sequence) complexity which is the total length of different sequences in a genome or component thereof; it is the information or coding potential. The kinetic complexity is the complexity calculated from reassociation rate measurements and is equal to A × C for the fast component. This value is then compared with that determined under the same experimental conditions for a bacterial or phage DNA whose complexity (in NTP) is known from independent methods.

The concept of complexity is often misunderstood. A given complexity does not necessarily mean that the number of nucleotides specified exist in a contiguous array. In contrast to the simple animal satellites, fast components of the intermediately repetitive type exist in blocks of sequences much shorter than the length indicated by the complexity of these components. In a recent example (Taylor, 1976; Taylor and Bendich, manuscript in preparation) component I of DNA from the broad bean *Vicia faba* has a kinetic complexity of 2×10^5 NTP but exists in contiguous lengths averaging 400 NTP. These 400 NTP lengths are interspersed with sequences from a second repetitive sequence component (one with fewer "copies" per genome) and with "single-copy" sequences.

We now come to the concept of families of repeated sequences, also a frequent source of confusion. A family is defined as a group of sequences of sufficient similarity that they may reassociate with one another at a given criterion of stringency set by the temperature and salt conditions under which reassociation took place. However, the assignment of the number of families present in a DNA component is currently a matter of subjective judgement since we do not know for any intermediately repetitive component what length of sequence constitutes a functional unit. For example, the number of families in component I of *Vicia faba* DNA could be taken as 500, $[(2 \times 10^5) \div 400]$, if we consider the length which is not interspersed with sequences from other components as likely to be important. Alternatively the number of families could be considered as one, if the complexity is deemed the basic unit or as a number greater than 500, if subdivisions of the 400 NTP length are chosen as units. This length, as determined by agarose gel electrophoresis, represents the average of a distribution of lengths. The corresponding length for all other organisms thus far examined (Efstratiadis *et al.*, 1976) is similarly a modal value of a distribution which is often quite broad. This fact becomes important when postulating coordinately functional properties for families of related sequences since the number of families assigned depends on the length of the member sequences. A 300 NTP sequence might have membership in a family which also includes related sequences 400 NTP in length, or these two lengths may be functionally segregated into different families. Yet another possibility is that one sequence may have no functional relationship to a second despite their obvious chemical similarity.

On the other hand, whereas the number of families and length of sequence per family member are not uniquely determined by reassociation rate measurements, the number of members per family (the copyness) is fixed by the ratio of B/A in Fig. 1. It should be emphasized that the sequences in a fast component are each present the same number of times per genome only when the DNA of that component reassociates as a single second order component.

HOW DO WE MEASURE DNA BASE SEQUENCE REPETITION?

Surprisingly little is known about the relationship among "repeated" sequences, despite a decade of research since the discovery of the phenomenon by Britten and his coworkers (Bolton *et al.*, 1965). Let us examine the commonly used operational definition of repeated sequences and the difficulties associated with this definition.

The procedure normally used to assess base sequence similarity in almost all work is to allow single DNA strands to form a duplex structure (reassociate) at 25°C below the T_m of the native DNA. According to Ullman and McCarthy (1973), this permits sequences to reassociate if their average complementarity is at least 85%. They estimate that 0.6% mispaired bases lowers the T_m of reassociated DNA by about 1°C. An estimate of 1% mispairing lowering the T_m by 1°C has also been made (Britten *et al.*, 1974). Consequently, those sequences less than 85% (or 75%) related are classified as "non-repeated." Furthermore, most studies utilize hydroxyapatite (HA) chromatography to separate single from double stranded structures. The difficulties of interpretation which have been associated with HA-based measurements include the following: (1) Short repeated sequence elements may go undetected. In the usual assay conditions (0.12 M sodium phosphate buffer at 60°C) the minimal stable length of duplex can be calculated to be about 20 NTP, but a run of at least 40 NTP is required for quantitative retention by HA (Wilson and Thomas, 1973). In bacteria and viruses, operator sites and sequences in promoters responsible for the formation of a tight binary complex with RNA polymerase are about 20 NTP or less (Gilbert *et al.*, 1974; Maniatis *et al.*, 1975; Pribnow, 1975). Important sequences such as these might not be detected using HA. (2) Unreassociated sequences contiguous to reassociated ones will be retained by HA, altering the shape of the kinetic curve and increasing the apparent extent of reassociation. This effect is striking and is a function of strand length (Grouse *et al.*, 1972; Pays and Ronsse, 1975; Goldberg *et al.*, 1975). Since we may expect the interspersion pattern of repeated and non-repeated sequences to vary with the organism in question, HA-based species comparisons of percent repetitious DNA are difficult to interpret. (3) At any convenient strand length, unpaired loops and ends of strands may be present in duplex structures which are retained by HA. Thus we do not know--for most of the organisms reported--what fraction of a duplex structure participates in base-pairing. (4) Low-melting regions in a duplex-containing structure would not elute from HA until the most stable region melts. (5) Molecules eluted from HA at high temperature under certain conditions may not be melted (Martinson and Wagenaar, 1974)--a possibility not discussed in most reports which do include melting profiles.

Thus, for the eucaryotes investigated, we know that there exist families of sequences related by at least 85% (or 75%) complementarity. We know also that in some organisms a larger fraction of the DNA meets this criterion than in others. But we know almost nothing of less closely related families of sequences. Perhaps most important of all, we do not know for any eucaryote what degree of complementarity constitutes a functional repeat (the "biological criterion" for repeated sequences). Consequently the fraction of any eucaryotic genome which is functionally repeated is an unknown quantity.

HOMOGENEOUS FAMILIES OF SEQUENCES

Two parameters used in assessing the reassociation of DNA strands are the fraction of bases involved in the duplex and the degree of mispairing within the duplex. The first may be determined by measuring the hypochromic effect as strands come together or the hyperchromic effect as they dissociate. The second is determined by a thermal stability measurement. To avoid some of the difficulties mentioned above, reassociation kinetics were measured at several temperatures in the spectrophotometer.

In our approach to the characterization of repeated sequences, a survey was made of 20 vascular plants. All of these exhibited DNA which reassociated with kinetics characteristic of intermediately repetitive sequences when assayed at T_m - 25°C (25°C below the T_m of native DNA). Four species were chosen for further analysis because each lacked sequences that reassociated extremely fast (less than a few percent) but did contain a DNA component whose reassociation followed a single ideal second order curve. I will designate such a component as "early." These early components comprised a sufficient fraction of the total DNA to permit their analysis on unfractionated DNA. The presence of a single early component is usually interpreted to indicate a single family of related sequences, or possibly several different families with the same number of members. Two types of families which may exist are heterogeneous families and homogeneous families. In heterogeneous families (Fig. 2a) the many family members would be of varying similarity since the size of the family decreases with increasing stringency used for reassociation. However, a single early component may also indicate the presence of several or many families of the homogeneous type (Fig. 2c). These families would differ from one another in degree of similarity within their membership, but each family would contain members of the same similarity. In one family the members are all related by, for example, 90% sequence homology. Thus, for homogeneous families, as the stringency for reassociation is raised (*e.g.*, by increasing the reassociation temperature), entire families are operationally moved to the non-repeated sequence class, since their members are no longer similar enough to reassociate at the higher temperature. The size of each family would have to be the same to show a single early

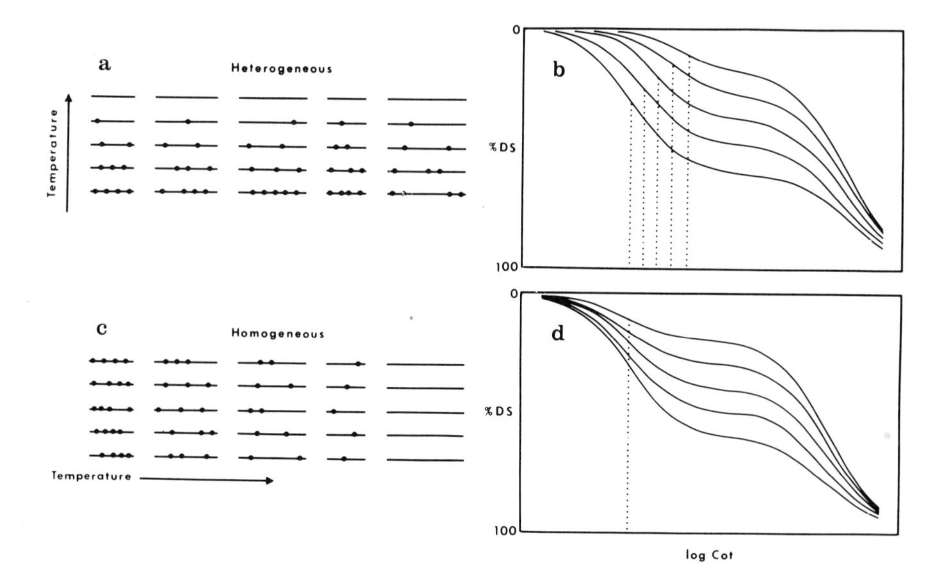

Fig. 2. The difference between heterogeneous and homogeneous families of repeated DNA sequences. Each sequence represented by a line is a member of a vertically arrayed family. Members of one family are sufficiently similar to reassociate with each other but not with members of other families. Dots on the lines represent base changes. As the temperature of reassociation is increased, the more divergent sequences can no longer reassociate and the fraction of rapidly reassociating DNA decreases. For heterogeneous families (a) increasing temperature decreases the size of each family which decreases the rate of reassociation of the early (repeated sequence) component. Thus, as the size of the early component decreases with increasing temperature, its $C_0t_{\frac{1}{2}}$ increases (b). For homogeneous families (c) increasing temperature eliminates entire families but the size and rate of reassociation of the remaining families is unaltered. Thus as the size of the early component decreases with increasing temperature, its $C_0t_{\frac{1}{2}}$ remains constant (d). Rate effects due to mispairing and suboptimal temperature (Fig. 3, legend) have been disregarded for the sake of simplicity.

component. Were the size to differ among the several families, no single early component would be observable. The two alternatives, heterogeneous *versus* homogeneous families, may be distinguished experimentally as shown below.

For heterogeneous families, the observed rate of reassociation should decrease as the family size decreases (Fig. 2b). Thus if the $C_ot_{\frac{1}{2}}$ were, say, equal to 1 for a 60% component at T_m - 25°C, it would increase to 3 at that higher temperature at which the component size becomes 20% of the total DNA. This follows since there are now only 1/3 the number of family members which can interact. For homogeneous families, the observed rate of reassociation should remain constant as the families comprised of less well-matched members are eliminated (Fig. 2d). The $C_ot_{\frac{1}{2}}$ of the 60% component would still be equal to 1 at the higher temperature at which the component size becomes 20% of the total DNA. This follows because the size, and therefore the reassociation rate, of each family is independent of the presence of other families with dissimilar sequences.

To determine whether families are homogeneous or heterogeneous, reassociation kinetics were measured as a function of increasingly stringent conditions from T_m - 35°C to about T_m - 5°C (Figs. 3, 4). For all 4 species of vascular plants the size of the repeated component of their DNA is unchanged between

Fig. 3. Reassociation kinetics of plant DNAs. Hypochromicity was measured at from 35.1 to 5.1°C (parsley fern and barley) or 34.6 to 7.6°C (daffodil and deer fern) below the T_m in 1 M $NaClO_4$, 0.03 M Tris buffer pH 8.0 with DNA at 77-100 μg/ml using fragments of about 1100 nucleotide pairs. Data have been standardized against B. subtilis *DNA analyzed simultaneously to account for temperature effects on rate and for thermal damage, and have been corrected for the effect of base mispairing (ΔT_m)*

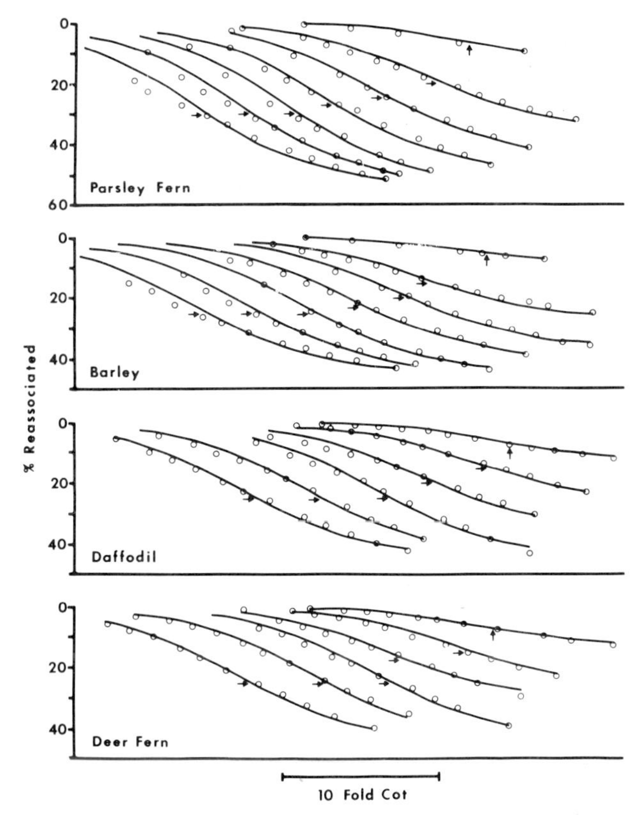

on rate (Marsh and McCarthy, 1974). Lines are ideal second order curves for components (to the nearest 5%) best fitting the data points which are arbitrarily displaced laterally. Arrows indicate the $C_o t_{1/2}$ *for each curve. The* $C_o t_{1/2}$ *for* B. subtilis *DNA was about 0.5 at* T_m *- 25°C. With DNA from* Bacillus megaterium, B. subtilis *and bacteriophages T4 and PS8, the decrease in absorbance at the completion of reassociation approaches only 90-91% of the hyperchromic effect of melted DNA. Therefore 90% hypochromicity was taken as the value for complete reassociation of DNA. The curves are "ideal" in the sense that they approximate ideal second order kinetics as closely as do reassociation data for the bacterial and phage DNAs. All data for each plant were obtained by serially melting and reassociating a single sample in a Gilford 2400 recording spectrophotometer equipped with the thermal programmer and reference compensator accessories.*

T_m - 35°C and T_m - 25°C and then decreases with further increasing stringency of reassociation. The $C_0t_{\frac{1}{2}}$ does not increase as the relative amount of the repeated component decreases. With DNA from parsley fern, for example, the heterogeneous family alternative predicts an increase in $C_0t_{\frac{1}{2}}$ of 4-fold (60/15) as the size of the early component is reduced from

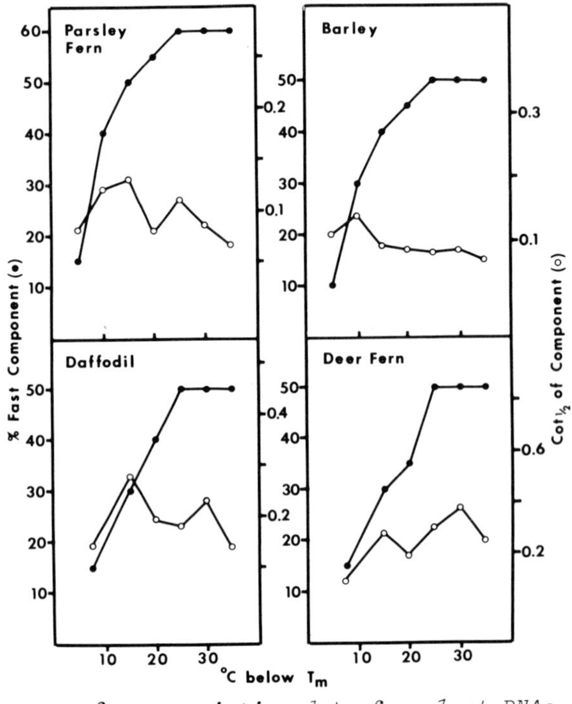

Fig. 4. Summary of reassociation data for plant DNAs. From the data in Fig. 3, the size of the second order component (to the nearest 5%) and the $C_0t_{\frac{1}{2}}$ of each component are plotted against degrees below the T_m for reassociation.

60% to 15% by increased stringency. The increases predicted for barley, daffodil and deer fern DNAs would be 5, 3.3 and 3.3-fold, respectively. The homogeneous families alternative predicts a constant $C_ot_{\frac{1}{2}}$ regardless of component size. Since the $C_ot_{\frac{1}{2}}$ does not increase appreciably for any of the four plant DNAs tested, these data favor the homogeneous family alternative.

A second type of experiment which supports the homogeneous family hypothesis assesses the sequence homology between closely similar and more distantly related classes of repetitive DNA. Tritium-labeled barley DNA was denatured and incubated at T_m - 35°C until all of its early component was reassociated. Duplex-containing structures were isolated on HA and then fractionated on the basis of thermal stability (Fig. 5). Five fractions "melting" at different temperatures were again reassociated at T_m - 35°C and remelted from HA. According to the heterogeneous family alternative the DNA which melted at a low (or high) temperature on the first fractionation should yield both low and high melting duplexes on the second cycle of reassociation since sequences in these melting classes are members of the same families. Conversely, the homogeneous family alternative predicts that the DNA that melted at the low and the high temperatures should reform DNA duplexes which are low and high melting, respectively, upon a second cycle of reassociation since sequences in these melting classes should not cross-react. The data in Fig. 5 conform to the latter expectation of homogeneous families making up the early component in barley DNA. Data consistent with the homogeneous family alternative may be found in the classic paper of Britten and Kohne (1968) in which salmon DNA was analyzed as in Fig. 5. These and other experiments with calf DNA indicated little sequence homology between the well-matched and less well-matched sets of repetitive DNA. However, such measurements could not themselves lead to the homogeneous family alternative because a single frequency class of sequences

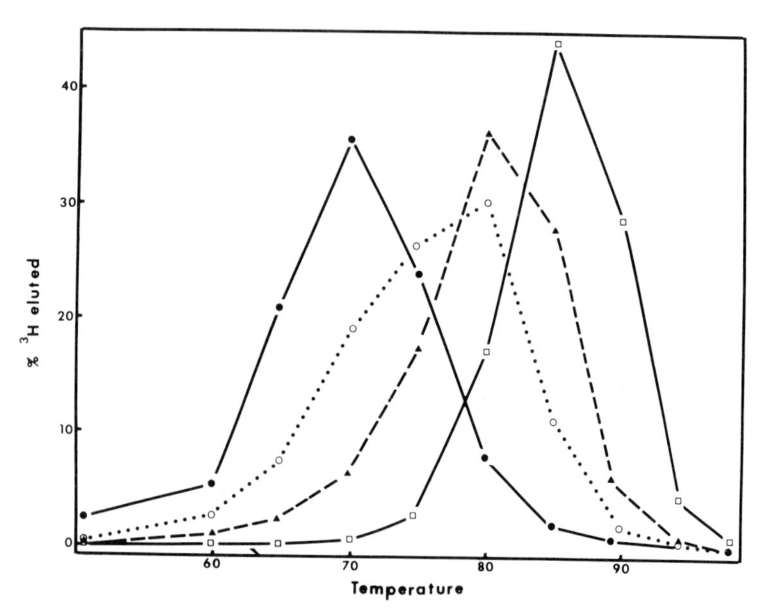

Fig. 5. Thermal fractionation on hydroxyapatite of reassociated barley DNA. Tritium-labeled barley DNA (about 350 nucleotide pairs) from which the foldback fraction (5% of the total) was removed at T_m - 35°C was denatured and incubated in 1 M $NaClO_4$, 0.03 M Tris buffer pH 8.0 at 53.5°C (T_m - 35°C) to a C_ot of 1.6 and loaded onto HA in 0.12 M potassium phosphate, pH 6.8 (KP) at 50.6°C (T_m - 35°C). The 63% of the ^{3}H which bound to the column was thermally eluted in 0.08 M KP (open circles). Fractions eluting at five temperatures were dialyzed, denatured and again reassociated in 1 M $NaClO_4$-Tris at 53.5°C to a C_ot of 0.5 [or a C_ot of 0.8 (0.5/0.63), if calculated on a total DNA basis]. A second melting profile was then constructed. For clarity only the remelts of the fractions eluting at the lowest (filled circles) and highest (filled triangles) of the five temperatures are shown, as is a profile for native DNA (open squares). The T_m values of the remelts were 67.8, 69.4, 72.5, 75.0 and 78.0°C for the fractions eluting at 64.9, 70.1, 74.8, 80.0 and 85.0°C, respectively. The T_m of the first melting profile was 73.7°C, while that for native DNA was 83.2°C.

was not used in that work.

The data in Figs. 3, 4 and 5 do not rule out the presence of a minor fraction of DNA in heterogeneous families which would cause a small increase in $C_ot_{\frac{1}{2}}$ for the lower temperature range. However, they do indicate that most, if not all, the sequences which reassociated at high temperature did not interact with more distantly related sequences at low temperature. Most or all of the families containing poorly matched sequences cease to exist at high temperature; their DNA is now recognized as "non-repeated."

As stated before, a second parameter for characterizing reassociated DNA is its thermal stability. When a DNA devoid of partially related base sequences (such as viral or bacterial DNA) is melted, reassociated and remelted, the difference (ΔT_m) between the first and second T_m is small. This, of course, is evidence for the absence of higher organism-type "repeated" sequences. Figure 6 shows the ΔT_m as a function of increasing stringency of the conditions used in reassociation; this represents a relatedness profile for the various homogenous families. The shape of the four curves is similar, indicating a similar, continuous gradation in the degree of relatedness among the various families of repeated DNA sequences in the four plants.

It should be clear that a more complete understanding of the relationships among related DNA sequences may be achieved when measurements of reassociation are made at several temperatures rather than at a single temperature.

SALTATION-TRANSLOCATION: MUSKMELON SATELLITE DNA SEQUENCES INTERSPERSED WITH MAIN BAND DNA

Muskmelon DNA contains two satellites (Bendich and Anderson, 1974; Bendich and Taylor, 1977). Satellites I and II represent about 10% and 20% of total DNA and have kinetic complexities of about 560 NTP and 1.7×10^6 NTP, respectively. Although both

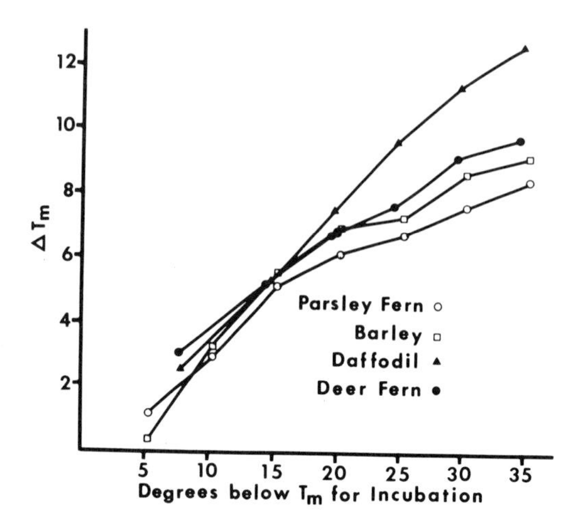

Fig. 6. Change in thermal stability of DNA reassociated at various temperatures below the T_m of native DNA. The difference between the T_m of native and reassociated DNA is plotted against the difference between the native T_m and the incubation temperature for reassociation.

satellites band at 1.706 g/cm^3 in neutral CsCl, satellite I melts about 8°C higher than satellite II and is not linked to it. Satellite I occurs in blocks of at least 67 tandemly repeating units of the 560 NTP sequence. There are 1.8×10^5 copies of satellite I and 72 copies of satellite II per haploid genome. The copies in each satellite are essentially identical replicas, for the ΔT_m of each is near 0°C.

Muskmelon DNA with main-band density contains short sequences that are homologous to satellite DNA (Bendich and

ey comprise a minor fraction of any given 4900-nucleotide long,
in-band fragment. Any main-band DNA fragment containing more
an a minor fraction of satellite DNA sequences would be found in
e dense side of the main-band peak which was discarded. The
st likely situation is that a short stretch of satellite I
quence is present on some 4900-nucleotide long, main-band frag-
nts which contain at least two repeated sequence elements unre-
ted to satellite I; the same can be inferred for satellite II
A. These conclusions were previously reached for satellite I
ing a mixture of I and II (not resolved in Ag^+ - Cs_2SO_4) with
in-band DNA that had been incubated to a C_ot of 58. In this
periment most of the ^3H in the aggregate would be from satellite
(Bendich and Taylor, 1977). We estimated that up to 1% of
gregated main-band DNA (half the DNA in that case) was homolo-
us to satellite I. This represented the equivalent of some
00 units of the 560 NTP sequence, or 3% of satellite I.
ughly the same degree of dispersal for satellite II can be
sumed since the amount of II in the aggregate approaches that
I (Table 1). Since the specific radioactivity of satellite
A in the present study is much greater than in the former
endich and Taylor, 1977), the fraction of main-band DNA homolo-
us to satellite DNA is even further below the saturation level
an in the former study, and thus we use the indirect estimate
r satellite II.

THE EVOLUTIONARY HISTORY OF HOMOGENEOUS FAMILIES

For the four plants surveyed here (Fig. 3) there is probably
re DNA which reassociates at T_m - 25°C with ideal second order
netics in the single early component than exists in the entire
mmalian genome (unpublished results). The single early compo-
nt implies that half or more of the genome is composed of
quences, each of which is represented the same number of times.

Taylor, 1977). This conclusion was reached from experiments in which denatured satellite DNA strands were incorporated into hyperpolymeric structures called networks (Bolton *et al.*, 1965; Waring and Britten, 1966) or aggregates (Bendich and Bolton, 1967; Thompson, 1976). Figure 7 shows an electron micrograph of a pea DNA aggregate. Both single and double stranded regions on inter-locking strands can be seen. This is a moderate sized aggregate; others contain more than 1000 strands. Britten first described the aggregation phenomenon with mouse DNA and suggested that to form such networks there probably is a requirement for an average of more than two regions per strand that are capable of pairing with complementary regions on other strands (Bolton *et al.*, 1965). When aggregates are formed with eucaryotic DNA incubated to low C_ot, it seems likely that the interacting strands contain an average of two or more repeated sequence elements.

Tritium-labeled satellites I and II were separated in a Ag^+ - Cs_2SO_4 gradient (Bendich and Anderson, 1974) and then sheared to about 1100 NTP. A trace of each satellite DNA was incubated with excess higher molecular weight main-band DNA and aggregation assays were performed (Table 1). The amount of satellite-density material in the main-band preparation used here was < 1%, a level of contamination which had no significant effect on the amount of ^3H in the aggregate pellet (Bendich and Taylor, 1977). After incubation to a C_ot of 32, 36% of satellite I but only 9% of satellite II is found in the main-band aggregate. Upon more extensive incubation, the amount of satellite II in the aggregate approaches that for satellite I. The more rapid aggregation for satellite I is consistent with its lower kinetic complexity. Neither satellite I nor satellite II DNA join the calf DNA aggre-gate and conversely, DNA of bacteriophage PS8 is not incorporated into the muskmelon DNA aggregate. Thus, short sequences comple-mentary to both satellites are interspersed in the main-band of muskmelon DNA. These sequences must be distributed such that

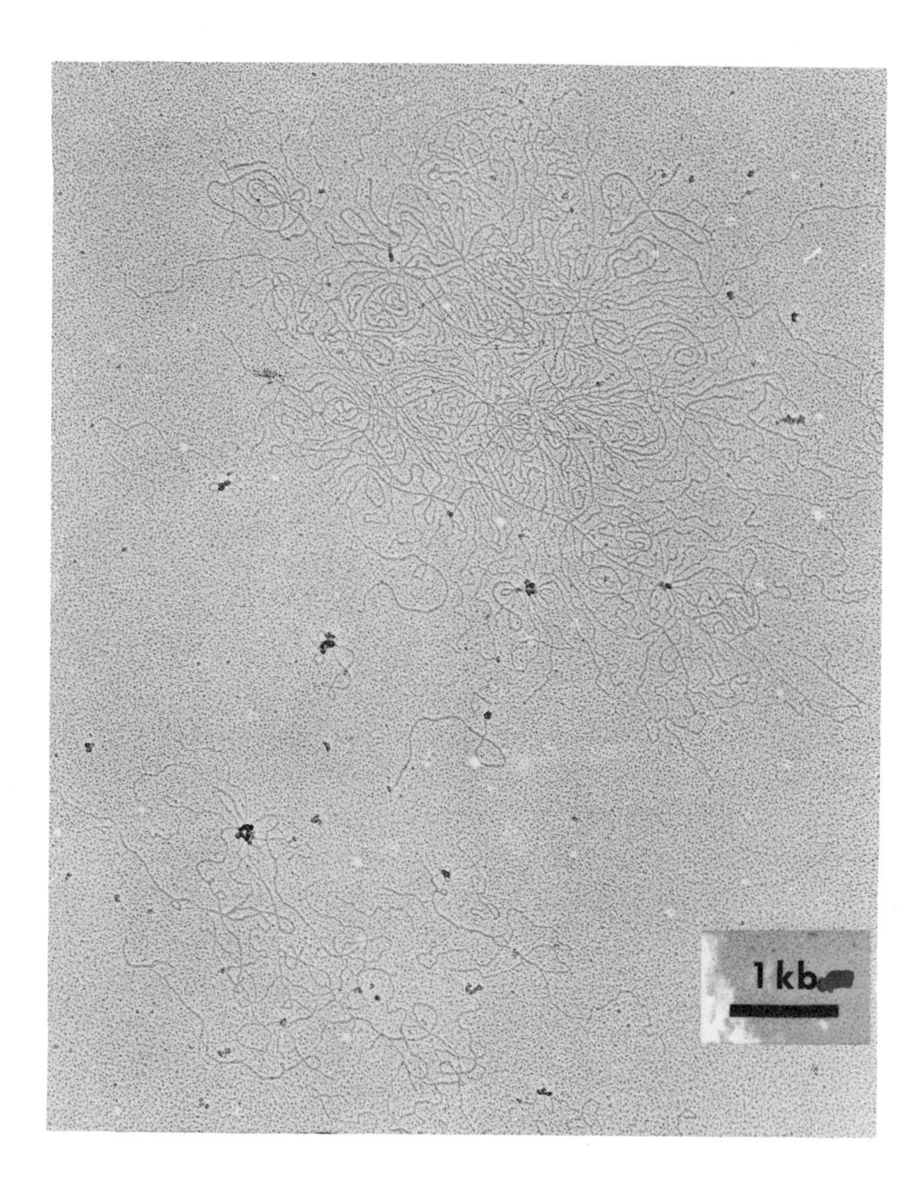

Fig. 7. Electron micrograph of a pea DNA aggregate. Fragments were sheared to an average length of 4900 nucleotide pairs, reassociated at 25°C below the T_m in 1 M $NaClO_4$, 0.03 M Tris buffer pH 8.0, to a C_ot of 1.8, and spread in 60% formamide by a modified Kleinschmidt method (Davis et al., *1971). The bar equals 1000 nucleotide pairs (1 kb).*

TABLE 1

Aggregation of sheared ^{3}H-DNA with excess long DNA

	Long DNA	^{3}H-DNA	$\%A_{260}$ in Agg.
Low C_ot	Main-band	I	41, 40
	Calf	I	55, 55
	Main-band	II	40, 40
	Calf	II	50, 53
	Main-band	PS8	46, 50
High C_ot	Main-band	I	48, 56
	Calf	I	70, 69
	Main-band	II	53, 54
	Calf	II	70, 71

Main-band DNA was incubated to C_ot of 32 or equivalen
Calf DNA was incubated to C_ot of 46 or equivalent C_ot

Tritium-labeled DNA (0.7-0.8 ng, 300-380 cpm for sate
II; 4.6 ng, 640 cpm for PS8) mixed with excess unlabe
μg main-band or 2.1 μg calf) was denatured at 103°C f
and incubated in 0.2 M NaCl, 5 mM Tris pH 8 at 56°C (
in 1 M $NaClO_4$, 0.03 M Tris pH 8 at 61°C (high, equiva
After incubation, the 10 μl samples were quickly cool
diluted to 0.2 ml with incubation buffer, reheated fo
at incubation temperature, again quickly cooled at 0°
centrifuged at 4°C for 30 minutes at 27,000 g. After
the supernatant, the aggregate pellet was resuspended
incubation buffer and both fractions were again heate
and quickly cooled to 0°C before absorbance and radio
were determined. The C_ot and equivalent C_ot values f
labeled satellites I and II were 0.017, 0.018 and 0.1
respectively; that for ^{3}H-PS8 was 0.10. Sheared and
were about 1100 and 4900 nucleotide pairs, respective
lite and main-band preparations were obtained by reba
dense and light portions from the satellite and main-
in CsCl, respectively. Satellites I and II were then
an Ag^+ - Cs_2SO_4 gradient (Bendich and Anderson, 1974)

If some sequences were represented more times than others, no single component would be observed. How did this come about? Let us consider an ancient event in which all DNA in the component arose from one sequence in a single saltatory replication or in saltations occurring over an evolutionarily short period of time. The length of the sequence taken for replication would be its kinetic complexity. It is likely, as Britten and Kohne (1968) have reasoned, that the nucleotide sequences of the members of the families are not conserved by severe selection. The members of all families may then change slowly and independently of each other leading, after a long period of time, to families with divergent members such as are observed. This process should result in heterogeneous families comprising the single early component. But the data indicate homogeneous families. To be consistent with the data and retain the idea of a single saltation, we must invoke an unusual selection. The thousands of members of each family must be restricted to a specific rate of change and members of different families must change at different and widely varying rates in order to generate homogeneous families. Furthermore, the changes must not alter the constant number of members in each family. Clearly this is an unlikely possibility and we should abandon the concept that a single event created the homogeneous families we observe.

Let us now assume a constant rate of accumulation of base changes in members of all families. The families composed of closely related members would then be identified as recently created families, while those with distantly related members would be older. Thus, the genesis of a set of homogeneous families is most likely an extended process involving many minor saltatory events over a long period of time. The length of sequence taken for replication could be variable and would be much shorter than the kinetic complexity measured for the single early component. This measurement represents the sum of the lengths of all sequences which have been copied the same number of times during the series of saltatory events. Therefore, we can only place an approximate upper limit on the total length of sequence taken for replication in the most recent saltations by using data in Fig. 4 at the most stringent criteria. The $C_ot_{\frac{1}{2}}$ for parsley fern is about 0.1 and the size of the component is 15% at T_m - 5°C. If not diluted in total DNA, its $C_ot_{\frac{1}{2}}$ would be 0.015, (0.1 × 0.15). Within a factor of 2, the corrected $C_ot_{\frac{1}{2}}$ values for the other species in Fig. 4 are the same. The DNA of *Bacillus subtilis* has a $C_ot_{\frac{1}{2}}$ of about 0.5 under the same conditions and has a kinetic complexity of 2 × 10^9 daltons or 3 × 10^6 NTP (Gillis and DeLey, 1975). A $C_ot_{\frac{1}{2}}$ of 0.015 would then represent a kinetic complexity of about 10^5 NTP. If the basic length of a family member in these plants is about 400 NTP, as in *Vicia faba*, the plant DNA components observed at high stringency would each contain about 250 families. However, the 400 NTP estimate was made from data obtained at T_m - 25°C and T_m - 10°C (the 300 NTP estimate for aminals was made from data obtained at only one condition of stringency: about T_m - 25°C). It is possible that a shorter repeat length may be observed at higher stringency. The fact that the length of the parental sequence of a family is not known precludes an accurate estimate of the number of families or saltations in the plants.

I suggest that the homogeneous families are the result of saltatory replications occurring at many different times with a constant rate of sequence change in all families. To account for the single early component we observe, one of two requirements must be met. Either the parental sequence was copied the same number of times in each saltation or family members were discarded subsequent to their synthesis until a constant family size remained. It is not obvious why such a sequence amplification to a constant level should exist. Nor is it easy to formulate a mechanism by which the constancy is maintained. Nevertheless, a phenomenon termed "gene compensation" has been described which appears relevant to the question of constancy of sequence amplification. In *Drosophila* the number of genes for ribosomal RNA (Tartof, 1971) and for 5S RNA (Procunier and Tartof, 1975) is sometimes returned to a wild-type level after individual flies with various deletions for some of these genes are crossed. Perhaps it is not unreasonable to accept a controlled amplification of sequences whose function we do not understand.

In my view the creation of each of the two muskmelon satellites represented the initial step in a process which leads to a family whose sequences are scattered throughout the genome. That the satellites were created recently can be inferred from the fact that the ΔT_m for each is near 0°C. The short sequences of main-band density which are homologous to the satellites represent the first translocation events which will ultimately convert each satellite into a single family of dispersed repetitive sequences. As the families grow older, their member sequences will gradually diverge from one another. It is the families composed of dispersed and divergent member sequences in animals that have been extensively studied as possible regulatory elements for the control of gene expression (Britten and Davidson, 1969; Davidson *et al.*, 1975). I shall consider the regulatory potential of such sequences in the following section.

If, as I suggest, each homogeneous family of dispersed sequences originated as a satellite (The term satellite is used here to include sequences present in multiple copies which may be detected as a discrete component in density gradients using DNA of moderate to high molecular weight.), why are the complexities of the animal satellites generally believed to be less than 20 NTP while the repeat length of the dispersed repeats is at least 300 NTP? Several points are relevant to this question. (1) There have been reports of complex animal satellites. The reassociation kinetics of a satellite from the wood mouse, *Apodemus agrarius*, indicate its complexity may approach that of a bacterial genome (Walker, 1971). The major component of human satellite A reassociates with a $C_ot_{\frac{1}{2}}$ approximately that of a bacterial genome (Chuang and Saunders, 1974). The kinetic complexity of satellite III of *Drosophila nasutoides* is also as great as that of a bacterium (Cordeiro-Stone and Lee, 1976). (2) Complex animal satellites may be common but present in such low proportions that they are ordinarily overlooked. Human satellite A represents only 0.5-1% of total DNA. (3) The greater the sequence complexity the less the chance that the sequence will differ in base composition from bulk DNA so as to be separable in neutral CsCl gradients. Human satellite A could be isolated in Ag^+ - Cs_2SO_4 but not CsCl and thus is a "cryptic" satellite. Calf DNA showed no obvious complex density pattern in CsCl but was fractionated into at least 7 discrete density components when Ag^+ - Cs_2SO_4 gradients were employed (Filipski *et al.*, 1973). Several of these components behaved as complex satellites, when density shifts in CsCl before and after reassociation of denatured DNA were used as assays. It is therefore possible that complex satellites, including cryptic ones, occur as frequently as do simple satellites, but the latter have been more easily identified by the commonly used methods.

ON THE VALUE OF SEQUENCE REPETITION FOR THE INVESTIGATOR AND FOR THE CELL

In this section I will consider the recent direction of inquiry concerning sequence repetition, and whether it may lead us to an understanding of the biological role of repetitive DNA.

Let us assume for the moment that there are DNA sequences--other than those which specify the stable RNA's--that can function in concert because of their similarity and that such sequences are needed for orderly functioning of the organism. What degree of sequence homology would the cell require for such functional equivalence? Would the cell consider two stretches of DNA to be unrelated if they differ in sequence by 10%? By 1%? I have seen no discussion of this issue in the literature which deals with sequence repetition in eucaryotes. Investigators in this area usually set an operational definition for repetition by choosing T_m - 25°C as the single criterion for their reassociation, although there have been some notable exceptions (Walker and McLaren, 1965; Bendich and McCarthy, 1970; Marsh, 1974). However, we do not know what level of discrimination is used by any eucaryote to distinguish repeat from non-repeat. It would therefore be fortuitous if the properties of sequences classified as repetitious at T_m - 25°C were to be representative of functionally repetitious DNA.

Our data indicate that several to many homogeneous families comprise the single rapidly reassociating components in the plants studied. Which of the families might be utilized for function? Since there is such a large proportion and amount of DNA in these families, it seems unlikely that all this DNA would function in a manner requiring sequence similarity. The families composed of more divergent members might be useful, but those with closely similar members should afford the advantage of more precise control over a postulated network of functions. I will assume for this argument that sequences must be nearly perfect replicas to be considered as reiterated by a cell. The short, dispersed sequences homologous to the muskmelon satellites might be examples of such sequences. Just how similar the sequences need be is conjectural, but there are examples in procaryotic systems in which a single base change can alter the recognition of a regulatory protein for its binding sequence (Gilbert *et al.*, 1974; Maniatis *et al.*, 1975). In accord with my assumption, nearly all the families would be unsuitable for purposes which require functional repetition. Perhaps only some of the sequences which reassociated at T_m - 5°C (0-1°ΔT_m) would be sufficiently similar. Data for many organisms have been gathered from reassociation experiments conducted at T_m - 25°C (the criterion of stringency adopted as standard by most investigators). These data have led to estimates of the length, spacing, number and complexity of DNA sequences which appear to be repetitious at the standard criterion, as well as to assignments of the fraction of total DNA in such sequences. However, these properties and the generalizations derived from them may not be characteristic of functionally repetitious sequences. If functional repeats exist as a small number of nearly perfect replicas, and if their properties are different from those for the much larger amount of DNA which reassociates rapidly at the standard criterion, then the properties of functional repeats have been obscured by those of sequences regarded as reiterated by the investigator but not by the cell. The current of research may, in fact, be carrying us away from our goal of understanding the biological utility of repeated DNA sequences.

At present the manner in which families of repeated sequences are important to the phenotype remains elusive. A solution to this central problem will require a more quantitative description of sequence repetition than the one usually provided. For example, properties such as length, spacing and distribution should be derived from measurements employing several criteria, rather than a single criterion. We should determine whether a single kinetic component is comprised of heterogeneous or homogeneous families. After we obtain such information for a variety of organisms, we will be better able to formulate generalizations concerning sequence organization. Without this basic analysis it seems unlikely that we will understand the putative role of repeated sequences in gene regulation, chromosome structure, evolution, or any other process.

SUMMARY

Reassociation kinetics were measured at several temperatures for DNA from barley, daffodil, deer fern, and parsley fern. The data indicate that several to many families of related DNA base sequences comprise half or more of the genome in these plants. The various families are not related to one another. They are termed homogeneous because each family contains member sequences related by the same degree of similarity. In one family the members would all be related by, for example, 95% sequence homology. Sequences homologous to each of two satellite DNAs from muskmelon have been detected in its main-band DNA. The observations provide the first experimental evidence to support the idea that families of repeated sequences are created as clustered repeats which are subsequently dispersed throughout the genome. Thus, sequences recognized as "intermediately repetitive" may exist in homogeneous families which originated as satellites or as cryptic satellites.

The question is considered: are sequences designated by investigators as "repeated," in fact, accepted by the cell as repetitious in some functional sense? I conclude that before we can understand the functional and phenotypic significance of repeated sequences, we may need to ascertain the physical parameters of sequence organization by methods different from those in common use.

ACKNOWLEDGEMENTS

I thank G.L.G. Miklos and W.C. Taylor for their helpful consultation, and W.C. Taylor for the electron microscopy. This work was supported by National Science Foundation Grant GB41179 and National Institutes of Health Grant 1 R01 GM22870-01.

REFERENCES

Bendich, A.J., and Anderson, R.S. (1974). Novel properties of satellite DNA from muskmelon. *Proc. Nat. Acad. Sci. U.S. 71*, 1511-1515.

Bendich, A.J., and Bolton, E.T. (1967). Relatedness among plants as measured by the DNA-agar technique. *Plant Physiol. 42*, 959-967.

Bendich, A.J., and McCarthy, B.J. (1970). DNA comparisons among barley, oats, rye, and wheat. *Genetics 65*, 545-565.

Bendich, A.J., and Taylor, W.C. (1977). Sequence arrangement in satellite DNA from the muskmelon. *Plant Physiol. 59*, 604-609.

Bolton, E.T., Britten, R.J., Cowie, D.B., Roberts, R.B., Szafranski, P., and Waring, M.J. (1965). "Renaturation" of the DNA of higher organisms. *Carnegie Inst. Wash. Yrbk. 64*, 316-333.

Britten, R.J., and Davidson, E.H. (1969). Gene regulation for higher cells: a theory. *Science 165*, 349-357.

Britten, R.J., Graham, D.E., and Neufeld, B.R. (1974). Analysis of repeating DNA sequences by reassociation. In *Methods In Enzymology, 29E*, (Grossman, L., and Moldave, K., eds.), pp. 363-418. Academic Press, New York.

Britten, R.J., and Kohne, D.E. (1968). Repeated sequences in DNA. *Science 161*, 529-540.

Cech, T.R., Rosenfeld, A., and Hearst, J.E. (1973). Characterization of the most rapidly renaturing sequences in mouse main-band DNA. *J. Mol. Biol. 81*, 299-325.

Chuang, C.R., and Saunders, G.F. (1974). Complexity of human satellite A DNA. *Biochem. Biophys. Res. Commun. 57*, 1221-1229.

Cordeiro-Stone, M., and Lee, C.S. (1976). Studies on the satellite DNAs of *Drosophila nasutoides*: their buoyant densities, melting temperatures, reassociation rates, and localizations in polytene chromosomes. *J. Mol. Biol. 104*, 1-24.

Davidson, E.H., Galau, G.A., Angerer, R.C., and Britten, R.J. (1975). Comparative aspects of DNA organization in Metazoa. *Chromosoma 51*, 253-255.

Davis, R.W., Simon, M., and Davidson, N. (1971). Electron microscopy heteroduplex methods for mapping regions of base sequence homology in nucleic acids. In *Methods In Enzymology, 21D*, (Grossman, L., and Moldave, K., eds.), pp. 413-428. Academic Press, New York.

Efstratiadis, A., Crain, W.R., Britten, R.J., Davidson, E.H., and Kafatos, F.C. (1976). DNA sequence organization in the lepidopteran *Antheraea pernyi*. *Proc. Nat. Acad. Sci. U.S.* *73*, 2289-2293.

Filipski, J., Thiery, J.-P., and Bernardi, G. (1973). An analysis of the bovine genome by Cs_2SO_4-Ag^+ density gradient centrifugation. *J. Mol. Biol.* *80*, 177-197.

Flamm, W.G., Walker, P.M.B., and McCallum, M. (1969). Some properties of the single strands isolated from the DNA of the nuclear satellite of the mouse (*Mus musculus*). *J. Mol. Biol.* *40*, 423-443.

Gilbert, W., Maizels, N., and Maxam, A. (1974). Sequences of controlling regions of the lactose operon. *Cold Spring Harbor Symp. Quant. Biol.* *38*, 845-855.

Gillis, M., and DeLey, J. (1975). Determination of the molecular complexity of double-stranded phage genome DNA from initial renaturation rates. The effect of DNA base composition. *J. Mol. Biol.* *98*, 447-464.

Goldberg, R.B., Crain, W.R., Ruderman, J.V., Moore, G.P., Barnett, T.R., Higgens, R.C., Gelfand, R.A., Galau, G.A., Britten, R.J., and Davidson, E.H. (1975). DNA sequence organization in the genomes of five marine invertebrates. *Chromosoma* *51*, 225-251.

Grouse, L., Chilton, M.-D., and McCarthy, B.J. (1972). Hybridization of ribonucleic acid with unique sequences of mouse deoxyribonucleic acid. *Biochem.* *11*, 798-805.

Jones, K.W. (1970). Chromosomal and nuclear location of mouse satellite DNA in individual cells. *Nature* *225*, 912-915.

Maniatis, T., Ptashne, M., Backman, K., Kleid, D., Flashman, S., Jeffrey, A., and Mauer, R. (1975). Recognition sequences of repressor and polymerase in the operators of bacteriophage lambda. *Cell* *5*, 109-113.

Marsh, J.L. (1974). Characterization of repeated sequences in eukaryotic genomes. Ph.D. thesis. Univ. of Washington, Seattle.

Marsh, J.L., and McCarthy, B.J. (1974). Effect of reaction conditions on the reassociation of divergent deoxyribonucleic acid sequences. *Biochem.* *13*, 3382-3388.

Martinson, H.G., and Wagenaar, E.B. (1974). Thermal elution chromatography and the resolution of nucleic acids on hydroxyapatite. *Anal. Biochem.* *61*, 144-154.

Pays, E., and Ronsse, A. (1975). Interspersion of repetitive sequences in rat liver DNA. *Biochem. Biophys. Res. Commun. 62*, 862-867.

Pribnow, D. (1975). Nucleotide sequence of an RNA polymerase binding site at an early T7 promoter. *Proc. Nat. Acad. Sci. U.S. 72*, 784-788.

Procunier, J.D., and Tartof, K.D. (1975). Genetic analysis of the 5S RNA genes in *Drosophila melanogaster. Genetics 81*, 515-523.

Tartof, K.D. (1971). Increasing the multiplicity of ribosomal RNA genes in *Drosophila melanogaster. Science 171*, 294-297.

Taylor, W.C. (1976). DNA sequence organization in *Vicia faba.* Ph.D. thesis. Univ. of Washington, Seattle.

Thompson, W.F. (1976). Aggregate formation from short fragments of plant DNA. *Plant Physiol. 57*, 617-622.

Ullman, J.S., and McCarthy, B.J. (1973). The relationship between mismatched base pairs and the thermal stability of DNA duplexes. II. Effects of deamination of cytosine. *Biochim. Biophys. Acta 294*, 416-424.

Walker, P.M.B. (1971). "Repetitive" DNA in higher organisms. In *Progress in Biophysics and Molecular Biology, 23*, (Butler, J.A.V., and Nobel, D., eds.), pp. 147-190. Academic Press, New York.

Walker, P.M.B., and McLaren, A. (1965). Specific duplex formation *in vitro* of mammalian DNA. *J. Mol. Biol. 12*, 394-409.

Waring, M.J., and Britten, R.J. (1966). Nucleotide sequence repetition: a rapidly reassociating fraction of mouse DNA. *Science 154*, 791-794.

Wilson, D.A., and Thomas, C.A., Jr. (1973). Hydroxyapatite chromatography of short double-helical DNA. *Biochim. Biophys. Acta 331*, 333-340.

GENOME ORGANIZATION IN PLANTS

Virginia Walbot

Department of Biology
Washington University
St. Louis, Mo., 63130

The divergence between the expressed genetic content of eukaryotic organisms and the large amount of DNA found in the eukaryotic cell has engendered a great deal of research to determine the function and organization of eukaryotic DNA. This "C" paradox between genetic content and physical content of DNA is not yet solved. In plants where some of the more extreme "C" paradox cases exist there are fern species with 630 chromosomes (Brown and Bertke, 1974), and DNA content does not correlate with complexity nor evolutionary age (Sparrow *et al.*, 1972). Certain closely related genera can exhibit differences of 75-fold in the amount of DNA per cell (Rothfels and Heimburger, 1968). Plant cytogeneticists also have recognized a number of chromosome elimination and translocation phenomena which must to some extent depend on the organization of the genetic material within the chromosomes. These phenomena have not yet been explained in molecular terms. An excellent summary of the existing literature on plant DNA content recently has been prepared by Price (1976).

During the 1960's several important discoveries were made about the organization and composition of DNA from eukaryotes. Britten and associates (Britten and Kohne, 1967, 1968, 1969; Waring and Britten, 1966) in examining the reassociation kinetics of DNA from eukaryotic and prokaryotic species demonstrated that although prokaryotic DNA reassociates as if each sequence were present once per genome (first shown by Doty *et al.*, 1960), the kinetics of reassociation of eukaryotic DNA were more complex. DNA from eukaryotes contains sequences repeated many thousands or millions of times per genome (highly repetitive sequences), sequences repeated a moderate number of times (re-

ISBN 012-601950-9

petitive sequences), and sequences present once per haploid genome (unique copy DNA).

The existence of these kinetic classes narrowed the possible number of genes since it is assumed that Mendelian traits are present only once per genome. In organisms with a low "C" value the complexity of unique copy DNA is sufficient to specify 1×10^4 to 5×10^4 genes of 1,000 nucleotides each.

However, even after correction for the existence of repetitive DNA and non-transcribed specialized sequence DNA most higher eukaryotes still contain sufficient unique copy DNA to code for 10^5 to 10^6 or more different polypeptides. This number is far in excess of the number of proteins identified in eukaryotes and is several orders of magnitude greater than the approximately 5,000 genes in *Escherichia coli*. If all of these potential unique copy genes existed, the genetic load, *i.e.* the accumulation of deleterious mutations due to mistakes in DNA replication and repair, would be exceedingly high in most eukaryotes (Ohno, 1971).

The number of messinger RNA (mRNA) sequences transcribed in sea urchin gastrulae has been estimated at 10,000-15,000 different sequences (Galau *et al.*, 1974). About 90% of the polysomal mRNA, the "prevalent message class," consists of several hundred distinct species. The remaining 10% of the total mRNA population consists of approximately 10^4 sequences, the "complex class." The complex class mRNAs are present on the average only 1-10 times per cell. Both classes are composed primarily of mRNA transcribed from non-repetitive DNA (Goldberg *et al.*, 1975; McColl and Aronson, 1974). The existence of prevalent and infrequent classes of mRNA have been reported in friend cells (Birnie *et al.*, 1974), mouse fibroblasts (Williams and Penman, 1975), HeLa cells (Bishop *et al.*, 1974), *Drosophila* larvae (Levy and McCarthy, 1975), and several mouse organs and cell types (Ryffel and McCarthy, 1975).

Initial investigation of the proportion of the genome transcribed in various cell types indicated that differences in the extent of the genome utilized did vary considerably (reviewed by Lewin, 1974 and 1975). The most elegant demonstration of the extent of genome usage is a recent study by Galau *et al.* (1976) in which the mRNA populations of sea urchin embryonic and adult tissue types are compared to sea urchin gastrula mRNA. Only a small set of mRNAs were common to most tissue types, these 1,000 to 1,500 genes may be the "housekeeping genes" required for all cells. Oocyte mRNA, the most complex mRNA, consisted of transcripts of 3×10^4 to 4×10^4 different sequences. Each embryonic stage and adult contains a different although partially overlapping set of structural gene transcripts. The total number of sequences represented at all developmental stages of the sea urchin has not been determined but it seems likely that at least 5×10^4 sequences will be found and perhaps 10^5. There is enough unique copy DNA in sea urchin to specify approximately 5×10^5 mRNAs. Thus, despite the arguments of genetic load it is likely that 10^5 genes is a reasonable number for a complex eukaryote; a total minimum complexity of some 10^8 nucleotide pairs would be required.

Although a large number of genes are active in eukaryotes, genes are not transcribed with the same frequency in a given cell type or in different cell types. Gene activity is precisely regulated. It is the goal of sequence organization studies to provide a structural framework from which models of gene regulation can be derived and tested. In the following section the basic features of eukaryotic genome organization will be described.

CHARACTERISTICS OF THE MAJOR KINETIC CLASSES OF EUKARYOTIC DNA

Highly Repetitive Sequences

The most highly reiterated sequences of the genome are the subject of the paper by Bendich in this volume; consequently, I will only briefly discuss a few aspects of this sequence component relevant to our understanding of genome organization. These sequences are unlikely to code for proteins or be transcribed into RNA, but they constitute a significant percentage of the genome in many eukaryotes. These highly reiterated sequences are purified either by virtue of their buoyant density or of their rapid reassociation.

Satellite DNAs have been separated from the bulk (main-band) of DNA by centrifugation to equilibrium in neutral CsCl gradients of DNA from a variety of plants; the proportion of satellite(s) in the genome may be as high as 44% (Ingle *et al.*, 1973). Satellite DNAs may be a heavier density (higher G+C content) or a lighter density than the main-band DNA. Some species have no detectable satellite DNA in CsCl gradients; however, when the main-band DNA is complexed with Ag^+, Hg^+ or both metals and centrifuged to equilibrium in Cs_2SO_4 the resolution of known satellites is improved and cryptic satellites are sometimes found in animals (Corneo *et al.*, 1970; Skinner and Beattie, 1973) and plants (Bendich and Anderson, 1974). Satellite DNA in mouse (Pardue and Gall, 1970; Jones, 1970) and insects (Eckhardt and Gall, 1971; Gall *et al.*, 1971; Botchan *et al.*, 1971) hybridizes *in situ* to the centromeric heterochromatin. Satellite DNA of guinea pig (Yunis and Yasmineh, 1970), crab (Duerksen and McCarthy, 1971), and calf

(Yasmineh and Yunis, 1971) is localized *in situ* primarily to heterochromatin. The satellite DNAs that have been sequenced have a simple oligonucleotide repetitive sequence unlikely to code for protein [*Drosophila virilis* (Gall and Atherton, 1974), guinea pig alpha-satellite (Southern, 1971); kangaroo rat (Fry *et al.*, 1973)]. The satellite DNAs are apparently not transcribed into RNA. Due to their heterochromatic and centromeric localization it has been postulated that these sequences may be involved in chromosome pairing or chromosome superstructure (Walker, 1971; Pardue, 1975).

Several interesting observations have been made on plant satellite DNAs. Ingle *et al.* (1973) surveyed 59 dicotyledonous species for satellite DNA and found buoyant density satellites in 27 species; however, they detected no satellite DNA in the 11 monocotyledonous species examined. Although buoyant density satellites have not been detected in monocots, approximately 10% of the DNA of barley, wheat and rye reassociates very rapidly suggesting that highly repetitive (possibly simple sequence) DNA does exist in monocotyledonous plants (Bendich and McCarthy, 1970); each sequence forms a kinetic satellite class. Chromosomal localization by *in situ* hybridization of these buoyant density and kinetic satellite sequences is required before speculation on their role in chromosome structure; however, it seems likely that all chromosomes contain some specialized DNA sequences involved in chromosome recognition and pairing. Plant species lacking clear centromeres, "homocentric chromosomes," such as *Luzula purpurea* (Braselton, 1971) might be especially interesting to examine for the presence and localization of satellite DNA. The neocentric chromosomes of particular genotypes of corn (Rhoades and Vilkomerson, 1942) and the supernumerary chromosomes found in many plant species, *i.e.* the "B" chromosomes of corn, might also be interesting systems in which to look for novel distributions of satellite DNA sequences. The "B" chromosomes do not consist solely of satellite DNA, however, but contain DNA sequences found in the normal chromosomal complement (Chilton and McCarthy, 1973; Rimpau and Flavell, 1975; Dover, 1975).

Pearson *et al.* (1974) have reported that the percentage of satellite DNA as measured by neutral CsCl equilibrium centrifugation varies during development in subspecies of melon. Satellite DNA constitutes a higher percentage of total DNA in meristematic tissues (seed, root tip) than in mature, differentiated tissue (fruit, leaf, cotyledon). The variation in percentage of satellite is correlated with the amount of heterochromatic DNA found in the various cell types. These observations suggest that the satellite sequences are under-replicated in some cell types. Other studies indicate that endoreduplication of the genome of *Phaseolus* during polytenization

(Brady, 1973) or during maturation of *Vicia* (Millerd and Whitfeld, 1973) or cotton cotyledons (Walbot and Dure, 1976) does not involve the selective under-replication of any sequences. It is not clear whether there is a minimum amount of satellite DNA compatible with satellite DNA function. A systematic investigation within a genus (with species having widely divergent amounts of satellite DNA) might provide some evidence as to both the function of the satellite DNA and the amount of satellite DNA required to perform such functions.

The ribosomal cistrons are a special case of satellite DNA. These tandemly linked genes of high G+C content have a density of approximately 1.71 gm/cm^3 in neutral CsCl gradients. Although insufficient DNA mass is involved to reliably detect this satellite band in many species, hybridization of radioactively labeled rRNA to fractions taken across the CsCl gradient indicate that the rDNA does form a satellite in all species examined. The paper by Ingle (this volume) summarizes what is known about rDNA cistron number and distribution in a variety of higher plants.

Intermediately Repetitive and Unique Copy DNA

All plant species examined have a large component of intermediately repetitive DNA. Flavell *et al.* (1974) in a study of 23 species of higher plants demonstrated that plants with genomes of 5 to 98 pg DNA per 2C nucleus contained on the average 80% repetitive sequences; plants with 1 to 4 pg of DNA per 2C nuclei had only 62% repetitive sequences. Repetitive DNA is a feature of all eucaryotic DNA.

Do these repetitive sequences function as genes? Structural RNAs such as ribosomal and transfer RNAs and the mRNA for histone proteins have been conclusively demonstrated to be transcribed from repetitive sequence DNA (reviewed by Pardue, 1975). The repetitive sequences that are transcribed to make structural or mRNA have a unique organization. These genes are arranged as gene clusters in one or more chromosomal locations. For example, the ribosomal RNA cistrons are found in the nucleolus organizer region of plants [*Zea* (Ramirez and Sinclair, 1975), *Happlopappus* (Stahle *et al.*, 1975), *Phaseolus* (Brady and Clutter, 1972)] and animal genomes (Lewin, 1974; Pardue, 1975). The rDNA is clustered, and the genes are separated by transcribed as well as non-transcribed spacer DNA; the transcribed spacer is discarded during precursor rRNA processing (reviewed by Lewin, 1974). The 5S rDNA and 4S tRNA sequences are also reiterated. The 5S rDNA of *Xenopus laevis* are found on the extreme ends of the long arms of most *X. laevis* chromosomes; there are approximately 24,000 copies of the 5S gene per haploid chromosome complement. Each cluster contains a large number of 5S genes. As with the large rRNA genes, the 5S rRNA genes are separated by spacer sequences. The 5S rRNA genes of *Zea* have been localized by *in situ* hybridization to the long arm of chromosome 2 (Wimber *et al.*, 1974). A clustered arrangement is also found for tRNA genes in *Xenopus* (Clarkson *et al.*, 1973). The histone genes of *Drosophila* and a sea urchin are transcribed from repetitive DNA; in *Drosophila melanogaster* the genes for histones are clustered in bands 39D-39E of chromosome 2 of the polytene cells (reviewed by Pardue, 1975).

All genetic traits that show Mendelian inheritance are presumed to be present only once per haploid genome and thus should be contained within the unique copy fraction of the genome. Most specific mRNAs have been shown to be transcribed from DNA sequences present only once or a few times per genome. Silk fibroin mRNA (Suzuki *et al.*, 1972), ovalbumin mRNA (Harris *et al.*, 1973) and globin mRNA (Bishop and Freeman, 1973; Bishop and Rosbash, 1973; Harrison *et al.*, 1974) are transcribed from non-repetitive DNA sequences. By extrapolation from what is known about the repetitive gene families, it seems likely that mRNA will be proven to be derived from a larger initial transcript and that the mRNA genes will be separated by nontranscribed spacers. Heterogeneous nuclear RNA (HnRNA), a higher molecular weight class of molecules than mRNA, is the most likely mRNA precursor. The literature on mRNA transcription and processing has been summarized by Lewin (1974, 1975).

Britten and Kohne (1968) demonstrated that intermediately repetitive and unique sequence elements of plant and animal genomes are interspersed. This interspersion was tested by reassociating DNA having to different average fragment lengths to a low C_ot, a C_ot at which only repetitive DNA would reassociate. The reassociation of the longer fragment length DNA resulted in a larger proportion of the mass of DNA in duplexes containing molecules than was found with shorter fragment length DNA. This experiment demonstrated that the unique copy sequences were contiguous to repetitive sequence.

The model of gene regulation proposed by Britten and Davidson (1969) and since updated (Davidson and Britten, 1973) is an attempt to integrate what is known of DNA sequence organization and gene regulation in eucaryotes. This model asserts that structural genes (unique copy DNA) are contiguous to repetitive sequence DNA that serves to regulate transcription of the adjacent gene. The model provides an explanation for the coordinate regulation of structural genes dispersed throughout the genome. This is the key feature of the model since in contrast to procaryotes in which the genes of a particular biochemical pathway are often physically linked and coordinately regulated, genes of a particular biochemical in eucaryotes typically have different chromosomal locations.

The hierarchical organization of the genome as proposed by Britten and Davidson (1969) provides a series of hypotheses amenable to experimental investigation. These hypotheses include:

1. unique and repetitive genome elements are interspersed
2. unique genome regions must be of sufficient length to code for proteins
3. mRNA should hybridize primarily to unique DNA that is contiguous to a repetitive sequence element
4. mRNAs peculiar to a particular developmental stage or biochemical pathway should share a common repetitive sequence.

The discovery of HnRNA, the putative initial gene transcript processed by endonucleolytic cleavage to shorter mRNA molecules, adds another set of predictions:

5. HnRNA should be drawn from the same group of unique sequences as mRNA
6. if HnRNA is the initial transcript, it may contain some transcript from repetitive DNA sequences and these sequences should be selectively lost during RNA processing.

In this paper I would like to describe a set of experiments which sought to demonstrate that hypotheses one and two are true for a higher plant. At the time these experiments were initiated, detailed interspersion studies were available for *Xenopus* (Davidson *et al.*, 1973) and sea urchin (Graham *et al.*, 1974). These studies demonstrated that there is extensive interspersion of the unique and intermediately repetitive sequence classes in these two animal genomes; a basic unit of approximate length-300 nucleotides repetitive DNA:800 nucleotides unique copy DNA-explained the sequence organization of about fifty percent of the genome. A longer period interspersion pattern of the same short repetitive element contiguous to a longer (up to 4,000 nucleotides) unique copy sequence comprises an additional large fraction of each genome.

Subsequent studies have described the interspersion pattern in a variety of marine invertebrates (Goldberg *et al.*, 1975), in *Dictyostelium* (Firtel and Kindle, 1976), in *Drosophila* (Manning *et al.*, 1975) and in angiosperms (Walbot and Dure, 1976; Zimmerman and Goldberg, 1977; Gurley *et al.*, 1976). Studies of DNA:DNA reassociation to low Cot values in which the increased length of initial fragments resulted in more reassociation indicate that interspersion exists in many additional plant species although the lengths of the interspersed elements has not been determined (Britten and Kohne, 1967; Sivolap and Bonner, 1971; Millerd and Whitfeld, 1973; Flavell *et al.*, 1974; Cullis and Schweizer, 1974; Nze-Ekekang *et al.*,

1974; Ranjerkar *et al.*, 1974). With the exception of *Drosophila*, the interspersed repetitive and unique copy sequence elements of each species are easily grouped into one or several size classes; in *Drosophila*, however, there is a continuum of both repetitive and unique copy lengths.

The hypothesis that structural genes are contiguous to interspersed repetitive DNA sequences has been tested by Davidson *et al.* (1975). Total mRNA was recovered from gastrula stage sea urchin polyribosomes by puromycin release. The kinetics of mRNA reassociation with whole DNA and with repeat-continuous unique copy DNA (single copy sequences contiguous to an interspersed repetitive element) were measured. The data demonstrated that 80 to 100% of the mRNA is transcribed from the repeat-contiguous unique copy component. Bishop and Freeman (1974) have shown that a specific mRNA, that for hemoglobin, is near repetitive sequence elements in the duck genome. Therefore, interspersion of repetitive and unique copy DNA in eucaryotes does contribute in a significant way to the organization of the structural genes.

ORGANIZATION OF THE GENOME OF *GOSSYPIUM HIRSUTUM*

Methodology

Since an excellent summary of the theory and practice of DNA reassociation is available (Britten *et al.*, 1975), no general description of methodology will be given, but rather a brief discussion of some problems peculiar to the preparation of plant DNA. The details of experiments illustrated in this paper are given in a previous publication (Walbot and Dure, 1976).

DNA Extraction: DNA used in reassociation experiments must be free of all impurities that affect the rate of reassociation, integrity of DNA, or measurement of DNA concentration, i.e. metal ions, protein, RNA. Depending on the size and complexity of the genome up to 10 mg of purified DNA may be required to characterize the reassociation kinetics and gain some insight into the sequence organization of a particular genome. The choice of starting material is, therefore, very important since the yield of DNA per gram fresh weight may vary from 1 ug/gm for roots to 1 mg/gm for seeds. Within a genus, diploid or low chromosome number species will generally provide more favorable starting material than species of higher ploidy level or DNA content per cell.

Standard extraction and alcohol precipitation methods (Britten *et al.*, 1975; Walbot and Dure, 1976) will be most successful when DNA concentration is high. DNA binding columns (MUP procedure described by Britten *et al.*, 1975) in which DNA can be specifically recovered and concentrated from a large volume of starting material are successful with many plant species (R. Goldberg, personal communication). The only apparent disadvantage to this method is that the DNA is fragmented to approximately 4,000 base pairs in length (2.5×10^6 daltons). Such DNA is suitable for all our studies except those examining the arrangement of sequences more than several thousand bases apart. DNA samples are routinely purified from RNA and carbohydrates by preparative equilibrium density centrifugation in CsCl, a procedure which will also indicate the presence of buoyant density satellite DNA.

Methods for Generating DNA of Precise Length and Length Determination: The average length of a mixture of DNA fragments can be determined by: (1) electron microscopy (Davis *et al.*, 1971), (2) alkaline sucrose gradient sedimentation analysis relative to standards (Noll, 1967), (3) determination of $S_{20,w}$ by analytical ultracentrifugation (Studier, 1965), or (4) gel filtration (Davidson and Britten, 1973).

Single stranded length is used since single stranded DNA is the reactive species in reassociation experiments. Electron microscopy will give a number average length of DNA molecules while the latter three methods give a weight average molecular weight. Most experimental procedures depend on detection of DNA mass (A_{260}, isotope content), and it is preferable to express starting lengths in terms of weight average lengths. Length and weight average measurements are readily interconverted.

A variety of methods for fragmenting DNA have been employed. The most reproducible method is fragmenting by shearing at high speed in a Virtis homogenizer, a method which is more independent of DNA concentration than sonication or extrusion. *In vitro* DNA labeling procedures, HAP fractionation and other procedures also fragment DNA. Table 1 shows the effects of iodination on initial fragment length; 9,000 nucleotide fragments were shortened approximately one-third during iodination; shorter fragments are less affected. HAP binding and elution reduces the average length of most preparations to approximately 4,000 to 5,000 nucleotides.

Methods for Labeling DNA *In Vitro*: *In vivo* labeled plant DNA of high specific activity is often difficult and/or expensive to obtain. Consequently, various *in vitro* labeling procedures have been used including methylation, iodination, mercuration, and nick translation. Methylation using ^{3}H-sodium borohydride typically gives a low specific activity of $\cong$ 20,000 cpm/ug. Iodinated DNA (Commerford, 1971; Orosz and Wetmur, 1974) and RNA (Getz *et al.*, 1972; Prensky *et al.*, 1973) of

TABLE 1
Characterization of DNA Fragments Before and After Iodination

Shearing Conditions	Before Iodination					After Iodination		
	Analytical Ultracentrifugation		Alkaline Sucrose Gradients[a]			Analytical Ultracentrifugation		
	S°_{20w}	Single[b] stranded length	S°_{20w}	Single[b] stranded length	Average[b] single stranded length	S°_{20w}	Single[b] stranded length	Average[b] single stranded length
10 min. Sonication	4.7	260	4.8	280	250	4.5	280	250
	4.6	230	4.6	240		4.4	200	
6 min. Sonication	6.6	520	6.7	540	530	-	-	-
	6.6	520	6.4	500				
	6.9	590						
3 min. Sonication	9.2	1150	8.2	850	900	8.7	1070	900
	8.1	820	8.6	960		7.5	730	
	7.6	750						
1 min. Sonication	9.9	1450	11.2	1840	1650	10.0	1500	1500
	10.4	1650				10.0	1500	
15 sec. Sonication	11.5	2100			2050	10.4	1040	1800
	11.3	2000				10.8	1960	

Virtis, 25 min.	13.3	2900			3000	11.3	2100	2650
8,100 revs/min	13.7	3100				14.1	3200	
Virtis, 25 min.	17.1	5000		1550	4360	14.6	3600	3600
5,400 revs/min.	14.5	3500				14.1	3200	
	15.2	4000				15.2	4000	
25 gauge syringe	22.5	11400	17.8	6320	9210	18.7	7120	6320
10 passages	19.8	8270				17.6	6330	
	22.0	10800				16.9	5500	

[a] *Markers for alkaline sucrose gradients were unlabeled 530 nucleotide cotton DNA fragments and* ^{32}P *labeled 165 and 1400 nuc leotid* Euglena *DNA fragments (determined by analytical ultracentrifugation).*

[b] *Fragment length in nucleotides.*

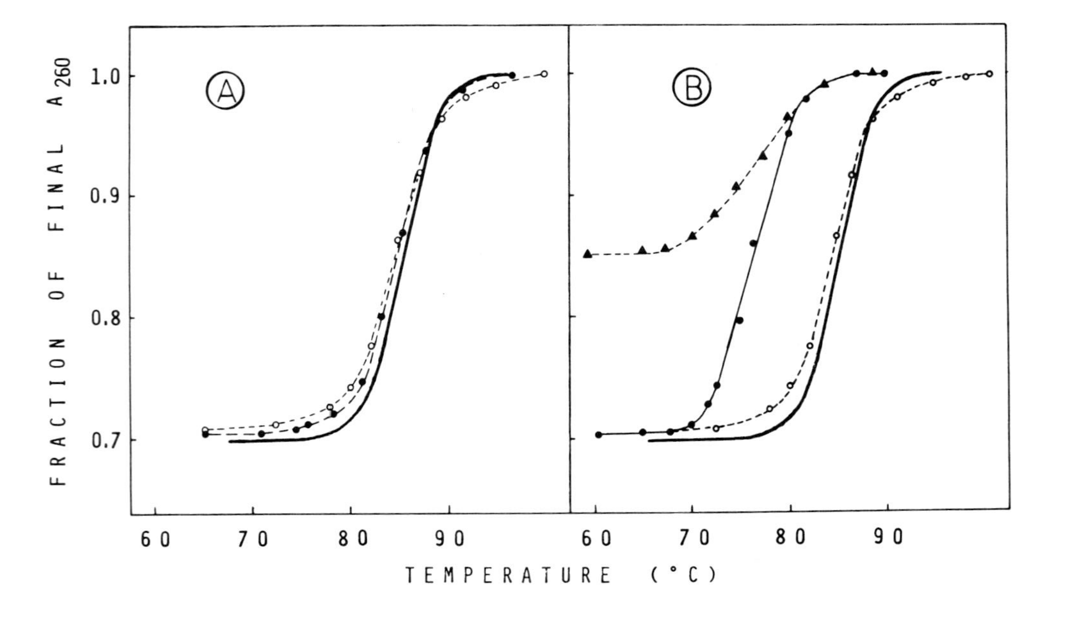

FIGURE 1. Thermal denaturation of native, sheared, and reassociated cotton DNA plotted relative to A_{260} at 97°C after correction for thermal expansion.

(A) Native cotton DNA (solid line) and sheared fragments of 530 base pairs (open circles) and 3,000 base pairs (closed circles) thermally denatured in 1 × SSC.

(B) Native cotton DNA (solid line) and 530 nucleotide single stranded length non-repetitive DNA (not bound to HAP at C_ot 100, bound at C_ot 5,000) thermally denatured in 1 × SSC (Standard Saline Citrate) (open circles). S_1 nuclease resistant duplexes obtained from 12,000 base pair DNA denatured and reassociated to Cot 10 were melted in 0.25 × SSC and the data normalized to 1 × SSC using the equations of Mandel and Marmur (1968) (closed circles). The HAP bound fraction from 1650 base pair DNA denatured and reassociated to Cot 50 was melted in 0.25 × SSC and the data normalized to the conditions for 1 × SSC as above (solid triangles).

high specific activity (10^6-10^8 cpm/μg) have been used in reassociation and *in situ* hybridization experiments with considerable success. Sufficient iodine can be introduced to cause a shift in the buoyant density of DNA, a feature useful in the collection of heteroduplexes.

Mercurated polynucleotides which are prepared using a double stranded DNA substrate reacted at low temperature provide isotopically labeled nucleic acid probes of high specific activity, more uniform labeling of each kinetic class, and less degradation than is possible with iodination (Dale and Ward, 1975; Dale *et al.*, 1975). Mercurated DNA also exhibits readily detectable buoyant density shifts; the half life of ^{203}Hg is 46 days, and this radionucleotide is very inexpensive. The primary disadvantage of mercurated polynucleotides is the lability of the carbon-mercury bond to cleavage by strong electrophiles and reducing agents; this property can, however, be put to advantage for the tritiation or iodination of mercurated polynucleotides under extremely mild reaction conditions. If mercurated polynucleotides are used in reassociation experiments low temperatures are required.

DNA polymerase I (Boehringer Co.) can be used to incorporated α-^{32}P, ^{3}H, or ^{14}C-labeled nucleotides into existing or induced nicks in a polynucleotide; this labeling process is called nick translation. DNA of high specific activity can be obtained; ^{32}P-labeled DNA is convenient for autoradiography and because of its high counting efficiency less material is required. Some investigators using simple sequence DNA have reported problems such as the formation of triple helices using this method. High molecular weight DNA can also be degraded during the labeling by metal ions or nuclease action; however, mitochondrial DNA of *Zea* has been labeled with α-^{32}P-deoxyATP, and the product was of the same size as the initial DNA, approximately 50×10^6 (Smith and Walbot, unpublished observations).

Whatever method is used to label the DNA, the following criteria should be used to evaluate the product: (1) size of single stranded DNA after labeling, (2) T_m-a reduction in T_m indicates generation of nicks or single stranded ends or extensive shearing (Fig. 1), (3) uniform labeling of kinetic classes, (4) stability of label during reassociation reaction and processing of product (this is a problem with iodination in which heat labile pyrimidine hydrates are formed in addition to the stable 5'iodocytidine derivative), and (5) specific activity of product sufficient for experiment

Reassociation Reactions: DNA:DNA reassociation reactions are used to measure the extent of homology within populations of molecules and the proportion of molecules of sufficient reiteration frequently to form stable hybrids at a particular

Cot value. These determinations are dependent on the criterion conditions set for an experiment. The stringency of the criteria set will determine how perfectly matched a duplex must be to be scored as a hybrid. Temperature, ionic strength, viscosity, base composition of the DNA, initial fragment length of DNA, as well as DNA concentration (Co) and time (t) of reaction determine the rate at which reassociated duplexes will form hybrids of sufficient stability to be counted as doubled stranded; these factors are discussed in detail by Britten *et al.* (1975).

The rate and extent of reassociation can be measured in several ways including: (1) electron microscopy, (2) spectrophotometry, taking advantage of the increase in absorbance (hyperchromicity) exhibited by single vs. double stranded DNA, (3) HAP (hydroxylapatite) chromatography in which molecules containing a duplex of approximately 50 base pairs will bind to the column, and (4) S_1 nuclease digestion of non-reassociated single stranded DNA leaving only well matched duplexes.

A combination of these methods is often desirable for confirmation of results, and there is generally good agreement in comparison of two techniques. Recently, Smith *et al.* (1975) and Britten and Davidson (1976) have reported on anomaly in the rate of reassociation as measured by S_1 nuclease when large and small initial fragments are mixed due to an inhibition of hybridization of single stranded ends in duplex-containing molecules.

Ploidy Level Considerations: Polyploidization is common in many plant taxa. If the larger genomes evolved from the smallest by exact duplication of all DNA then no new genetic loci nor additional sequence complexity should be found in the larger genomes. This hypothesis can be tested using DNA reassociation kinetics. An increase in ploidy levels affects the frequency of each sequence equally and consequently the concentration of each sequence is unchanged. DNA:DNA reassociation is concentration dependent and if concentration is unchanged, the rate constant (k) and $C_0t_{\frac{1}{2}}$ of each kinetic class will remain unchanged. In comparing related genomes of unequal size, an increase in the rate of reassociation of a component in the larger genome relative to the smaller genome indicates there has been preferential amplification of some sequences; conversely, a decrease in the rate of reassociation of a component indicates that the sequences of that kinetic class are now relatively more rare.

Autotetraploids in which there has been an exact duplication of all sequences will have a "unique copy" fraction that reassociates as expected for sequences present once per haploid genome. The reassociated duplexes should have a Tm close to that of native DNA since the unique copy fraction will find exact or near exact matches. The extent of depression in Tm will give a measure of the divergence of sequences during species evolution; that is, it will reflect the species age. A complication to this analysis is the existence of chromatid as well as chromosome segregation in autotetraploids; chromatid segregation is one of a number of phenomena that can reduce the measurable divergence in a genome. Extensive inbreeding combined with strong selection might also result in the production of individuals with a much lower measurable divergence than the actual evolutionary age of the species.

Allotetraploids, new species produced by the hybridation of two related species, contain the chromosome sets from both parental types. In this case chromatid segregation due to exclusive intragenomic pairing can result in a complete lack of segregation for any particular marker. If intergenomic pairing occurs genetic markers will segregate. Grant (1975) has summarized the existing data on marker segregation in a variety of plant species. Segregation ratios range from 0 to the expected tetrasomic ratios (35:1 F_2 if chromatid segregants, 21:1 if F_2 chromosomal segregants). Hypothetical cases could arise in which pairing will result in gametes containing only one chromatid type.

Gerstel and Phillips (1957) have examined chromosome pairing in cotton species by constructing synthetic allohexploids. They found that segregation ratios are correlated with the average degree of chromosome pairing between genomes. The two genomes of *G. hirsutum* represented as AADD can be tested for their degree of divergence from AA or DD bearing diploids. Closely related A genomes in *G. hirsutum* (AADD) × *G. arboreum* (AA) show chromosomal segregation; whereas *G. hirsutum* (AADD) × *G. raimonaii* (DD) genes in the D genome show much less segregation and hence more divergence of the chromatids as evidenced by reduced intergenomic pairing. The evolutionary divergence or convergence of the AA and DD chromosome sets have not been directly tested but since two species must be closely related to from a fertile allopolyploid the new species will contain many homologous genes. The homology may be nearly as complete as that predicted for autotetraploids since allelic heterozygosity may nearly equal the species differences. Thus, definition of evolutionary age and determination of origin by allo-*vs.* auto-tetraploidization in higher plants may be a difficult problem to solve using DNA reassociation experiments. This point will be discussed in more detail when the organization of *Nicotiana tabacum* is compared to *G. hirsutum*.

Experimental Results

The data discussed in this section are presented in detail in Walbot and Dure (1975); the summary reported here does not include all of the experimental detail upon which specific conclusions are based.

Reassociation Kinetics: The reassociation kinetics of short fragment length DNA of cotton are shown in Fig. 2 and summarized in Table 2. The genome consists of three kinetic classes: highly repetitive, repetitive, and single copy DNA. The reassociation of repetitive DNA is complete by C_0t 50; few single copy sequences will have reassociated at this C_0t value. The proportion of repetitive DNA is approximately 35% with 8% of the genome composed of highly repetitive sequences and 27% of intermediately repetitive sequences.

A more precise measurement of the proportion of repetitive DNA can be obtained by measuring the reassociation kinetics of DNA fragments at several lengths as is shown in Fig. 3(b)(c). Following the data for C_0t 50 we can extrapolate to the Y-axis to determine that 35% of the genome would reassociate at C_0t 50 at an infinitely short fragment length. In actual experiments we are constrained to use DNA of at least 100 or more bases since a duplex of approximately 50 base pairs is required for HAP binding (Wilson and Thomas, 1973). However, since interspersion of repetitive and unique copy DNA is likely, the extent of HAP binding at longer fragment lengths reflects both the reassociation of repetitive DNA as well as the presence of unreacted single copy DNA tails contiguous to the repetitive sequence. An absolute measure of the percent duplex using S_1 nuclease which cleaves single stranded but not well matched duplex DNA would show that the same amount of duplex is present at all fragment lengths.

A plot of percent HAP binding *vs*. fragment length (Fig. 3) can be used to measure the average spacing between nearby repetitive sequences (Graham *et al.*, 1974). As fragment length increases more single copy sequence contiguous to a repetitive element will be HAP bound up to the fragment length within which two repetitive sequences are likely to occur. In cotton there is a sharp change in slope of HAP binding *vs*. fragment length at about 1,800 nucleotides; this length is the average length of unique copy DNA between neighboring repeats.

About 60% of the genome is HAP bound at C_0t 50 at 1,800 nucleotides indicating that this proportion of the genome is found in a short period interspersion pattern. Since HAP binding continues to increase even at the longest fragment length measured, there must also be a long period interspersion pattern in which the distance between neighboring repeats exceeds 1800 nucleotides.

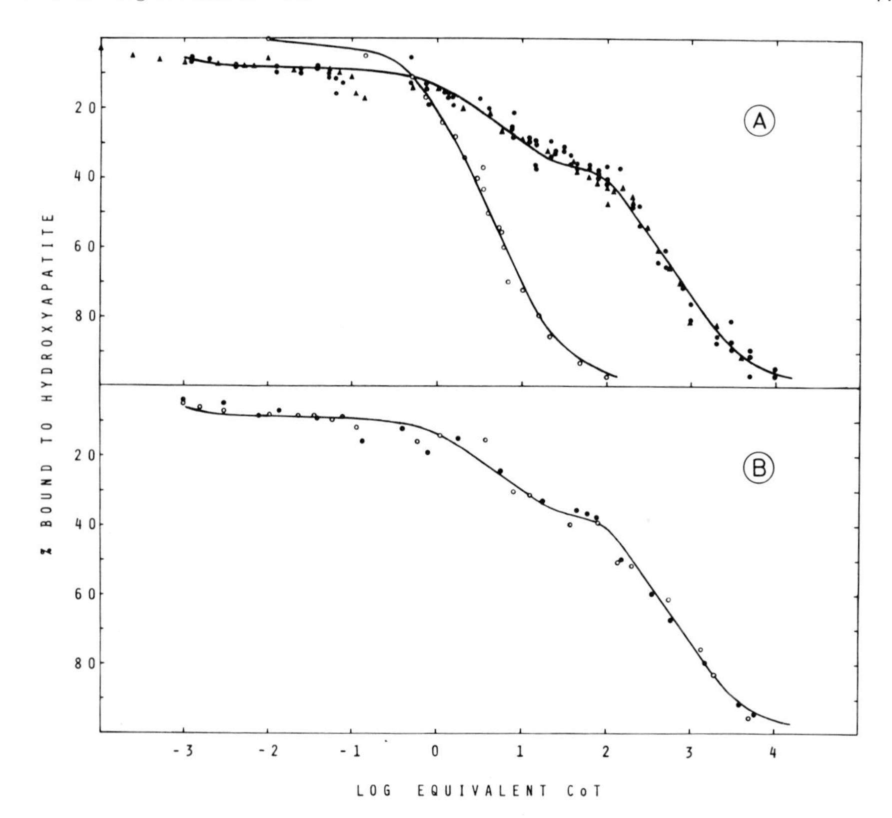

FIGURE 2. Reassociation kinetics of cotton DNA from various developmental stages analyzed by a computer curve fitting program.

(A) Optical reassociation data from HAP fractionation of 350 nucleotide fragment length cotton DNA samples isolated from 30-40 mg cotton cotyledons that had been reassociated to various C_0t values (solid circles); these data were used to generate the computer curve (solid line) using three second order kinetic components. Fractionation of radioactivity following HAP chromatography of ^{125}I-labeled 350 nucleotide fragment length DNA samples (triangles) was used to determine the lowest C_0t values. E. coli *nucleotide fragment length DNA was also denatured, reassociated, and fractionated on HAP (open circles) and a computer curve fit to the data was plotted using a single second order kinetic component (solid line).*

(B) 350 nucleotide fragment length DNA samples from root (closed circles) and mature cotyledons after endoreduplication of the genome (open circles) were reassociated to various C_0t values and the percent HAP bound plotted for comparison to the computer curve fitted to data obtained using DNA isolated from immature cotyledons (solid line as in (A)).

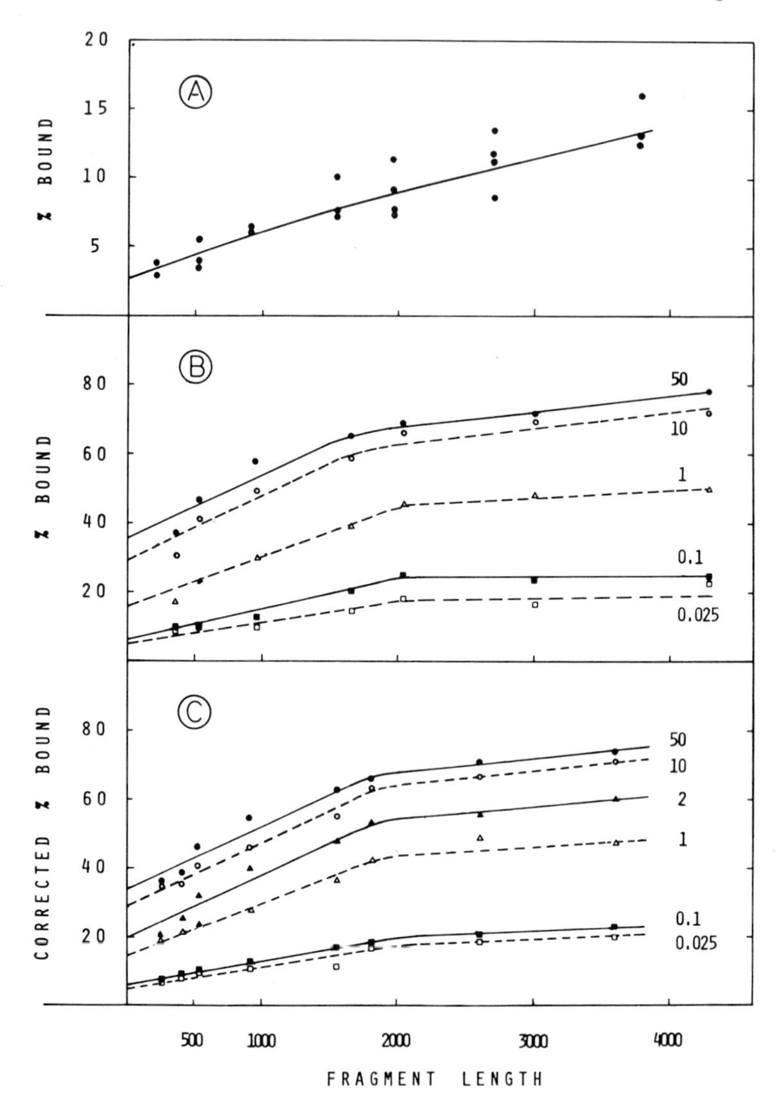

FIGURE 3. The percent HAP binding vs. *fragment length for zero time binding, self reassociation, and tracer reassociation experiments.*

(A) Zero time binding of ^{125}I-labeled cotton DNA plotted vs. *fragment length. The ^{125}I-labeled DNA preparations described in Table 1 were denatured by boiling for 7 min. and then loaded directly onto the standard HAP column with 50 ug heat denatured* E. coli *DNA (effective C_ot < 10^{-5} for cotton DNA) and the percent of radioactivity bound determined.*

The length of the repetitive elements can be estimated by determining the length of S_1 nuclease resistant duplexes formed by C_ot 50 by gel filtration, electron microscopy, or sedimentation coefficient (data in next section). Alternatively the length of the duplexes can be determined from optical melting profiles. Native DNA from cotton shows 30% hyperchromicity whereas 1,650 nucleotide long DNA that is HAP bound at C_ot 50 (60% DNA bound) shows 15% hyperchromicity. Therefore the length of the duplex within each 1,650 length fragment is $\frac{15}{30} \times 1,650$ or approximately 850 nucleotides.

Zero time binding shown in Fig. 3(a) is that component of the genome that reassociates by C_ot 10^{-4}. This component is about 2 to 3% of the cotton genome and consists of intramolecular duplexes often called palindromes. The palindromic sequences in the cotton genome are clustered (Walbot and Dure, 1976) which explains the unexpected low percentage of DNA that bonds to HAP at zero time at the longest fragments lengths examined. In sea urchin (Graham *et al.*, 1974) zero time binding increases more nearly proportional to the fragment length. Clustering of palindromes has been noted in several animal genomes (Wilson and Thomas, 1974) and in soybean DNA (Gurley *et al.*, 1976).

The increase in HAP binding at C_ot 50 as a function of fragment length is due to the reassociation of mildly repetitive rather than highly repetitive or intramolecular duplex containing molecules. This is shown in Fig. 4 in which the data of Fig. 3 have been replotted as a function of C_ot. It is clear that the major increases in HAP binding occurs between Cot 0.1 and 10, the domain within which the intermediately repetitive sequences of cotton DNA reassociate.

S_1 Nuclease and Electron Microscopic Studies: S_1 nuclease can be used to measure both length of duplexes and the fidelity

---

(B) Self reassociation: the unlabeled DNA fragments described in Table 1 were reassociated to various C_ot values and the percent binding to HAP determined by the fractionation of A_{260} units. No correction was made for zero time binding. The symbols represent C_ot values of 0.025 (□), 0.1 (■), 1 (△), 10 (○) and 50 (●).

(C) Tracer reassociation: tracer quantities of the ^{125}I-labeled DNA fragments described in Table 1 were mixed individually with an 8,000 to 12,000 fold excess of 530 base pair long driver DNA, denatured, reassociated to various C_ot values, and the fractionation by HAP chromatography of radioactivity monitored. The HAP binding of tracer was corrected for zero time binding (Fig. 2A) as described by Davidson et al. *(1973). The symbols represent C_ot values of 0.25 (□), 0.1 (■), 1 (△), 2 (▲), 10 (○) and 50 (●).*

TABLE 2
Reassociation Kinetics of Cotton DNA

Component	Fraction of Genome[a]	Observed $C_ot_{\frac{1}{2}}$	$C_ot_{\frac{1}{2}}$ pure[b]	Kinetic Complexity[c]	Mass per 1N genome (pg)
Unique	0.605	723	437	2.9×10^{11}	0.48
Repetitive	0.27	5.42	1.46	9.7×10^{8}	0.21
Highly[d] Repetitive	0.08	5.75×10^{-3}	4.6×10^{-4}	3.8×10^{6}	0.064

[a] *Calculated from Fig. 2 and ordinate intercepts of Fig. 3B and C.*

[b] *$C_ot_{\frac{1}{2}}$ pure = $C_ot_{\frac{1}{2}}$ observed × fraction of genome.*

[c] *Kinetic complexity (in daltons) calculated by comparison of the rate of reassociation of each component of cotton DNA to that of* E. coli *DNA ($C_ot_{\frac{1}{2}}$ 4.2, Fig. 2a; 2.8×10^9 daltons, Cairns, 1963).*

[d] *Highly repetitive sequences were defined as those sequences reassociating by C_ot 0.1, including the apparent zero time binding components; consequently the observed $C_ot_{\frac{1}{2}}$ and $C_ot_{\frac{1}{2}}$ pure reflect the inclusion of inter- as well as intramolecular duplex formation.*

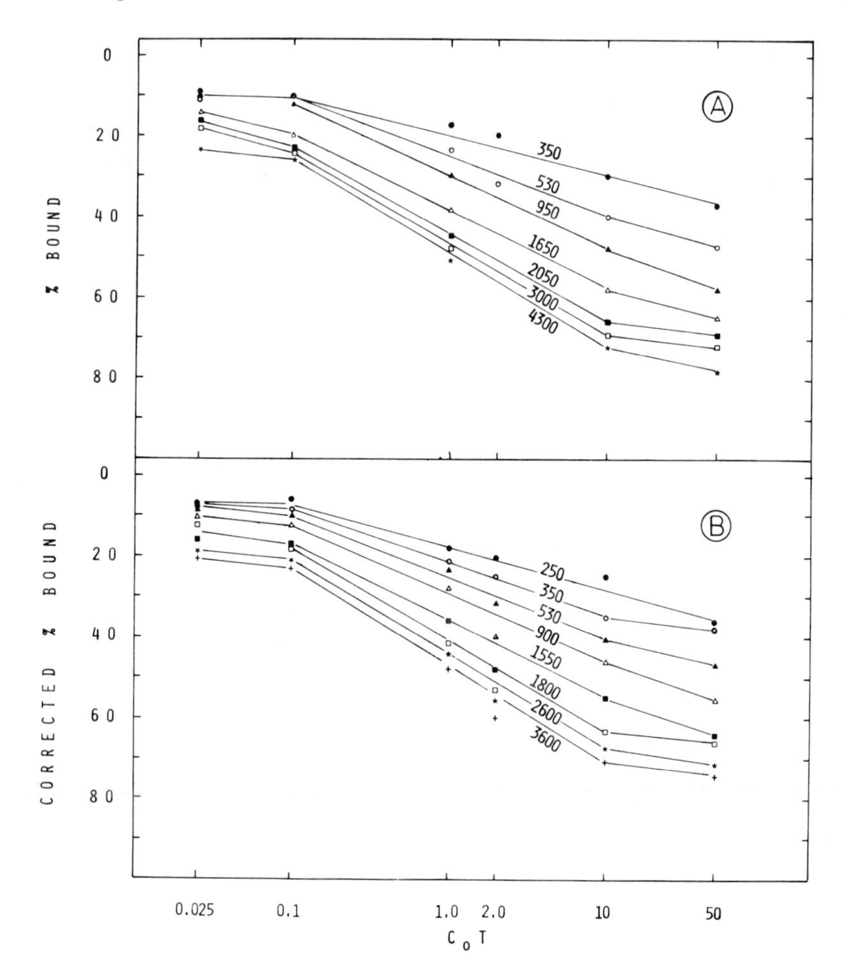

FIGURE 4. The percent HAP binding vs. *C_ot for various fragment length DNA in self and tracer reassociation experiments. The data shown in Fig. 3B and C have been replotted as a function of C_ot.*

(A) Self reassociation of various fragment length DNA. The symbols represent DNA fragment lengths of 350 (●), 530 (○), 950 (▲), 1650 (△), 2050 (■), 3000 (□), and 4300 (*) nucleotides.

(B) Tracer reassociation of various fragment length DNA. The symbols represent DNA fragment lengths of 250 (●), 350 (○), 530 (▲), 900 (△), 1550 (■), 1800 (□), 2600 (*) and 3600 (+) nucleotides.

of pairing. S_1 nuclease will recognize and cleave short, several base mismatched regions (Wilson and Thomas, 1974). In previous studies S_1 digests of duplexes formed by repetitive DNA have yielded a relatively homogeneous component of approximately 300 base pairs; these duplexes could be conveniently sized by agarose gel filtration (Davidson and Britten, 1973). However, the initial estimates from melting studies of the length of reassociated duplexes in cotton was 800 or more base pairs; duplexes of this size are excluded from gel filtration columns and must therefore be measured using electron microscopy.

A detailed electron microscopic analysis of duplex length before and after S_1 nuclease treatment (Walbot and Dure, 1976) indicated: (1) highly repetitive DNA forms long >3,600 base pair length duplexes, (2) intermediately repetitive DNA duplexes have a weight average length of 1,200 base pairs, and (3) non-reassociated DNA at C_0t 50, the single copy class, consists of a well defined class of molecules about 2,000 nucleotides in length (60% of the single copy component) with a second class that is a continuum of molecular lengths of from 2,500 to >6,000 bases.

The percent S_1 nuclease resistant duplex at low C_0t 0.1 and C_0t 50 was estimated by passage of the S_1 digest through a Sephadex G-25 column (Walbot and Dure, 1976) to demonstrate that the amount of duplex formed was a function of C_0t and not due to the fragment length.

Interspersion Pattern of the Cotton Genome: From the estimates of the average lengths of the repetitive and unique copy sequences which were shown to be interspersed and the proportion of the genome in each kinetic class, it is possible to describe a model interspersion pattern for the genome of cotton. Table 3 summarizes the data and model. About 80% of the genome is involved in a short and long interspersion pattern of 1250 base repetitive elements and 1800 or 4000 base non-repetitive elements. The rapidly renaturing sequences are long >3600 base sequences and also show a high degree of mismatching (Walbot and Dure, 1976). These sequences do not contribute significantly to the HAP binding of non-repetitive sequences at low C_0t and have, therefore, been grouped as a single non-interspersed component; however, these sequences are, in fact, interspersed in the genome but their contribution to reassociation is so small in total DNA preparation of cotton that only an incomplete description of this component is possible. If the DNA were fractionated to increase the proportion of highly reiterated sequence, it could be studied in more detail. The remainder of the genome (8%) consists of non-repetitive sequences more than 4,000 bases from the nearest repetitive element.

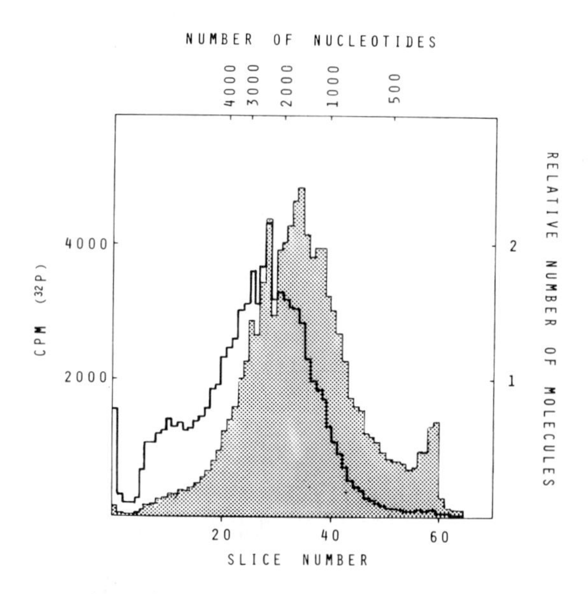

FIGURE 5. Size distribution of mRNA-poly (A) molecules from germinating cotton cotyledons. Radioactive mRNA-poly (A) was purified from this tissue, electrophoresed and the distribution of radioactivity in the gel determined by the procedures given in Harris and Dure (1974). The distribution of mass (open histogram) among mRNA-poly (A) molecules is shown by the distribution of radioactivity. The upper abscissa showing the number of nucleotides in RNA in various regions of the gel was determined from the mobility of 25S, 18S, and 5.8S rRNA markers. The relative number of molecules in each slice (shaded histogram) was determined from the mass distribution and the nucleotide length represented in each slice.

One requirement for any genome model is that it predict the size of genes. The cotton interspersion model predicts that most genes should be about 1,800 nucleotides in length (60%) but that a substantial portion of the genes may be considerably larger. The length of mRNA-poly(A) molecules in germinating cotton has been measured (Fig. 5; Harris and Dure, 1974) in acrylamide gels under fully denaturing conditions in 99% formamide. The mass average length of the mRNA minus poly(A) is 2,400 nucleotides. The mean length of mRNA molecules is 1,500 nucleotides; the class of mRNAs found at this length may reflect transcription from the short interspersion 1,800 nucleotide non-repetitive DNA.

TABLE 3

Model of the Cotton Genome

A. Population of Kinetic Components of Cotton DNA[a]

Component	Reiteration frequency	Number of individual sequence elements in component
Unique[b]	1	1.5×10^5
Repetitive[c]	134	1,184
Highly repetitive and palindromic[d]	---	---

B. Interspersion Pattern

Components[e]

A. 1,250 nucleotide repetitive element + 1,800 nucleotide non-repetitive element
B. 1,250 nucleotide repetitive element + 4,000 nucleotide non-repetitive element
C. Non-repetitive DNA more than 6,000 nucleotides from the nearest repetitive element
D. Palindromic and highly repetitive DNA reassociated by C_ot 0.1

Component	% of total genome[b]	Repetitive DNA % of genome[c]	Non-repetitive DNA % of genome
A	61	25	36
B	21	5	16
C	8	--	8
D	8	8	--

[a]Values are derived from an estimation of the haploid genome size as 0.795 pg DNA.

[b]This component is 60.5% of the genome or 0.48 pg DNA. This equals 4.4×10^8 base pairs. The mass average sequence element size (i.e. regions not interrupted by repeated sequences) is estimated to be 2,900 base pairs. Thus $4.4 \times 10^8/2{,}900$ yields 1.5×10^5 sequence elements.

[c]This component is 27% of the genome or 0.215 pg DNA, which equals 1.96×10^8 base pairs. Since the $C_0t_{\frac{1}{2}}$ (5.4) of this component is 134-fold less than the non-repetitive component, 1.48×10^6 base pairs are present as different sequence elements. The mass average sequence element size is estimated to be 1,250 base pairs, thus $1.48 \times 10^6/1{,}250$ yields 1184 different sequence elements.

[d]Data not amenable to computation as discussed in text.

[e]The components of the model were derived from: 1) The percent of the genome in each of the kinetic components given in Table 2. 2) The mass average size of repetitive and non-repetitive sequences as determined by electron microscopy. 3) The breakpoint in the slope in the plot of percent HAP bound vs. fragment length (Fig. 3B and C). The mildly repetitive sequence elements that average 1,250 base pairs are assumed to be adjacent to non-repetitive sequences, since there is no evidence for clustering.

IMPLICATIONS OF INTERSPERSION OF GENES FOR REGULATION

Analysis of Other Plant Genomes

The interspersion of repetitive and single copy sequence elements has been examined in detail for three additional angiosperms. R. Goldberg and his associates have examined tobacco (Zimmerman and Goldberg, 1977), pea, and soybean (Goldberg, 1978); the reassociation kinetics of these species are summarized in Tables 4, 5 and 6. Hepburn, Gurley and Key (personal communication) have also examined the reassociation of soybean DNA and have slightly different results as shown in Table 7. The two measurements of soybean DNA kinetic parameters agree as to genome size. However, the proportion of repetitive and single copy DNA, the description of the number of kinetic classes, and reiteration frequency of the intermediately repetitive sequence class(es) are different. This difference seems unlikely to be heterogeneity of the organization of the soybean genome. The summation of the reiteration frequency classes of 24 and 3550 of Goldberg's study (Table 6) yields a class with kinetic properties very similar to the single class of reiteration frequency reported by Hepburn and co-workers. This observation of a varying number of repetitive sequence classes reflects the existence of a number of repetitive families. Varying conditions of detection, *i.e.* changes in stringency and initial fragment length, yield different estimates of the frequency and dominance of a particular kinetic class. Since a sequence class must be a major component of a particular region of the reassociation curve to be detected, variable conditions may enhance or diminish the apparent contribution of a particular reiteration frequency class.

The extent of divergence of sequences within a genome is measured by the extent of mismatching detected by thermal denaturation or S_1 nuclease. This divergence presumably reflects the rate of evolution within a particular kinetic class as well as the evolutionary history of the species. *Nicotiana tabacum* and *Pisum sativum* both show a high degree of mismatching in the interspersed repetitive class while *Gossypium hirsutum* and *Glycine max* interspersed repetitive sequences are less divergent. The long, highly repetitive sequences of cotton are much more mismatched than the intermediately repetitive class; the opposite result was found for pea, tobacco and soybean (Table 8) and for melon (Bendich, This volume). This difference in the extent of divergence of the highly repetitive sequence is an interesting observation that may relate to the evolutionary history of each species.

TABLE 4

Characterization of the Tobacco *Nicotiana tabacum* Genome[f]

Component	[a]Fraction of 300 NT fragments bound to HAP	Kinetic[b] complexity	[c]Reiteration frequency	Average length	Number of elements	% Mismatching
Foldback	0.02	-	-	?	?	?
Fast	0.06	4.6×10^3 NTP	12,000	>2,000 NTP	?	5%[d]
Intermediate	0.59	2.5×10^6 NTP	250	>2,000 NTP	?	1-5%[e]
				300 NTP	3,700	15%[e]
Single Copy	0.23	2.8×10^8 NTP	1	1,500 NTP	370,000	8%[d]
Genome Size	1.6 pg (by kinetics)					

[a]*From HAP fractionation and reassociation of specific kinetic fractions. Do not add to 1.00 as only the fraction of 300 NT fragments containing sequences of each component which were identified by HAP* fractionation and reassociation *were used.*

[b]*Based on K_{pure} from self-reassociation of kinetic components and* E. coli *K of 0.22 $M^{-1}sec^{-1}$ for 300 NT fragments. Expressed as nucleotide pairs.*

[c]*Derived by ration of* $\frac{K_{whole\ repetitive\ component}}{K_{whole\ single\ copy}}$ *per 1 C genome.*

[d]*Melt of reassociated isolated component.*

[e]*Melt of S1 resistant repetitive duplexes.*

[f]*Zimmerman and Goldberg (1977).*

TABLE 5

Characterization of the Pea *Pisum Sativum* Genome[c]

Component	[a]Fraction of 300 NT fragments bound to HAP	Kinetic[b] complexity	Reiteration frequency	Average length	Number of elements	% Mismatching
Foldback	0.01	–	–	–	?	?
Fast	0.16	1.8×10^4 NTP	775	?	?	?
Intermediate	0.58	2.4×10^6 NTP	90	>2,000 NTP	?	3%
				300 NTP	?	13%
Single copy	0.18	6.8×10^7 NTP	1	<2,700 NTP	?	?
Genome size	0.42 pg (from kinetics)					

[a]*Based on 3 component least squares analysis of total HAP pea data. No minicots or slave minicot curve data. 60°C, 0.18 M Na^+ criterion.*

[b]*Determined similarly to soybean data.*

[c]*Goldberg, R. B., personal communication.*

TABLE 6

Characterization of the Soybean *Glycine max* Genome[c]

Component	[a]Fraction of 250 NT fragments bound to HAP	Kinetic[b] complexity	Reiteration frequency	Average length	Number of elements	% Mismatching
Foldback	0.05	-	-	?	?	?
Very fast	0.03	187 NTP	375,000	>2,000 NTP	?	1%
Fast	0.33	2.1×10^5 NTP	3,550	>2,000 NTP	?	1%
				200-400 NTP	?	6%
Slow	0.33	3.2×10^7 NTP	24	>2,000 NTP	?	1%
				200-400 NTP	?	6%
Single copy	0.22	4.9×10^8 NTP	1	<1,900 NTP	?	?
Genome size	2.5 pg (by kinetics)					

[a]*Based on Four Component Least Squares Fit of* Total *HAP Reassociation Data. Rates of repetitive components determined by reassociation of pure tracer repetitive fractions (prepared by HAP fractionation) in whole soybean DNA.*

[b]*Based on Total Reassociation curve; fractions given in column 1 and* E. coli *rate of* $0.22M^{-1}sec^{-1}$ *corrected for fragment length.*

[c]*Goldberg (1978).*

TABLE 7
Characterization of the Soybean *Glycine max* Genome[g]

Component	Fraction of genome	Kinetic[a] complexity	Reiteration frequency	Average length	Number of[c] different families	% Mismatching
Foldback	0.03	--	--	--	--	--
Highly repetitive (cryptic satellite)	0.26	8.705×10^4 [d]	3,857	350[b]	248	14
Repetitive	0.095	1.894×10^6 [d]	153	350[b]	2,284	8
Single copy	0.615	5.164×10^8 [d]	1	1,500	4.478×10^5 [e]	[f]

[a]*Total chemical complexity from kinetics = 1.286×10^9; haploid in a tetraploid species such as soybean is 2× the basic complement.*

[b]*From reassociation vs. fragment length and electron microscopic data.*

[c]*From chemical complexity/reiteration frequency × 1/average length.*

[d]*Derived from total C_ot curve observed K_s corrected to pure K_s by adjusting for observed fractions.*

[e]*Size is from reassociation vs. fragment length data--only 85% of unique is interspersed at this size.*

[f]*Not determined.*

[g]*Hepburn, A. G., Gurley, J. L., and Key, J. L., personal communication.*

Note: If rates of reassociation are corrected for mis match the highly repetitive sequences are then calculated to contain 37 families with an average reiteration frequency of 26,190 (and the repetitive sequences a reiteration frequency of 459 in 764 families.

If highly repetitive sequences are involved in chromosome pairing, rapid divergence of these sequences might preclude certain chromosome pairings. For example, in the consideration of polyploidization the phenomena of chromosome *vs.* chromatid segregation was discussed. Rapid divergence of the highly repetitive sequences within an allotetraploid genome set might preclude chromatid segregation in that species despite the homologous nature of the gene elements as well as of the general organization of the chromosomes as is found in autotetraploids. The large divergence noted in cotton highly-repetitive DNA may be a reflection of the divergence between the progenitor AA and DD genomes noted by Gerstel and Phillips (1957), as a lack of segregation of A and D markers.

There is a very interesting difference in the extent of mismatching in the unique copy fraction of tobacco and cotton. Both *N. tabacum* and *G. hirsutum* are allotetraploids, however, reassociated cotton unique copy DNA has a T_m and per cent hyperchromicity close to that of native DNA (Fig. 1) while reassociated tobacco single copy DNA shows 8% mismatching by these same criteria (Zimmerman and Goldberg, 1976). Mismatching in the mildly repetitive DNA is quite similar. There are several possible explanations for these observations. One line of reasoning would be that the two diploid genomes constituting *N. tabacum* were either somewhat diverged prior to polyploidization or have diverged "rapidly" within the same cellular environment. Alternatively, *G. hirsutum* can be viewed as the amalgamation of two nearly homologous progenitors; subsequently these evolved "slowly" in the unique copy fraction but have maintained and extended the divergence of the highly repetitive DNA responsible for chromosome pairing and hence the cytogenetic and segregational analysis identification of the AA and DD genomes.

Analysis of evolutionary history by determining the extent of mismatching of kinetic classes will be a difficult problem unless both the extant and progenitor species are examined. For example, the AA and DD genomes of cotton should be examined separately as should the extent of divergence of *G. hirsutum* kinetic classes from the progenitor types. These data would allow an assessment of the apparent anomaly of closely matched unique copy *vs.* highly divergent repetitive sequences.

The complexities of relating reassociation kinetic data to evolution do not preclude the conclusion that single copy and repetitive DNA interspersion appears to be a stable and general feature of the genomes of all higher plants examined so far. In Table 8 are presented some data on the extent and pattern of interspersion in the tobacco, pea and soybean genomes. It is clear that each genome has a high (0.8 to 0.9) proportion of its single copy DNA interspersed with repetitive DNA when

TABLE 8
Characteristics of the Genomes of Three Higher Plant Species

Organism	1C genome size (pg)	Genomic fraction of single copy DNA	Genomic complexity (nucleotide pairs)	Repetitive sequence classes (number of copies per 1C genome)	Fragment length used for genome organization studies (nucleotides)
Nicotiana tabacum (tobacco)	1.60	0.45	6.2×10^8	250 15,000	2,400
Pisum sativum (pea)	0.42	0.40	1.6×10^8	90 800	2,700
Glycine max[a] (soybean)	2.50	0.40	9.2×10^8	25 3,500 350,000	2,700
Glycine max[b])soybean)	2.58	0.61	1.29×10^9	153 3,857 ?	several

Organism	Are repetitive sequences interspersed at this fragment length?	Fraction of repetitive DNA 200-400 Nucleotides in length	Fraction of single copy DNA interspersed at this fragment length	Average % mismatching of repetitive sequences	
				Long (1500 NTP)	Short (300 NTP)
Nicotiana tabacum (tobacco)	Yes No	0.35	0.80	1-2%	10-15%
Pisum sativum (pea)	Yes ?	0.55	0.90	1-2%	10-15%
Glycine max[a] (soybean)	Yes ?	0.55	0.90	1-2%	5-10%
Glycine max[b] (soybean)	No: cryptic satellite Yes: other repetitive	0.36	0.85	1	8%

[a] *From R. Goldberg (1978).*

[b] *From Hepburn, A. G., Gurley, W. B., and Key, J. L., personal communication.*

examined at a moderate fragment length. The more highly repetitive DNA elements of tobacco and of soybean are not generally interspersed; the highly repetitive DNA elements of soybean can be isolated as a cryptic buoyant density satellite in Ag^+-CsCl gradients (Hepburn *et al.*, unpublished results). Interspersion of the unique and repetitive sequence elements of angiosperm genomes would seem to be a general character of the group.

Repetitive Sequence DNA and Regulation

The differential transcription of genes contiguous to different repeats during development is the central hypothesis of the Britten-Davidson model. Plant tissues such as seeds which contain stored mRNAs may be excellent systems in which to follow the coordinate transcription and processing of a class of messages utilized during a particular developmental stage. If this class of mRNAs is coordinately regulated then each gene should be contiguous to the same or a low diversity of repetitive sequences. Other major shifts in development program such as flowering, spore germination in lower plants, and senescence may also contain a large set of specialized mRNAs. The advantages of using seeds, spores, or another developmentally arrested system is that the experimentor can control progression to the next developmental stage. Previous studies with sea urchin (Goldberg *et al.*, 1975) demonstrated the feasibility of isolating DNA sequences adjacent to functional transcribed genes for further study. It may also be possible to isolate pre-mRNA apparently stored in seeds for analysis (Walbot *et al.*, 1975).

The possible existence of sub-repeats within long repetitive and highly repetitive sequence elements is another potentially interesting line of investigation. The Britten-Davidson model predicts that several regulatory sites exist preceding each gene; regulatory sequences such as operators, promotors, and repressor binding sites in procaryotes are short sequences of 20 nucleotides or loss. Such short sequences are not amenable to study by DNA:DNA reassociation experiments. However, it is possible that each regulatory sequence is contained within a longer element in which case the unit could be studied.

Evolution of DNA Sequence Organization in Eucaryotes

Deuterostomic and protostomic animals (reviewed by Davidson *et al.*, 1975), with the exception of *Drosophila* (Manning *et al.*, 1975), a slime mold (Firtel and Kindle, 1976), and the angiosperm species discussed previously all have similar sequence organization of interspersed nonrepetitive and repetitive DNA. In most of these organisms the interspersed repetitive elements are short 300 to 500 nucleotide sequences. The total complexity of non-repetitive sequence varies over two orders of magnitude; however, measurement of the transcribed sequences is required in several species in addition to sea urchins (Galau *et al.*, 1976) to determine the number of functional genes in a species.

Sequence interspersion raises several questions about the evolution of genes. Do genes and adjacent repetitive sequences coevolve? Do regulators change chromosomal location during crossing-over? Are neighboring repetitive sequences related and derived from a progenitor sequence, and have these sequences evolved more slowly than the contiguous non-repetitive sequences? Do related species retain more or less homology between their repetitive or non-repetitive DNA classes?

Plants offer favorable material for an examination of these questions and others raised in the text. The existence of different ploidy of individual chromosomes and tolerance of aneuploidy and alloploidization suggest that a wide variety of artificial chromosome sets could be constructed.

Plant cytogeneticists have described several unusual chromosome phenomena that deserve re-examination using new methodology. The existence of interspersion in so many diverse taxa suggests that discoveries and descriptions using plant material will be of general applicability in eucaryotic organisms.

ACKNOWLEDGMENTS

The experimental results described here were obtained while the author was a NIH postdoctoral fellow in the laboratory of Dr. Leon S. Dure III at the university of Georgia.

Special thanks are due to Drs. R. Goldberg, A. Hepburn, W. Gurley and J. L. Key for communication of their results prior to publication; the author apologizes to them and to the reader for the organization of the comparative tables which proved a difficult task since a variety of experimental approaches were used to answer the same questions. A more complete description of each genome and its organization will be available with the publication of each author's results.

REFERENCES

Bendich, A. J., and Anderson, R. S. (1974). Novel properties of satellite DNA from muskmelon. *Proc. Nat. Acad. Sci. U.S.* *71*, 1511-1515.

Bendich, A. J., and McCarthy, B. J. (1970). DNA comparisons among barley, oats, rye, and wheat. *Genetics 65*, 545-565.

Birnie, G. D., MacPhail, E., Young, B. D., Getz, M. J., and Paul, J. (1974). The diversity of the messenger RNA population in growing friend cells. *Cell Differentiation 3*, 221-232.

Bishop, J. O., and Freeman, K. B. (1973). DNA sequences neighboring the duck hemoglobin genes. *Cold Spring Harbor Symp. Quant. Biol. 38*, 707-716.

Bishop, J. O., and Rosbash, M. (1973). Reiteration frequency of Duck haemoglobin genes. *Nature, New Biology 241*, 204-207.

Bishop, J. O., Morton, J. G., Rosbash, M., and Richardson, M. (1974). Three abundance classes in HeLa cell messenger RNA. *Nature 250*, 199-204.

Botchan, M., Kram, R., Schmid, C. W., and Hearst, J. E. (1971). Isolation and chromosomal localization of highly repeated DNA sequences in *Drosophila melanogaster*. *Proc. Nat. Acad. Sci. U.S. 68*, 1125-1129.

Brady, T. (1973). Feulgen cytophotometric determination of the DNA content of the embryo proper and suspensor cells of *Phaseolus coccineus*. *Cell Differentiation 2*, 65-75.

Brady, T., and Clutter, M. F. (1973). Cytolocalization of ribosomal cistrons in plant polytene chromosomes. *J. Cell Biol. 53*, 827-832.

Braselton, J. P. (1971). The ultrastructure of the non-localized kinetochores of *Luzula* and *Cyperus*. *Chromosoma 36*, 89-99.

Britten, R. J., and Davidson, E. H. (1969). Gene regulation for higher cells: A theory. *Science 165*, 349-357.

Britten, R. J., and Davidson, E. H. (1976). Studies on nucleic acid reassociation kinetics: empirical equations describing DNA reassociation. *Proc. Nat. Acad. Sci. U.S. 73*, 415-419.

Britten, R. J., Graham, D. E., and Neufeld, B. R. (1974). Analysis of repeating DNA sequences by reassociation. In *Methods in Enzymology* (Grossman, L., and Moldave, K., eds.), *29*, 363-365. Academic Press, New York.

Britten, R. J., and Kohne, D. E. (1967). Nucleotide sequence repetition in DNA. *Carnegie Inst. Wash. Yrbk. 65*, 78-106.

Britten, R. J., and Kohne, D. E. (1968). Repeated sequences in DNA. *Science 161*, 529-540.

Britten, R. H., and Kohne, D. E. (1969). Repetition of nucleotide sequences in chromosomal DNA. In *Handbook of Molecular Cytology* (Lima-de-Faria, A., ed.), pp. 21-36. North Holland, Amsterdam.

Brown, W. V., and Bertke, E. M. (1974). In *Textbook of Cytology*, Second Edition, p. 361. Mosby, St. Louis.

Cairns, J. (1963). The chromosome of *Escherichia coli*. *Cold Spring Harbor Symp. Quant. Biol. 28*, 43-46.

Chilton, M.-D., and McCarthy, B. J. (1973). DNA from maize with and without B chromosome: a comparative study. *Genetics 74*, 605-614.

Clarkson, S. G., Birnstiel, M. L., and Purdom, I. F. (1973). Clustering of transfer RNA genes of *Xenopus laevis*. *J. Mol. Biol. 79*, 411-429.

Commerford, S. L. (1971). Iodination of nucleic acids *in vitro*. *Biochem. 10*, 1993-1999.

Corneo, G., Ginelli, E., and Polli, E. (1970). Different satellite deoxyribonucleic acids of guinea pig and ox. *Biochem. 9*, 1565-1570.

Cullis, C. A., and Schweizer, D. (1974). Repetitious DNA in some *Anemone* species. *Chromosoma 44*, 417-421.

Dale, R. M. K., Martin, E., Livingston, D. C., and Ward, C. D. (1975). Direct covalent mercuration of nucleotides and polynucleotides. *Biochem. 14*, 2447-2457.

Dale, R. M. K., and Ward, D. C. (1975). Mercurated polynucleotides: new probes for hybridization and selective polymer fractionation. *Biochem. 14*, 2458-2469.

Davidson, E. H., and Britten, R. J. (1973). Organization, transcription, and regulation in the animal genome. *Quart. Rev. Biol. 48*, 565-613.

Davidson, E. H., Hough, B. R., Amenson, C. S., and Britten, R. J. (1973). General interspersion of repetitive with non-repetitive sequence elements in the DNA of *Xenopus*. *J. Mol. Biol. 77*, 1-23.

Davidson, E. H., Hough, B. R., Klein, W. H., and Britten, R. J. (1975). Structural genes adjacent to interspersed repetitive DNA sequences. *Cell 4*, 217-238.

Davis, R. W., Simon, M., and Davidson, N. (1971). Electron microscope heteroduplex methods for mapping regions of base sequence homology in nucleic acids. In *Methods in Enzymology* (Grossman, L., and Moldave, K., eds.), *21*, 413-417. Academic Press, New York.

Doty, P., Marmur, J., Eigner, J., and Schildkraut, C. (1960). Strand separation and specific recombination in deoxyribonucleic acids: physical chemical studies. *Proc. Nat. Acad. Sci. U.S. 46*, 461-476.

Dover, G. A. (1975). The heterogeneity of b-chromosome DNA: No evidence for a b-chromosome specific repetitive DNA correlated with b-chromosome effects on meiotic pairing in the triticinae. *Chromosoma 53*, 153-173.

Duerksen, J. D., and McCarthy, B. J. (1971). Distribution of deoxyribonucleic acid sequences in fractionated chromatin. *Biochem. 10*, 1471-1477.

Eckhardt, R. A., and Gall, J. G. (1971). Satellite DNA associated with heterochromatin in *Rhynchosciara*. *Chromosoma 32*, 407-427.

Firtel, R. A., and Kindle, K. (1975). Structural organization of the genome of the cellular slime mold *Dictyostelium discoideum*: interspersion of repetitive and single-copy DNA sequences. *Cell 5*, 401-411.

Flavell, R. B., Bennet, M. D., Smith, J. B., and Smith, D. B. (1974). Genome size and the proportion of repeated nucleotide sequence DNA in plants. *Biochem. Genet. 12*, 257-269.

Fry, K., Poon, R., Whitcome, P., Idriss, J., Salser, W., Mazrimas, J., and Hatch, F. (1973). Nucleotide sequence of HS-β satellite DNA from Kangaroo Rat *Dipodomys ordii*. *Proc. Nat. Acad. Sci. U.S. 70*, 2642-2646.

Galau, G. A., Britten, R. J., and Davidson, E. H. (1974). A measurement of the sequence complexity of polysomal messenger RNA in sea urchin embryos. *Cell 2*, 9-20.

Galau, G. A., Klein, W. H., Davis, M. M., Wold, B. J., Britten, R. J., and Davidson, E. H. (1976). Structural gene sets active in embryos and adult tissues of the sea urchin. *Cell 7*, 487-505.

Gall, J. G., and Atherton, D. D. (1974). Satellite DNA sequences in *Drosophila virilis*. *J. Mol. Biol. 85*, 633-664.

Gall, J. G., Cohen, E. H., and Polan, M. L. (1971). Repetitive DNA sequences in *Drosophila*. *Chromosoma 33*, 319-344.

Gerstel, D. U., and Phillips, L. L. (1957). Segregation in new allopolyploids of gossypium. II. Tetraploid combinations. *Genetics 42*783-797.

Getz, M. J., Altenburg, L. C., and Saunders, G. F. (1972). The use of RNA labeled *in vitro* with iodine -125 in molecular hybridization experiments. *Biochim. Biophys. Acta 287*, 485-494.

Goldberg, R. B. (1978). DNA sequence organization in the soybean plant. *Biochem. Genet. 16*, 45-68.

Goldberg, R. B., Galau, G. A., Britten, R. J., and Davidson, E. H. (1973). Nonrepetitive DNA sequence representation in sea urchin embryo messenger RNA. *Proc. Nat. Acad. Sci. U.S. 70*, 3516-3520.

Goldberg, R. B., Crain, W. R., Ruderman, J. V., Moore, G. P., Barnett, T. R., Higgins, R. C., Gelfand, R. A., Galau, G. A., Britten, R. J., and Davidson, E. H. (1975). DNA sequence organization in the genomes of five marine invertebrates. *Chromosoma 51*, 225-251.

Graham, D. E., Neufeld, B. R., Davidson, E. H., and Britten, R. J. (1974). Interspersion of repetitive and non-repetitive DNA sequences in the sea urchin genome. *Cell 1*, 127-137.

Grant, V. (1975). *Genetics of Flowering Plants*. Columbia University Press, New York.

Gurley, W. B., Angus, G., Hepburn, A. G., and Key, J. L. (1976). Organization of soybean DNA. III. Interspersion of unique and repetitive sequences. *Plant Physiol. 57*, 14 (abst.).

Harris, B., and Dure, L. S., III. (1974). Differential effects of 3'-deoxynucleosides on RNA synthesis in cotton cotyledons. *Biochem. 13*, 5463-5467.

Harris, S. E., Means, A. R., Mitchell, W. M., and O'Malley, B. W. (1973). Synthesis of [^{3}H]DNA complementary to ovalbumin messenger RNA: evidence for limited copies of the ovalbumin gene in chick oviduct. *Proc. Nat. Acad. Sci. U.S. 70*, 3776-3780.

Harrison, P. R., Birnie, G. D., Hell, A., Humphries, S., Young, B. D., and Paul, J. (1974). Kinetic studies of gene frequency. *J. Mol Biol. 84*, 539-554.

Hepburn, A. G., Gurley, W. B., and Key, J. L. (1976). Organization of soybean DNA. I. Kinetic analysis of total and Cs_2SO_4/Hg^2 + fractionated DNA. *Plant Physiol. 57*, 13 (abst.).

Ingle, J., Pearson, G. G., and Sinclair, J. (1973). Species distribution and properties of nuclear satellite DNA in higher plants. *Nature, New Biology 242*, 193-197.

Jones, K. W. (1970). Chromosomal and nuclear location of mouse satellite DNA in individual cells. *Nature 225*, 912-915.

Levy, W. B., and McCarthy, B. J. (1975). Messenger RNA complexity in *Drosophila melanogaster*. *Biochem. 14*, 2440-2446.

Lewin, B. (1974). *Gene Expression*, Vol. 2. Wiley, New York.

Lewin, B. (1975). Units of transcription and translation: sequence components of heterogeneous nuclear RNA and messenger RNA. *Cell 4*, 77-93.

Mandel, M., and Marmur, J. (1968). Use of ultraviolet absorbance-temperature profile for determining the guanine plus cytocine content of DNA. In *Methods in Enzymology* (Grossman, L., and Moldave, K., eds.), *21B*, pp. 195-206. Academic Press, New York.

Manning, J. E., Schmid, C. W., and Davidson, N. (1975). Interspersion of repetitive and nonrepetitive DNA sequences in the *Drosophila melanogaster* genome. *Cell 4*, 141-155.

McColl, R. S., and Aronson, A. I. (1974). Transcription from unique and redundant DNA sequences in sea urchin embryos. *Biochem. Biophys. Res. Commun. 56*, 47-51.

Millerd, A., and Whitfeld, P. R. (1973). Deoxyribonucleic acid and ribonucleic acid synthesis during the cell expansion phase of cotyledon development in *Viscia faba* L. *Plant Physiol. 51*, 1005-1010.

Noll, H. (1967). Characterization of macromolecules by constant velocity sedimentation. *Nature 215*, 360-363.

Nze-Ekekang, L., Patillon, M., Schafer, A., and Kovoor, A. (1974). Repetitive DNA of higher plants. *J. Exp. Bot 25*, 320-329.

Ohno, S. (1971). Simplicity of mammalian regulatory systems inferred by single gene determination of sex phenotypes. *Nature 234*, 134-137.

Orosz, J. M., and Wetmur, J. G. (1974). *In vitro* iodination of DNA. Maximizing iodination while minimizing degradation; use of buoyant density shifts for DNA-DNA hybrid isolation. *Biochem. 13*, 5467-5473.

Pardue, M. L. (1975). Evolution of DNA content in higher plants. *Genetics 79*, 159-170.

Pardue, M. L. and Gall, J. G. (1970). Chromosomal localization of mouse satellite DNA. *Science 168*, 1356-1358.

Pearson, G. G., Timmis, J. N., and Ingle, J. (1974). The differential replication of DNA during plant development. *Chromosoma 45*, 281-294.

Prensky, W., Steffensen, D. M., and Hughes, W. L. (1973). The use of iodinated RNA for gene localization. *Proc. Nat. Acad. Sci. U.S. 70*, 1860-1864.

Price, H. J. (1976). Evolution of DNA content in higher plants. *Botanical Rev. 42*, 27-52.

Ramirez, S. A., and Sinclair, J. H. (1975). Ribosomal gene localization and distribution (arrangement) within the nucleolar organizer region of *Zea mays*. *Genetics 80*, 505-518.

Ranjekar, P. K., Lafontaine, J. G., and Pallotta, D. (1974). Characterization of repetitive DNA in rye (*Secale cereale*). *Chromosoma 48*, 427-440.

Rhoades, M. M., and Vilkomerson, H. (1942). On the anaphase movement of chromosomes. *Proc. Nat. Acad. Sci. U.S. 28*, 433-436.

Rimpau, J., and Flavell, R. B. (1975). Characterization of rye b chromosome DNA by DNA/DNA hybridization. *Chromosoma 25*, 96-103.

Rothfels, K., and Heimburger, M. (1968). Chromosome size and DNA values in sundews (*Droseraceae*). *Chromosoma 25*, 96-103.

Ryffel, G. U., and McCarthy, B. J. (1975). Complexity of cytoplasmic RNA in different mouse tissues measured by hybridization of polyadenated RNA to complementary DNA. *Biochem. 14*, 1379-1385.

Sivolap, Y. M., and Bonner, J. (1971). Association of chromosomal RNA and repetitive DNA. *Proc. Nat. Acad. Sci. U.S. 68*, 387-389.

Skinner, D. M., and Beattie, W. G. (1973). $CsSO_4$ gradients containing both Hg^{2+} and Ag^+ effect the complete separation of satellite deoxyribonucleic acids having identical densities in neutral CsCl gradients. *Proc. Nat. Acad. Sci. U.S. 70*, 3108-3110.

Smith, M. J., Britten, R. J., and Davidson, E. H. (1975). Studies on nucleic acid reassociation kinetics: reactivity of single-stranded tails in DNA-DNA renaturation. *Proc. Nat. Acad. Sci. U.S. 72*, 4805-4809.

Southern, E. M. (1970). Base sequence and evolution of guinea-pig α-satellite DNA. *Nature 227*, 794-798.

Sparrow, A. H., Price, H. J., and Underbrink, A. G. (1972). A survey of DNA content per cell and per chromosome of prokaryotic and eukaryotic organisms: some evolutionary considerations. *Brookhaven Nt. Lab. Symp. Biol. 23*, 451-494.

Ståhle, U., Lima-de-Faria, A., Chatnekar, R., Jaworska, H., and Janley, M. (1975). Satellite DNA, localization of ribosomal cistrons and heterochromatin in *Haplopappus gracilis*. *Hereditas 79*, 21-28.

Studier, F. W. (1965). Sedimentation studies of the size and shape of DNA. *J. Mol. Biol. 11*, 373-390.

Suzuki, Y., Gage, L. P., and Brown, D. D. (1972). The genes for silk fibrion in *Bombyxmori*. *J. Mol. Biol. 70*: 637-649.

Tartof, K. D. (1975). Redundant genes. *Ann. Rev. Genetics 9*, 355-385.

Walbot, V., and Dure, L. S. (1976). Developmental biochemistry of cotton seed embryogenesis and germination. *J. Mol. Biol. 101*, 503-536.

Walker, P. M. B. (1971). Repetitive DNA in higher organisms. *Progr. Biophys. Mol. Biol. 23*, 145-190.

Waring, M., and Britten, R. J. (1966). Nucleotide sequence repetition: a rapidly reassociating fraction of mouse DNA. *Science 154*, 791-794.

Williams, J. G., and Penman, S. (1975). The messenger RNA sequences in growing and resting mouse fibroblasts. *Cell 6*, 197-206.

Wilson, D. A., and Thomas, C. A., Jr. (1973). Hydroxyapatite chromatography of short double-helical DNA. *Biochem. Biophys. Acta 331*, 333-340.

Wilson, D. A., and Thomas, C. A., Jr. (1974). Palindromes in chromosomes. *J. Mol. Biol. 84*, 115-144.

Wimber, D. E., Duffey, P. A., Steffenson, D. M., and Prensky, W. (1974). Localization of the 5S RNA genes in *Zea mays* by RNA-DNA hybridization *in situ*. *Chromosoma 47*, 353-359.
Yasmineh, W. G., and Yunis, J. J. (1971). Satellite DNA in calf heterochromatin. *Exp. Cell Res. 64*, 41-48.
Yunis, J. J., and Yasmineh, W. G. (1970). Satellite DNA in constitutive heterochromatin of the guinea pig. *Science 168*, 263-265.
Zimmerman, J. L., and Goldberg, R. B. (1977). DNA sequence organization in the genome of *Nicotiana tabacum*. *Chromosoma 59*, 227-252.

REPLICATION OF CHROMOSOMAL DNA FIBERS OF ROOT MERISTEM CELLS OF HIGHER PLANTS

J. Van't Hof
C. A. Bjerknes
J. H. Clinton

Biology Department
Brookhaven National Laboratory
Upton, New York

The successful completion of a normal cell cycle simultaneously marks the doubling of the cell number and the equal distribution of genetic information to each of the recipient daughter cells. Success is by no means left to chance. The parental cells when born at some earlier mitosis progress through an ordered sequence of physiological and biochemical states that assures an adequate inheritance to their progeny. This progression is well defined temporally and has two positions of reference that can be measured experimentally. The references are the functions of DNA synthesis and mitosis which are bracketed by two other time periods that are called G1 and G2 (Howard and Pelc, 1953). The former is the period before and the latter the period after completion of chromosomal DNA replication. The sequence of states through which more eucaryotic cells proceed as they copy, distribute, and perpetuate genetic information can be symbolized as M→G1→S→G2→M, etc., where M and S refer to the periods of mitosis and DNA synthesis respectively. While many factors must function in a cooperative manner to assure the birth of normal cells, none is more important than the genetic code itself. During S the information of each strand of the DNA duplex of the parental chromatid is copied in a complementary daughter strand and when completed a chromosome results. Each chromatid of the chromosome has a DNA duplex consisting of one parental and one newly synthesized polynucleotide strand and each daughter cell at mitosis is destined to receive one of the chromatids. The

ISBN 012-601950-9

focal point of this paper is the replication unit that is responsible for the formation or growth of the DNA duplex in the chromatid.

The term *replicon* was used by Taylor (1963) to describe segments of DNA that replicate as units in a manner similar to that proposed by Jacob and Monod (1961 as quoted by Taylor 1963) for bacteria. Autoradiography of DNA fibers of *Escherichia coli* (Prescott and Kuempel, 1972) and *Bacillus subtilis* (Gyurasits and Wake, 1973; Wake, 1975) showed that the chromosome replicated via two moving forks that diverged in opposite directions from a single position on the circular DNA duplex. The position where replication began was designated the *origin* and the point at which the two forks converge and finally fuse was called hypothetically the *terminus*. The term *fork* is descriptive of the spatial arrangement often assumed by the parental and two newly replicated daughter DNA fibers. The pattern on an autoradiograph of isolated DNA molecules often resembles that of the letter Y with the direction of fork movement being indicated by the apex of the crotch. The apex is also the site of action of the multi-enzyme replication complex that is responsible for DNA fiber growth (Kornberg, 1974). In this paper, the origin and terminus of a replicon will refer to the positions on the DNA molecule where replication begins and ends as suggested by Edenberg and Huberman (1975). Replicon size is defined as the distance between adjacent origins on a given DNA molecule. In procaryotes, *E. coli* and *B. subtilis*, one replicon equals one chromosome but this 1:1 relation does not apply to eucaryotes.

Cells of all higher organisms examined to date have long chromosomal DNA fibers that are replicated in tandemly joined sections each of which has an initiation point or origin. In 1963 Taylor postulated that unlike bacteria, eucaryotes replicate chromosomes by "...a tandem-linked series of 'replicons', segments of DNA which replicate as units..." While Taylor's early notion was based on the premise that DNA duplex segments were linked by 3' R and 5'R units, the idea that the chromatid had asynchronous, and therefore multiple, replication sites was firmly grounded on cytological observations (Taylor, 1958; Lima-de-Faria, 1959; Wimber, 1961; Keyl and Pelling, 1963). When Cairns applied his autoradiographic technique (Cairns, 1963) to replicating DNA of HeLa cells, he saw evidence of multiple replication activity tandemly arranged on the molecules (Cairns, 1966). This early observation of Cairns was extended and expanded by Huberman and Riggs (1968) in their experiments with Chinese hamster cells. They showed that mammalian chromosomal DNA fibers had multiple, tandemly arranged replication sites, that DNA chain growth occurred in a bidirectional manner, and that not all forks were active simultaneously. In the opinion of Huberman and Riggs, DNA fiber replication was initiated at the origin of a replicon by two forks that diverged, moving in opposite directions, toward different termini. Thus, unlike *E. coli* or *B. subtilis*, the replication forks of mammalian DNA share an origin but not a terminus. The terminus, in fact, is shared by forks of adjacent replicons.

The size of replicons of eucaryotes is much less than that of bacteria. For example, the replicon and chromosome of *B. subtilis* is 900-1100 μm (Wake, 1974, 1975) but in Chinese hamster chromosomes the replicon can be as small as 14 and as large as 106 μm (Huberman and Riggs, 1968). Most replicons, however, were 15-60 μm and the modal size was 30-40 μm. Very small replicons are observed in the cleavage nuclei of dipteran eggs. Wolstenholme (1973) estimated a mean size of 3.87 μm in *Drosophila* DNA, while Lee and Pavan (1974) reported an average of 2.3 μm in the DNA of *Cochliomyia* cleaving eggs. Judging from these observations alone, it is evident that replicons of eucaryotes vary in size. In fact, Blumenthal *et al.* (1974) measured a change in size of over 6-fold in *Drosophila*. In cleaving eggs the mean replicon size was 2.6 μm but in cultured somatic cells there were two groups, one of 9 and another of 19 μm. Likewise, Callan (1972) also observed a change in replicon size in the DNA of cells of different tissues of *Triturus viridescens*. Given the observation that replicon size varies, it follows that not all potential origins are active and that only under special conditions, such as those that occur in the cleaving egg, do they all function. If the origins function at one time and not at another, one cannot help but question the specificity of an origin and a terminus in the eucaryotic chromosome. Unlike bacteria, where an inactive or impaired origin results in cell sterility, in higher organisms the segment of DNA alloted to an inactive replicon is replicated by a neighboring fork. The result of inactivity is neither impaired DNA replication nor cell sterility but a larger, measurable replicon.

The rate at which a fork moves along the parental DNA duplex may vary among different replicons of the chromosomes in eucaryotes and even during replication in bacteria. Wake (1975) using a thymine requiring strain, 168 of *B. subtilis*, noted that the fork rate increased by a factor of 2-3 during the first half of the round of replication. In mammalian cells the situation is reversed. Both HeLa and Chinese hamster cells have faster moving forks in the latter portion of S (Painter and Schaefer, 1971; Housman and Huberman, 1975).

The experiments performed with eucaryotic DNA have disclosed a number of characteristics that apply to chromosomal fiber replication. However, none of these was shown to be the property of higher plants until recently. The list below summarizes the characteristics and is used as an outline for the discussion to follow. Other topics covered concern the multiplicity of replicon sizes of plant DNA and the average replicon size of unrelated species with different amounts of nuclear DNA.

Some Characteristics of Eucaryote Chromosomal DNA Fiber Replication as Visualized in Plants

The first characteristic is that chromosomal DNA is replicated semi-conservatively. This characteristic was demonstrated over 20 years ago by Taylor, Woods, and Hughes (1957) in their classical experiments with *Vicia faba* root-tip cells. Their experiment heralded modern molecular cytogenetics and provided strong evidence for the uninemic structure of the chromatid.

The second characteristic is that chromosomal DNA fibers replicate by a multitude of replication units (replicons) tandemly spaced along the molecule. The early work of Taylor (1958) with *Crepis capillaris*, Lima-de-Faria (1959) with *Secale* and Wimber (1961) with *Tradescantia paludosa*, provided broad evidence for multiple replication sites on the chromosome. The size and quantity of these sites, however, awaited the application of DNA fiber autoradiography to plant material. In 1975b Van't Hof described a technique for radioactive labeling and isolation of DNA molecules from dividing cells of the root-tip of *Pisum sativum* (pea) that made possible the visualization of replicating plant DNA. The autoradiographs showed that the chromosomal DNA fibers of pea were replicated by a multitude of tandem forks.

In brief, the experiments involved a pulse of tritiated thymidine that was sufficient to flood the precursor pool with isotope. As soon as the pulse ended the nuclei were isolated, enzymatically treated with either pronase or trypsin, and lysed with detergent. The nuclear DNA was spread over the surface of a microscope slide and dried. After washing with trichloroacetic acid and ethanol, the slides were covered with liquid photographic emulsion. On the developed autoradiographs, the segment of DNA that replicated during the pulse is registered by an array of contiguous silver grains that are located along the stretched DNA duplex. In Fig. 1 and 3 are photomicrographs of autoradiographs of replicating DNA of pea, and those of Fig. 2 show arrays of grains produced by labeled DNA isolated from *Helianthus annuus* (sunflower) nuclei. The DNA of both species, as in the case of all other plants examined to date, show that replication occurs at multiple sites along the DNA duplex.

The third characteristic of eucaryotic DNA fiber replication is that it is bidirectional. The direction of fiber growth can be demonstrated unambiguously by either a step-down or a step-up pulse label protocol (Huberman and Riggs, 1968; Huberman and Tsai, 1973; Callan, 1972). So far only the step-down protocol has been successful with higher plants. The reason for this is unknown, but it probably involves DNA precursor pool sizes. The term step-down refers to the use of two sequential pulses of tritiated thymidine, the first being at least three times the specific activity of the second (Huberman and Tsai, 1973). The DNA that is replicated during the first pulse is more radioactive and causes a higher density of silver grains on the autoradiograph than that produced by that portion of the molecule replicated during the pulse with the lower specific activity isotope. The relative positions of the higher and lower grain densities on an array indicates the direction of fork movement during the pulses. If an origin is activated during the first pulse and if the forks diverge in opposite directions, the array will have a high grain density in the middle with tails of decreasing density on either side. Also, if initiation occurred just before the first pulse, the dense middle portion will be split by an area with very few or no grains. However, each dense portion will have a tail that leads in opposite directions. The autoradiographs shown in Fig. 1 are examples of grain density changes produced by a step-down protocol of 45 minutes high and 90 minutes low specific activity pulses. Upon examination of the arrays of grains, one will note that in addition to the density patterns described above there are two others. These have no grain density gradients and are either very dense or light. The dense arrays are interpreted to result from DNA replicated exclusively during the high specific activity pulse and those of a lesser density without gradients, the DNA synthesized during the lower specific activity pulse.

If the autoradiographs of pea DNA shown in Fig. 1 are interpreted correctly, then a comparison of grain array patterns produced by a step-down protocol should differ from those observed after a simple pulse (Huberman and Riggs, 1968). The autoradiographs of replicating DNA of sunflower shown in the photomicrographs of Fig. 2 demonstrate this difference. The grain arrays of Fig. 2a were produced by a step-down protocol of 30 minutes high, 60 minutes low specific activity pulses. Those shown in Fig. 2b were produced by a simple pulse of 90 minutes with high specific-activity isotope. The arrays produced by a step-down pulse have grain density gradients with the expected pattern, if fork movement were bidirectional, *i.e.*, the central dense portions are flanked by tails of lesser grain density. The simple pulse, in contrast, produced arrays

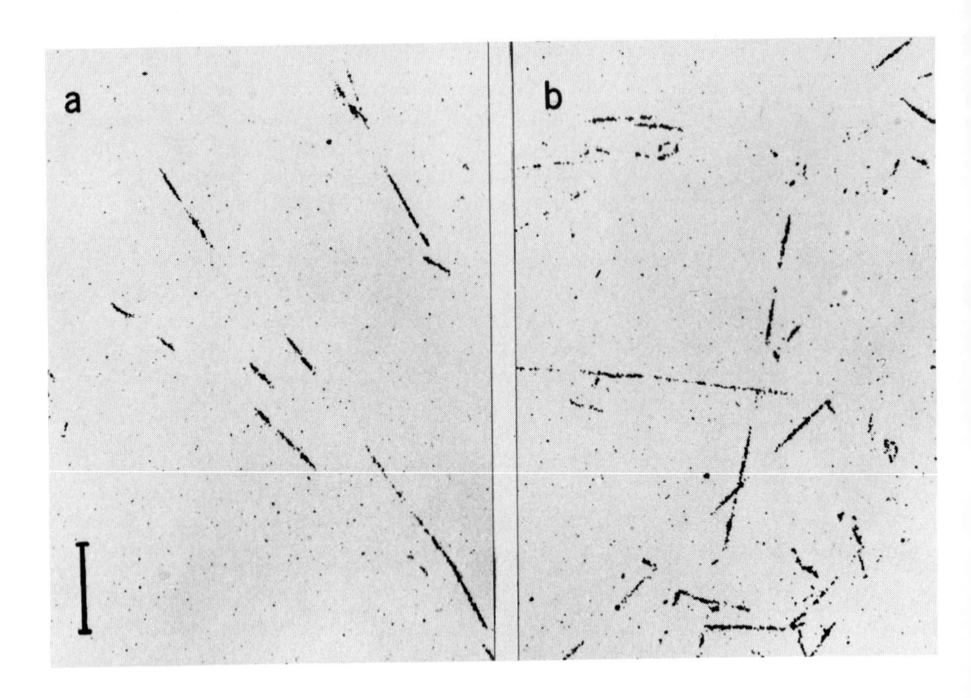

FIGURE 1. Autoradiographs of labeled chromosomal DNA fibers isolated from cells of cultured pea roots that were pulsed with tritiated thymidine by a step-down protocol of 45 minutes high, 90 minutes low specific activity. The cells were in early S when labeled. The bar scale is 50 μm.

of nearly uniform high density. The grain density is less only when the daughter duplex molecules are separated as shown in the lower right hand portion of Fig. 2b. The autoradiographic data of both pea and sunflower DNA show that much of the chromosomal fiber of these species replicated bidirectionally.

The fourth characteristic of chromosomal fiber replication is that the rate of fork movement of different replicons may not be equal. Because DNA replication occurs at multiple sites and because the replicons that are active early in S may not function later on, this fourth characteristic must be viewed in two ways. The first comparison is between contemporary replicons that are active within a given portion of S; the second, is between replicons that are active at different times in S. To compare replicons at a particular time in S, it is essential to selectively label DNA in cells that are positioned temporally in the portion of interest. In higher plants, tissue with no cells in S can be obtained by the cul-

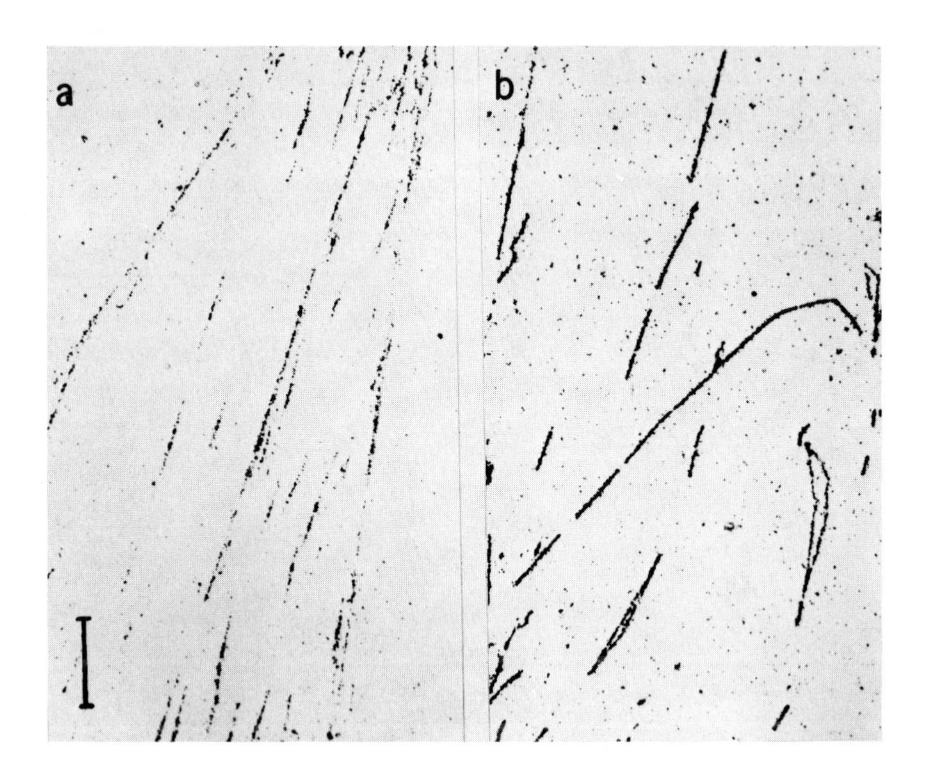

FIGURE 2. Autoradiographs of labeled DNA fibers isolated from asynchronous cells of cultured sunflower roots. (a) Tandem arrays of grains with density gradients produced by the step-down labeling protocol of 30 minutes high, 60 minutes low specific-activity tritiated thymidine, (b) tandem arrays produced by a single 90 minute pulse with high specific-activity tritiated thymidine. The bar scale is 50 μm.

ture of excised root-rips in medium lacking carbohydrate (Van't Hof, 1966, 1975a). When fed sugar the starved cells leave G1 and enter S. If tritiated thymidine is present when the cells cross the G1/S boundary, the DNA fibers replicated in early S can be selectively labeled. The autoradiographs shown in Figs. 1 and 3 are of DNA fibers isolated from cells of cultured pea roots that were labeled in early S. The relative rate of movement between forks of different replicons is seen in the length of the grain arrays: the longer the array, the faster the fork. Absolute fork rates may be estimated by dividing the length of the array by the duration of the tritiated thymidine pulse. The value determined, however, is an estimate because of inherent errors in the technique such as

TABLE 1
Replication Size of Cells of *Pisum Sativum* in Different Portions of S Measured by Autoradiography of Isolated Chromosomal DNA Fibers

Portion of S	Replicon Size (μm)			
	Range	Mode	Mean	% >50 μm
Late[a]	15-75	30-35	38	11
Early[b]	5-95	30-35	40	19
Early to middle[b]	15-125	30-40	49	40
Asynchronous[c]	20-105	50-55	55	61

[a] *Van't Hof (1976a)*
[b] *Van't Hof (1976b)*
[c] *Van't Hof (1975b)*

the time interval between immersion in isotope and the incorporation of label in the DNA duplex. Nevertheless, the relative lengths of grain arrays on a given fiber (Table 1) differ by a factor of 1.5 or more (Van't Hof, 1975b, 1976a, 1976b).

Even if the contemporary replicons that function in early S have forks that move at different rates, it is possible that, on the average, the rate is dissimilar from that of forks that are active in mid- to late S. In the case of pea root cells there is a strong hint that the more rapid forks are activated in the middle and late portions of S. The source of this evidence is the four curves shown in Fig. 4 where the average length of arrays of grains on DNA fibers is expressed as a function of the duration of the tritiated thymidine pulse. These curves will be discussed in order, moving from left to right. The first curve follows the open squares and the data shown were those previously published for asynchronously dividing cells (Van't Hof, 1975b). The rule to follow for such cell populations is that the shorter pulses produce the more reliable measurements. They minimize the errors introduced by the fusion of adjacent forks and the lack of space between forks that initiated movement during the pulse. In these instances a single array is produced by the movement of two forks. The bend in the curve at 30 minutes is an effect of pulses of excessive duration. If the isotope is available too long, there is an increased number of fused forks, a higher proportion of completely labeled fibers, and fewer scorable tandem arrays. The initial slope of the curve is useful to

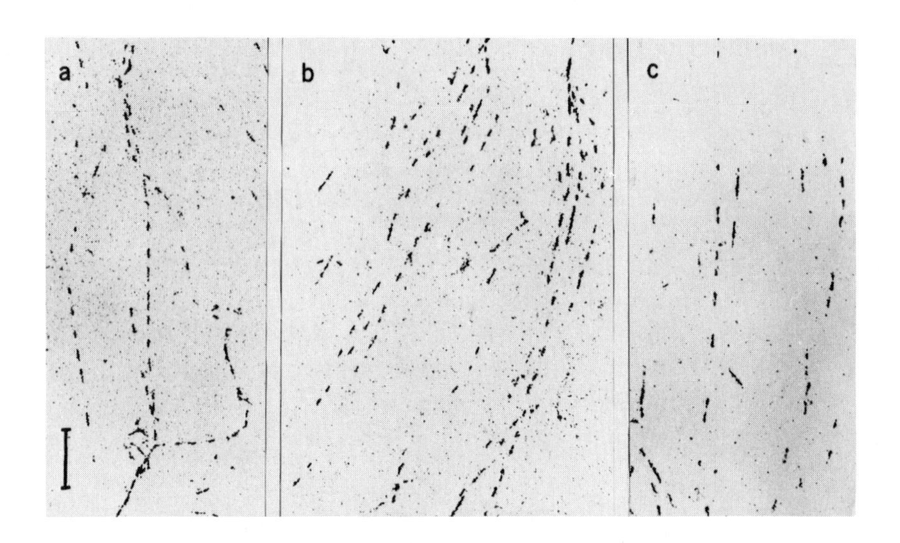

FIGURE 3. Autoradiographs of labeled DNA fibers isolated from pea root-tip cells in early S. (a) DNA of cells labeled by 90 minute pulse with high specific-activity tritiated thymidine; (b and c) DNA of cells labeled by a step-down protocol of 45 minutes high, 90 minutes low specific-activity isotope. Note in b and c the absence of grain density gradients on many arrays indicative of a slow fork rate. The bar scale is 50 μm.

estimate the average fork rate for asynchronously dividing cells and in this case its value is about 29 μm/hr.

The second curve follows the solid triangles and these data also were obtained from asynchronous cells. However, in this case the DNA was isolated during the first round of replication after 72 hours of carbohydrate starvation. The shape of the second curve is essentially the same as the first but the initial slope is slightly but not significantly less indicating an average fork rate of approximately 25 μm/hr.

Lower rates were observed in DNA isolated from cells that were in early S. The third curve in Fig. 4 traces the open circles that represent data from pea root-rip cells that were synchronized at the G1/S boundary by the technique of Kovacs and Van't Hof (1970, 1971). The cells were allowed to progress 30 minutes into S, and then pulsed with tritiated thymidine to label the DNA. Consequently, the points on the curve correspond to the portion of S that is equal to the sum of 30 minutes and the duration of the pulse. For example, the 30 minute point consists of data from cells that were

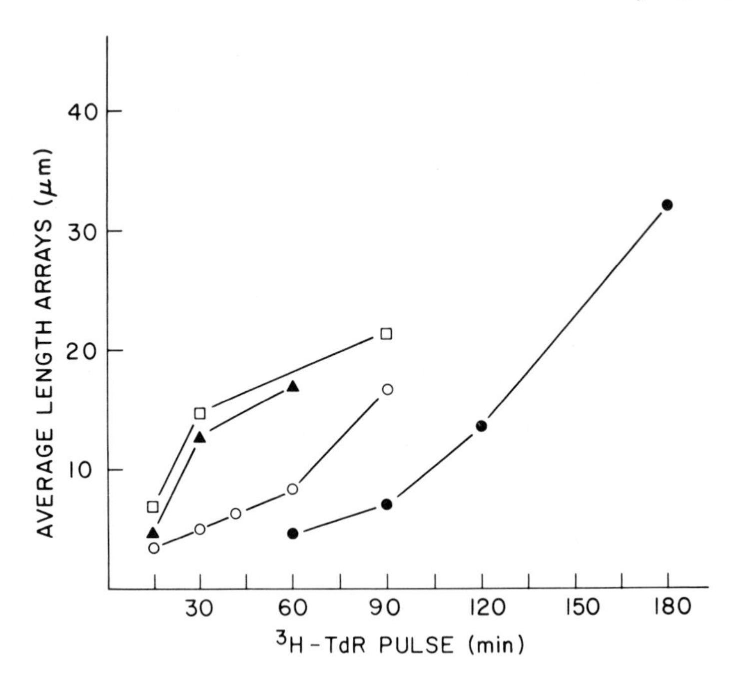

FIGURE 4. The average length of contiguous grain arrays on autoradiographs of DNA fibers from cells of cultured pea roots expressed as a function of the tritiated-thymidine pulse. Open squares, asynchronous cells of roots that were never starved of sugar; closed triangles, asynchronous cells first starved of sugar for 3 days then fed 2% sucrose for 18 hours; open circles, cells in early S after being synchronized at the G1/S boundary; closed circles, cells in early S positioned after 3 days of sugar starvation followed by 6 hours of feeding with 2% sucrose.

one hour into S. Like the previous two curves, the third one is complex and has two parts. The length of DNA replicated increased linearly at a rate of about 6 μm/hr in cells located in the early 45 to 90 minutes of S. Between 90 and 120 minutes into S, fork movement was more rapid and the observed rate was nearly 18 μm/hr.

The fourth and final curve displayed in Fig. 4 (solid circles) represents data obtained from cells that were in G1 and were pulsed as they entered S. The cells, previously starved of carbohydrate for 3 days, must synthesize the RNA and all the proteins needed for initiation of DNA synthesis before they enter S (Webster and Van't Hof, 1970). Consequently, the radioactive pulses were not performed until 6 hours after the tissue was fed sugar. At this time the cells were just beginning to initiate DNA replication (Van't Hof, 1976b). In early S the fork rate was at first slow. Radioactive pulses of 60 to 90 minutes gave a value of approximately 6 μm/hr. Concurrent with labeling, the cells continued to progress into and through S. As they did, the fork rate increased as indicated by the steeper slope of the curve generated between 120 and 180 minute pulses (Fig. 4). The rate at this time was 18 μm/hr and three times greater than that determined from the initial portion of the curve.

The average rate of fork movement increased as the cell population from which the DNA was isolated gradually filled the S period from G1 to G2. The rate in early S averaged 6 μm/hr, that of early to mid-S was 18 μm/hr, and finally a mixed asynchronous population that spanned all of S, had forks that averaged 25 and 29 μm/hr (Fig. 4). The change in DNA fiber growth with the distribution of cells in S offered a basis for the postulate that replicons activated in late S have rapid fork rates. The average rate of replication for DNA isolated from asynchronous cells is high, and that of the cells in early to mid-S, low. The faster forks compensated for the slower ones to produce a higher average replication length in late S; in an asynchronous population late replicating DNA is represented, but it is absent in the DNA of early S.

The fifth characteristic of eucaryotic chromosomal DNA fiber replication is quiescent replicons. As mentioned earlier, the DNA of cleaving eggs of *Drosophila* is estimated to average an activated origin every 3-4 μm but in the DNA of cultured somatic cells of the same species, origins are most frequently 9 to 19 μm apart (Blumenthal *et al.*, 1974). Also, in *Triturus* replicon size changes in DNA isolated from one tissue to the next by a factor of as high as 10 (Callan, 1972). Assuming that the molecular aspects of the genome does not change during cell development and differentiation, the altered replicon size implies that some origins are quiescent in tissues other than the most embryonic. While experiments are presently underway to test this characteristic in higher plants, no data are available currently.

Are There Major Size Groups of Replicons in Higher Plants

Hori and Lark (1974, 1976) described replicons in one species of kangaroo rat (*Dipidomys ordii*) that are absent in another (*D. desertii*). Autoradiographs of fibers first separated on a CsCl gradient showed the distinctive fibers to be associated

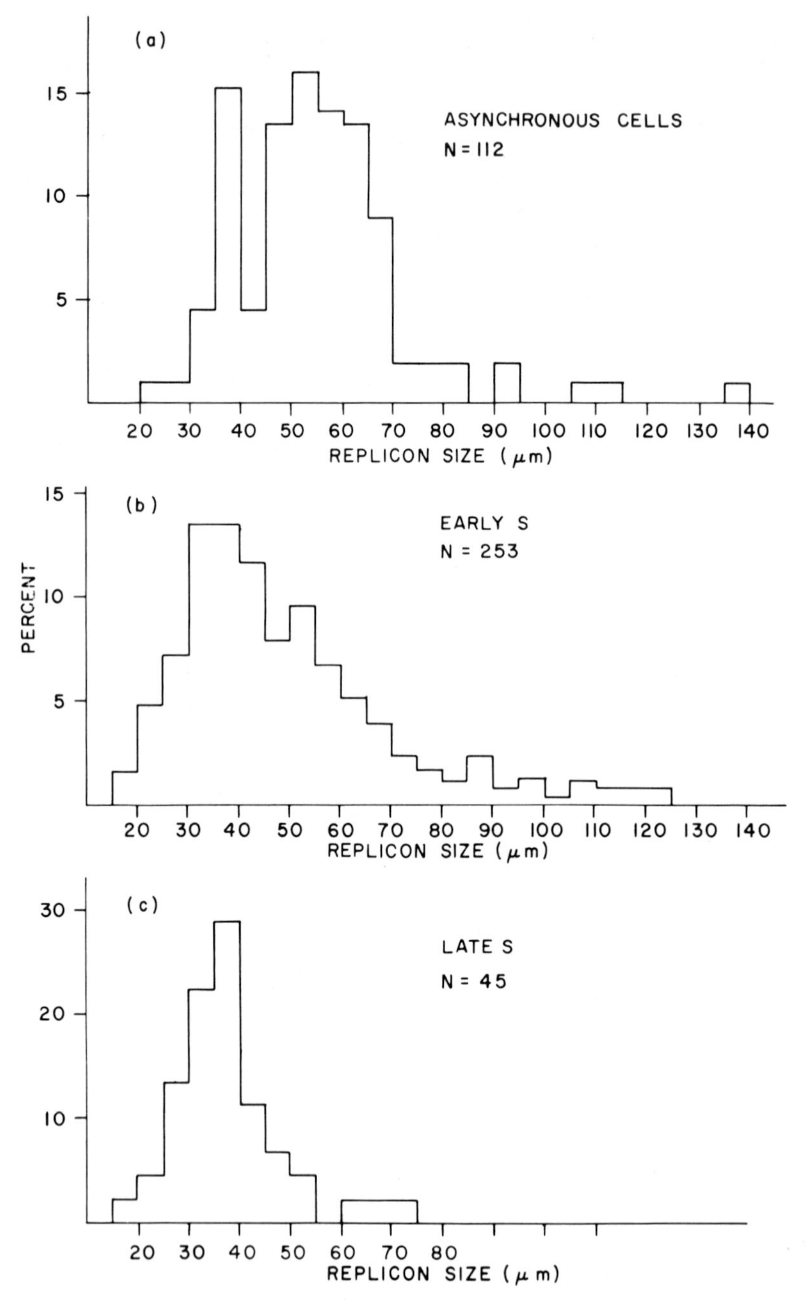

FIGURE 5. Histograms of replicon sizes of chromosomal DNA fibe of peas measured from center-center distances on tandem arrays of grains with density gradients produced by a step-down labeling protocol to (a) asynchronous cells; (b) cells in early to mid S; (c) cells at the end of S. Data from Van't Hof (1975b, 1976a, 1976b). N is the number of measurements.

with satellite DNA. Because the replicons on the unusual fibers are active in late S and because *D. ordii* has much more heterochromatin than *D. desertii*, Hori and Lark believe them to be replicons of heterochromatic DNA. Upon further analysis of their data, these authors found the replicated segments to be defined lengths and multiples of 7 (Hori and Lark, 1976). On a somewhat different scale, Blumenthal *et al.* (1974) detected two major size groups of 9 and 19 μm in the DNA of cultured somatic cells of *Drosophila*. Also, replicating DNA of pea may have multiple groups as indicated by the bimodal distribution of replicon sizes observed in DNA of asynchronous cells (Van't Hof, 1975b). In this case, as shown in Fig. 5a, the higher frequencies are seen between 35 and 40 μm and between 45 and 70 μm with the latter being the more frequent. However, in the DNA of cells in either early or late S (Figs. 5b and 5c), the smaller size is preponderant. Recent work with callus of pea-root explants from cortical and stelar tissue showed that these differentiated cells have replicons that averaged 18 μm (Van't Hof and Bjerknes, 1977). The two average size ranges of approximately 38 and 54 μm of replicons of DNA from the root-tip meristem cells are very close to multiples of 18 and it seems likely that the different size classes are singlets, doublets, and triplets of the 18 μm unit. The data listed in Table 1 summarizes the replicon size measurements of cells in different portions of S.

Does the Average Replicon Size Change With the Amount of Nuclear DNA

The preceding discussion pointed out that the origins of some replicons of dipteran cells are quiescent in somatic cells but activated in nuclei of cleaving eggs. A consequence of inactivity at one time and activation at another is a change in the length of the segment of DNA replicated by a given fork. For example, if two neighboring replicons were activated, their respective converging forks would travel one-half the distance between the origins. On the other hand, if every other replicon were activated, the distance traveled by the converging forks would be the total distance between the

TABLE 2
The 2C Amount of Nuclear DNA (pg)[a] and the Average Replicon Size (μm) of Chromosomal DNA Fibers of Several Unrelated Plant Species as Determined With Tritiated Thymidine By Either a Step-Down or a Single Pulse Protocol. Pulse Duration (min) is Noted in Parenthesis Below Replicon Size. Data from Root-Tip DNA Unless Noted Otherwise.

Species	2C DNA	Replicon size	
		Step-down	Single
Saccharomyces cerevisiae - (yeast)	0.0495[b]	---	30[c] (15)
Arabidopsis thaliana - seedling (var. Columbia)	0.4	---	24 (15)
Crepis capillaris - seedling	2.8	---	24 (20)
Lycopersicon esculentum - (tomato, var. Beefsteak)	3.0	---	24 (30)
Glycine max - (soybean)	4.0	45 (30 hi-30 lo)	25 (30)
Helianthus annuus - (sunflower var. Russian mammoth)	6.3	57 (30 hi-60 lo)	22 (15)
Pisum sativum - (pea, var. Alaska)	7.8	54 (45 hi-30 lo)	33 (20)
Vicia faba - (broad bean var. Longpod)	24.3	16[e] (30 hi-30 lo) 63 (45 hi-90 lo)	98[d] (30)
Allium cepa - (onion, var. Southport yellow globe)	33.0	---	29 (30)

[a]*Determined cytophotometrically and biochemically (Van't Hof, 1965; Bennett and Smith, 1976).*
[b]*Diploid value (Sober, 1970).* [c]*Petes and Williamson, 1975.*
[d]*Gaddipati and Sen, 1978.* [e]*Ikushima, 1976.*

origins. Even though replicon activation is not fully understood, it should be considered when comparisons are made between species. Two good rules to follow in comparative studies are that the DNA fibers be isolated from cells of the same organs grown under similar conditions, and that the duration of the tritiated thymidine pulse or combination of pulses is the shortest possible. As mentioned earlier in the text, the shorter the radioisotope pulse, the more reliable the data. The frequency of replication initiation and termination is more likely to be greater with longer pulses and each situation will exaggerate both the average fork rate and the average replicon size (Van't Hof *et al.*, 1978). The data listed in Table 2 exemplify the different replicon size estimates obtained from DNA that was labeled with either a step-down or a single pulse protocol. The higher plant species listed have 2C nuclear DNA amounts that range from 0.4 to 33 pg and except for one species (*V. faba*) the short, single pulse protocol gave values between 22 and 33 μm for the average replicon size.

The 20-40 μm Replicon, a Commonly Shared Unit Amongst Many Eucaryote Somatic Cells

In 1975 Petes and Williamson using an autoradiographic technique, estimated the replicon size of yeast to be 30.2 μm. This size agreed well with their earlier measurements made via electron microscopy. Because the yeast cell represents one of the simpler eucaryotes, it is plausible that a replicon size of 20-40 μm constitutes a conserved functional basic unit that was elaborated in more complex species. The reason for giving this notion some attention is derived from the fact that the somatic cells of many species either have an average size of 20-40 μm or subgroups that fall into this size range. *Arabidopsis, Crepis* and tomato, species with relatively small amounts of DNA, have average replicon sizes of 24 μm (Table 2). Also, sunflower, a species that has a moderate amount of DNA, has an average size of 22 μm. Other data from the animal kingdom offer further examples of the prevalency of the 20-40 μm unit. In Chinese hamster the average size is about 30 μm (Houseman and Huberman, 1975) and 30 μm is also the size reported for rat DNA (Priestly as mentioned by McFarlane and Callen, 1973). Hand and Tamm (1974) reported replicon sizes in somatic cell lines of five mammlain species (mouse, hamster, bovine, monkey, and human) that averaged respectively, 45, 30, 17, 42, and 22 μm. Because mammalian cells have 2C values of 5-7 pg of nuclear DNA, it is tempting to correlate smaller replicons with small genome sizes, particularly for cells with 2C values less than 8 pg. Callan (1972) reported that the DNA of *Xenopus*, which has a 2C amount of a little more than 6 pg, had an average replicon size of 58 μm, but the measurements also indicated a subgroup in the 20-40 μm range. In the case of chicken DNA, the average replicon size is a rather large 63 μm even though the 2C value is a low 2.8 pg (McFarlane and Callan, 1973). Nevertheless, in this species too, there is a small size subgroup of 20-40 μm.

Whether or not a 20-40 μm unit is conserved in the DNA of eucaryotes remains an open and intriguing question. The information available to date is just enough to motivate further work, particularly with higher plants. For they offer a high degree of DNA heterogeneity and a wide range of genome sizes that are adequate to resolve possible differences in the properties of the replicon.

ACKNOWLEDGMENTS

Research carried out under the auspices of the United States Energy Research and Development Administration. The authors wish to thank Dr. A. Kuniyki for his discussion and suggestions pertaining to the ideas conveyed in this manuscript.

REFERENCES

Bennett, M. D., and Smith, J. B. (1976). Nuclear DNA amounts in angiosperms. *Phil. Trans. Roy. Soc.* London (B) *247*, 227-274.

Blumenthal, A. B., Kriegstein, H. J., and Hogness, D. S. (1974). The units of DNA replication in *Drosophila melanogaster* chromosomes. *Cold Spring Harbor Symp. 38*, 205-223.

Callan, H. G. (1972). DNA replication in the chromosomes of eucaryotes. *Proc. Roy. Soc. Lond. B. 181*, 19-41.

Cairns, J. (1963). The bacterial chromosome and its manner of replications as seen by autoradiography. *J. Mol. Biol. 6*, 208-213.

Cairns, J. (1966). Autoradiography of HeLa cell DNA. *J. Mol. Biol. 15*, 372-373.

Edenberg, H. J., and Huberman, J. A. (1975). Eukaryotic chromosome replication. *Ann. Rev. Genetics 9*, 245-285.

Gaddipati, J. P. and Sen, S. K. (1978). DNA replication studies in genus *Vicia* through fibre radioautography. *J. Cell Sci. 29*, 85-91.

Gyurasits, E. B. and Wake, R. G. (1973). Bidirectional chromosome replication in *Bacillus subtilis*. *J. Mol. Biol. 73*, 55-63.
Hand, R., and Tamm, I. (1974). Initiation of DNA replication in mammalian cells and its inhibition by reovirus infection. *J. Mol. Biol. 82*, 175-183.
Hori, T., and Lark, K. G. (1974). Autoradiographic studies of the replication of satellite DNA in the kangaroo rat. *J. Mol. Biol. 88*, 221-232.
Hori, T., and Lark, K. G. (1976). Evidence for defined lengths of DNA replication units in *satellite* DNA from *D. ordii*. *Nature 259*, 504-505.
Housman, D., and Huberman, J. A. (1975). Changes in the rate of DNA replication fork movement during S phase in mammalian cells. *J. Mol. Biol. 94*, 173-181.
Howard, A., and Pelc, S. R. (1953). Synthesis of deoxyribonucleic acid in normal and irradiated cells and its relation to chromosome breakage. *Heredity 9 (Suppl.)*, 261-273.
Huberman, J. A., and Riggs, A. D. (1968). On the mechanism of DNA replication in mammalian chromosomes. *J. Mol. Biol. 32*, 327-341.
Huberman, J. A., and Tsai, A. (1973). Direction of DNA replication in mammalian cells. *J. Mol. Biol. 75*, 5-12.
Ikushima, T. (1976). Chromosomal DNA replication of *Vicia faba* cells. *Ann. Rep. Res. Reactor Inst. Kyoto Univ. 9*, 70-77.
Keyl, H. G., and Pelling, C. (1963). Differentielle DNS replikation in den Speicheldrüsen-Chromosomen von *Chironomus thummi*. *Chromosoma 14*, 347-359.
Kornberg, A. (1974). *DNA Synthesis*. W. H. Freeman and Co., San Francisco.
Kovacs, C. J., and Van't Hof, J. (1970). Synchronization of a proliferative population in a cultured plant tissue: kinetic evidence for a G_1/S population. *J. Cell Biol. 47*, 536-539.
Kovacs, C. J., and Van't Hof, J. (1971). Mitotic delay and the regulating events of plant cell proliferation: DNA replication by a G_1/S population. *Rad. Res. 48*, 95-106.
Lee, C. S., and Pavan, C. (1974). Replicating DNA molecules from fertilized eggs of *Cochliomyia hominivorex (Diptera)*. *Chromosoma 47*, 429-437.
Lima-de-Faria, A. (1959). Differential uptake of tritiated thymidine into hetero- and euchromatin in *Melanoplus* and *Secale*. *J. Biophys. Biochem. Cytol. 6*, 457-466.
McFarlane, P. W., and Callan, H. G. (1973). DNA replication in the chromosomes of the chicken, *Gallus domesticus*. *J. Cell Sci. 13*, 821-839.

Painter, R. B., and Schaefer, A. W. (1971). Variation in the rate of DNA chain growth through the S phase in HeLa cells. *J. Mol. Biol. 58*, 289-295.
Petes, Th. D., and Williamson, D. H. (1975). Fiber autoradiography of replicating yeast DNA. *Exp. Cell Res. 95*, 103-110.
Prescott, D. M., and Kuempel, P. L. (1972). Bidirectional replication of the chromosome in *Escherichia coli*. *Proc. Nat. Acad. Sci. U.S. 69*, 2842-2845.
Sober, H. A. (1970). *Handbook of Biochemistry*. Chemical Rubber Co., Cleveland.
Taylor, J. H. (1958). The mode of chromosome duplication in *Crepis capillaris*. *Exp. Cell Res. 15*, 350-357.
Taylor, J. H. (1963). DNA synthesis in relation to chromosome reproduction and the reunion of breaks. *J. Cell Comp. Physiol. 62*, 73-86.
Taylor, J. H., Woods, P. S., and Hughes, W. L. (1957). The organization and duplication of chromosomes as revealed by autoradiographic studies using tritium-labeled thymidine. *Proc. Nat. Acad. Sci. U.S. 43*, 122-127.
Van't Hof, J. (1965). Relationships between mitotic cycle duration, S period duration and the average rate of DNA synthesis in the root meristem cells of several plants. *Exp. Cell Res. 39*, 48-58.
Van't Hof, J. (1966). Experimental control of DNA synthesizing and dividing cells in excised root tips of *Pisum*. *Amer. J. Bot. 53*, 970-976.
Van't Hof, J. (1975a). The regulation of cell division in higher plants. *Brookhaven Symp. Biol. 25*, 152-165.
Van't Hof, J. (1975b). DNA fiber replication of chromosomes of a higher plant *(Pisum sativum)*. *Exp. Cell Res. 93*, 95-104.
Van't Hof, J. (1976a). DNA fiber replication of chromosomes of pea root cells terminating S. *Exp. Cell Res. 99*, 47-56.
Van't Hof, J. (1976b). Replicon size and rate of fork movement in early S of higher plant cells *(Pisum sativum)*. *Exp. Cell. Res. 103*, 395-403.
Van't Hof, J., and Bjerknes, C. A. (1977). 18 um Replication units of chromosomal DNA fibers of differentiated cells of pea *(Pisum sativum)*. *Chromosoma (Berl.) 64*, 287-294.
Van't Hof, J., Bjerknes, C. A., and Clinton, J. H. (1978). Replication properties of chromosomal DNA fibers and the duration of DNA synthesis of sunflower root-rip meristem cells at different temperatures. *Chromosoma* (in press).
Wake, R. G. (1974). Termination of *Bacillus subtilis* chromosome replication as visualized by autoradiography. *J. Mol. Biol. 86*, 223-231.
Wake, R. G. (1975). Bidirectional replication in *Bacillus subtilis*, In *DNA Synthesis and its Regulation* (Goulian, N., and Hanawalt, P., eds.) pp. 650-676. W. A. Benjamin, Menlo Park, California.

Webster, P. L., and Van't Hof, J. (1970). DNA synthesis and mitosis in meristems: requirements for RNA and protein synthesis. *Amer. J. Bot. 57*, 130-139.

Wimber, D. E. (1961). Asynchronous replication of deoxyribonucleic acid in root-tip chromosomes of *Tradescantia paludosa. Exp. Cell Res. 23*, 402-407.

Wolstenholme, D. R. (1973). Replicating DNA molecules from eggs of *Drosophila melanogaster. Chromosoma 43*, 1-18.

CHLOROPLAST DNA: STRUCTURE, INFORMATION CONTENT, AND REPLICATION

Krishna K. Tewari
Robert Meeker

Department of Molecular Biology and Biochemistry
University of California, Irvine
Irvine, California

Chloroplast (ct) DNA from higher plants has been found to exist in a circular conformation (Kolodner and Tewari, 1972a; 1972b; 1975a). About eighty percent of the total ctDNA has been obtained in circular molecules whether isolated from pea, bean, spinach, lettuce, corn, or oats. The covalently closed circular DNA molecules account for thirty to forty percent of the total ctDNA from these plants. The molecular sizes of ctDNAs relative to an internal standard (øX174 replicative Form II monomer) has been found to range from 25.9 øX units (139 kilobase pairs) for corn ctDNA to 29.0 øX units (155 kilobase pairs) for lettuce ctDNA. The ctDNAs from the monocots, corn and oats, have been found to be smaller than the ctDNAs from dicots; pea, spinach, and lettuce. However, all of the ctDNAs have been found to have differences in their molecular weights suggesting that ctDNAs have undergone significant changes in their molecular structure during evolution. The thermal denaturation studies with corn, pea, spinach, and lettuce ctDNA have also shown that the base sequences of the higher plant ctDNAs have diverged from each other.

The molecular weights of ctDNAs by renaturation kinetics have been found to range from about 85 to 95 $\times 10^6$ (Kolodner and Tewari, 1975a). The excellent agreement between the molecular weights of circular DNA molecules obtained by electron microscopy and by kinetic complexity suggests that the sequence of a circular ctDNA molecular represents the entire information content of the ctDNA. This suggestion is further confirmed by the denaturation mapping studies with the circular ctDNA. Denaturation maps of pea ctDNAs have shown that

ISBN 012-601950-9

all of the ctDNA molecules contain identical gross base sequences (Kolodner and Tewari, 1975b). These data again indicate that the information content of ctDNA is coded for by the base sequence of one circular DNA molecule. Furthermore, gross heterogeneity in base sequences or large repeating sequences were not detected in the denaturation maps of ctDNA. Using the denaturation mapping technique, circular dimers have been found which are formed by head to tail fusion of two circular monomers.

Circular ctDNAs from pea, lettuce, and spinach have been found to contain covalently linked ribonucleotides (Kolodner *et al.*, 1975). The pea and spinach ctDNAs contain a maximum of 18±2 ribonucleotides and lettuce ctDNA contains a maximum of 12±2 ribonucleotides. The electron microscopic studies of the alkali fragmented DNA and reannealed alkali fragmented DNA have made possible the localization of ribonucleotides in the circular map of ctDNA.

The ctDNAs from spinach, lettuce, and corn have been found to contain one long inverted repeat sequence of a molecular size of 15×10^6 (Tewari *et al.*, 1976). The molecular size and the location of the repeat sequence is approximately the same in all three ctDNAs. In contrast, pea ctDNA has not been found to contain any inverted repeat sequence. These observations again emphasize the divergence in base sequences of ctDNAs from higher plants.

The genetic information content in ctDNA has been studied by hybridizing the ribosomal (r) RNA, tRNA's and mRNAs with ctDNA. The DNA-RNA molecular hybridizations have shown that ctDNA from higher plants contain two rRNA genes (Tewari and Wildman, 1968; 1970; Thomas and Tewari, 1974a, 1974b). The base sequences of the ct-rRNA genes have been found to be quite similar in the various ctDNAs isolated from bean, lettuce, spinach, corn, and oats. This conclusion is based upon hybridization, competition, and thermal stability studies of DNA-rRNA hybrids involving RNA and DNA from homologous and heterologous systems. The hybridization studies between ctDNA from pea and tRNAs from pea chloroplasts have shown that there are 40±10 tRNA genes in ctDNA (Meeker and Tewari, in preparation). The competition hybridization experiments have shown that ct-tRNAs do not have any common base sequences with tRNAs from cytoplasm. Also, ct-tRNAs do not share base sequences with bacterial, yeast, or animal tRNAs. The ct-tRNAs have been charged with individually labeled amino acids with the aminoacyl tRNA synthetases from chloroplasts and/or *E. coli*. The hybridization of aminoacylated tRNAs with ctDNA has shown that the genes for 17 aminoacyl tRNAs are present in ctDNA. The thermal denaturation of the ctDNA-tRNA hybrids has confirmed the conclusions derived from hybridizations. The saturation hybridization studies with labeled total ctRNA have shown that practically all of the information contained in a single strand of ctDNA is transcribed *in vivo*. Some of the mRNAs coded by ctDNA contain poly A sequences at the 3' end (Tewari *et al.*, 1976).

The replication of ctDNA has been studied by electron microscopic studies of replicating intermediates (Kolodner and Tewari, 1975c; 1975d). The ctDNAs from higher plants have been found to replicate by the introduction of two displacement (D-) loops at a specific position in the DNA molecule. The two D loops are hydrogen bonded to the opposite parental strands and expand bidirectionally, resulting in a Cairn's type of DNA replicative intermediate. In addition to the replication of ctDNA by Cairns intermediates, the ctDNA has also been found to replicate by the rolling circle model of replication. In a growing leaf, the two types of replicative intermediates have been found to occur in almost equal proportions. The isolated chloroplasts have been found to have at least three proteins which polymerize deoxynucleoside triphosphates (Tewari and Wildman, 1967; Tewari *et al.*, 1976). The solubilized DNA polymerizing proteins carry out semiconservative replication of DNA *in vitro*. This complex mixture of proteins can also convert single stranded ϕX174 DNA into double stranded ϕX DNA.

CHARACTERIZATION OF ctDNAs

Buoyant Density of ctDNAs

The ctDNA from pea, lettuce, spinach, beans, corn, and oats have been found to have a buoyant density of 1.6980 ± 0.001 g/cm^3 in CsCl density gradients (Tewari, 1971). On denaturation, the buoyant density of ctDNAs increased by 0.013 g/cm^3 but incubation of the denatured ctDNA to a Cot (moles/litre/sec) of 2 resulted in the complete renaturation of the denatured ctDNAs. The renatured ctDNA banded at a buoyant density of 1.699 ± 0.001 g/cm^3. The buoyant density of the nuclear (n) DNAs from pea, lettuce, spinach, and bean were 1.695 ± 0.001 g/cm^3 and the nDNAs from corn and oats were 1.702 ± 0.001 g/cm^3. These nDNAs increased in their buoyant densities by 0.013 g/cm^3 on denaturation but incubation of the denatured nuclear DNA even at a Cot of 20 resulted in only about 20% renaturation. This characteristic renaturation property of ctDNA was utilized in establishing that the DNA isolated from higher plants is of the organelle origin and not derived from nuclear DNA (Tewari *et al.*, 1976; Tewari, 1971). Today, this property is routinely used for checking the purity of ctDNAs. It is interesting to note that all of the different higher plant ctDNAs have practically the same buoyant density whether the DNA is native, denatured, or renatured. In contrast, the buoyant density of nDNA has been found to differ in different plants (Tewari, 1971). In addition to the ctDNA of density 1.698 ± 0.000 g/ml, the higher plant cell contain mitochondrial (mt) DNA which bands at a density of 1.705 ± 0.001 g/cm^3 (Tewari, 1971). The mtDNAs from all higher plants have been found to band at a density of 1.705 ± 0.001 g/cm^3. The methods which we have published in the last few years routinely yield pure ctDNA without any contaminating nuclear or mitochondrial DNA.

Thermal Denaturation of ctDNAs

The T_m (thermal transition midpoint) of ctDNAs from higher plants was found to be distinctly different from each other (Fig. 1 and Table I). The most striking difference was between pea ctDNA and corn ctDNA; the T_m of corn ctDNA was 2.8° higher than the pea ctDNA (Kolodner and Tewari, 1975a). The derivatives of the melting profiles showed that pea, lettuce, spinach, and corn ctDNAs melted as a single component showing only one inflection point. The corn ctDNA melted considerably more broadly than the rest of the ctDNAs. None of the ctDNAs were as heterogeneous in base sequences as the ctDNAs from *Euglena, Chlamydomonas* and *Chlorella* (Stutz and Vandrey, 1971; Wells and Sanger, 1971; Bayen and Rode, 1973). It is obvious that the base sequences in any DNA would have to be heterogen-

TABLE I. Thermal Denaturation Characteristics of ctDNAs

DNA	T_m(°C)[a]	T_m(ct-T_m(pea-ct)(°C)	hmax(%)[b]	$\frac{2}{T3}$(°C)[c]
Pea	82.7 ± 0.7	-	36	9.3
Lettuce	85.1 ± 0.7	2.4 ± 0.05	37	8.3
Spinach	84.6 ± 0.5	1.9 ± 0.05	36	8.7
Corn	85.5 ± 0.7	2.8 ± 0.07	35	10

[a]*Transition point*
[b]*Maximum hyperchromicity*
[c]*Dispersion*

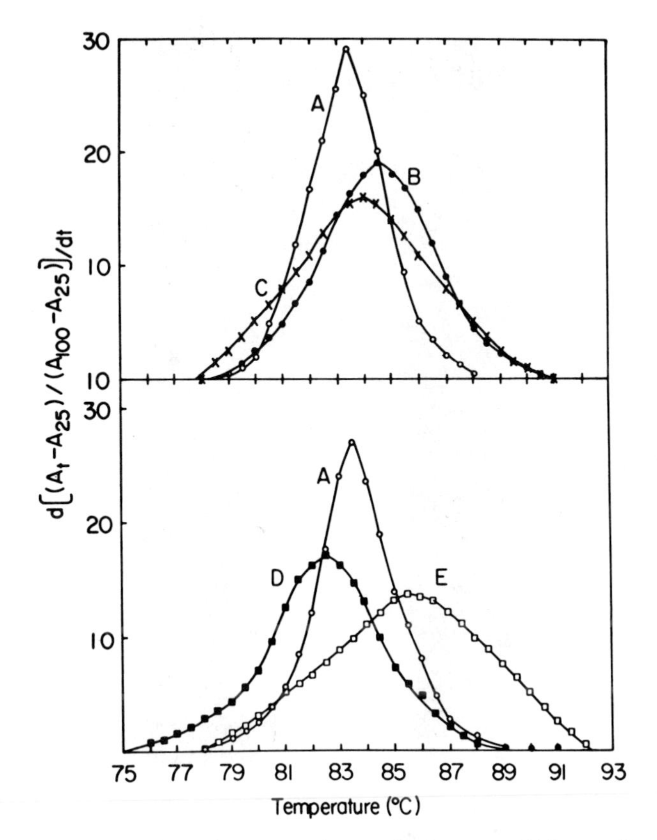

FIGURE 1. Derivative of the melting profiles of ctDNA. (A), T4 DNA; (B), lettuce ctDNA; (C), spinach ctDNA; (D), pea ctDNA; (E), corn ctDNA. The thermal denaturations (meltings) were carried out in 1 × SSC using a rate of temperature rise of 1° per 10 min. $d[(A_T-A_{25})/(A_{100}-A_{25})]/dt$ *is the derivative of* $(A_T-A_{25})/(A_{100}-A_{25})$ *with respect to temperature.* A_{25}, $A_{260\ nm}$ *of DNA at 25°;* A_{100}, $A_{260\ nm}$ *of DNA at 100°;* A_T, $A_{260\ nm}$ *of DNA at a given temperature.*

eous in order to have the genetic information for coding proteins. However, the detection of such a heterogeneity would require very sensitive techniques (see below).

Conformation of Circular DNA

Centrifugation of ctDNA in a propidium diiodide-CsCl gradient results in two fluorescent bands as shown in Fig. 2. The DNA present in the lower band was found to have supertwisted

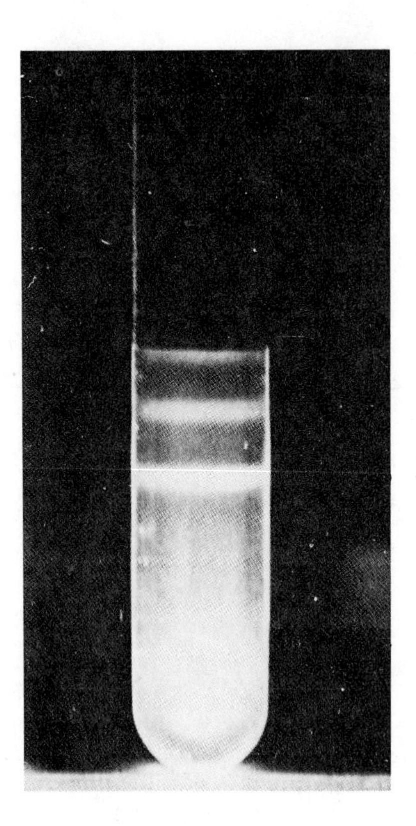

FIGURE 2. Propidium diiodide - cesium chloride density gradient of pea ctDNA. Twenty-five μg of covalently closed circular pea ctDNA was nicked with 750 rads of gamma rays. The resulting DNA solution was adjusted to a CsCl density of 1.57 g/cm^3, in 0.01 M Tris, 0.001 M EDTA, and 400 μg/ml propidium diiodide and was centrifuged for 36 hr at 32,000 rpm in a Spinco SW 41 rotor at 17°C. The resulting density gradient was photographed under long wavelength (365 nm) light.

molecules (Fig. 3) which accounted for 90% of the DNA. The remaining 10% of the molecules were relaxed circles (Fig. 3). The supertwisted molecules could be converted to relaxed open circular molecules by treatment with γ-rays which produces single strand breaks in DNA. In the upper band of the propidium diiodide-CsCl density gradients, 60% of the DNA molecules were relaxed circular molecules while the rest of the molecules were linear. No linear molecules were observed that were longer than the circular molecules. It has been possible to isolate 80% of the total ctDNA from pea, spinach, lettuce,

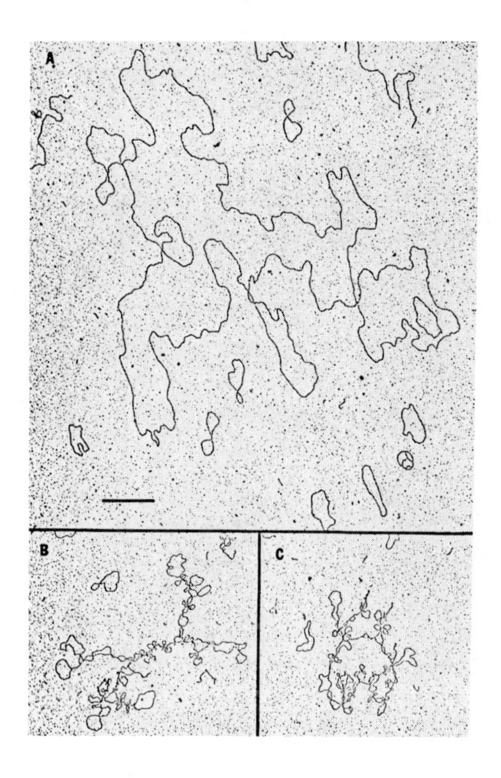

FIGURE 3. Relaxed circular and supertwisted ctDNA molecules. (A), relaxed circular lettuce ctDNA molecule. (B, C), supertwisted pea and oat ctDNA molecules, respectively.

corn, or oats in supertwisted and relaxed circular molecules. The ctDNA has also been found to occur in double length circular molecules (Fig. 4) and catenated dimers which consist of two topologically interlocked single length circular molecules (Fig. 5). These complex ctDNA molecules were found in practically all the higher plants (Table II).

Molecular Size of ctDNA

The molecular size of ctDNAs from higher plants has been determined by electron microscopy using ϕX174 RFII DNA as an internal standard. The distribution of the lengths of the circular monomers and dimers of pea, lettuce, spinach, corn, and oats ctDNA molecules is presented in Fig. 6. There is an excellent agreement between the single length and double

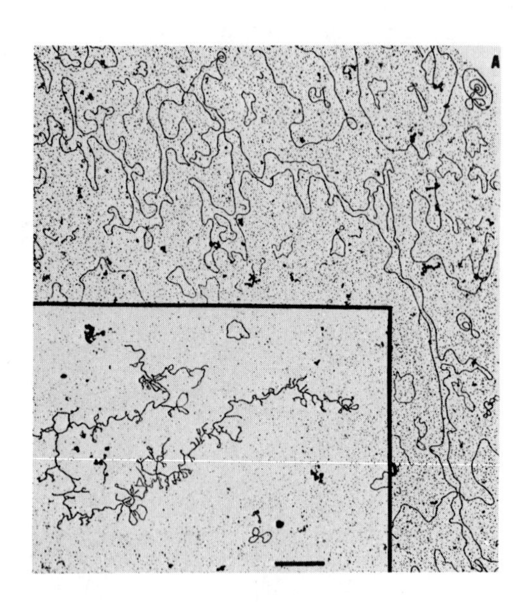

FIGURE 4. Dimer length circular ctDNA molecules. (A), relaxed double length circular pea ctDNA molecule; (B), supertwisted double length circular spinach ctDNA molecule. The small circular DNA molecules are ØX RF II DNA. The bar indicates 1 μm.

length circular ctDNAs from higher plants. It is clearly seen from Fig. 6 that corn and oat ctDNAs are much smaller in molecular size compared to the ctDNAs from peas, lettuce, and spinach. However, there are significant differences between the sizes of all of the ctDNAs that were examined.

The molecular weight of DNA can be determined by electron microscopy provided an internal standard of known molecular size is co-spread with the unknown DNA. The molecular weight of the unknown DNA is calculated from the ratio of its length to the length of the standard DNA. The molecular weights of ctDNAs calculated by using the molecular weight of ØX174 RFII DNA is given in Table 3. The molecular weights of ctDNAs are found to range from a low of 85×10^6 to a high of 96×10^6. The molecular size of ØXDNA by sequence analysis has been recently found to be about 5,370 nucleotides (Barrel *et al.*, 1976). Using this number, the ctDNAs from higher plants range from 139 kilobase pairs to 155 kilobase pairs. In such calculations, we are assuming that the mass per unit length of the ctDNAs and ØX RFII DNA is the same which may not be a valid assumption. It is obvious that absolute molecular weights

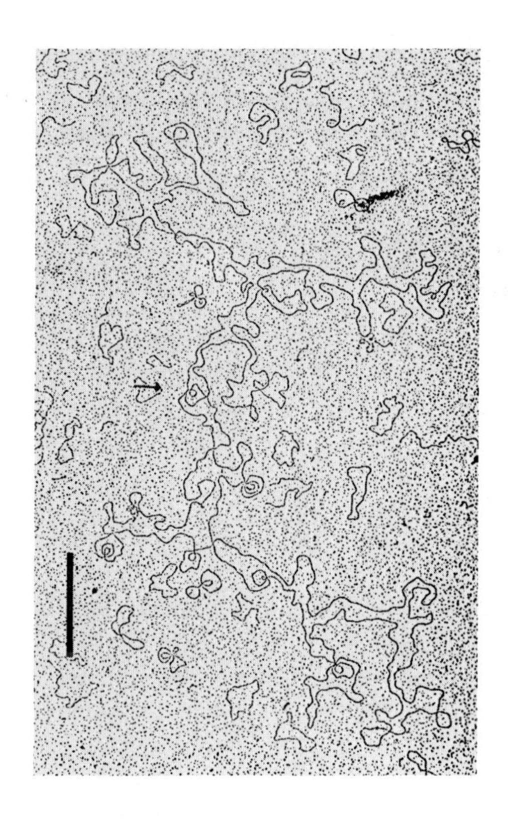

FIGURE 5. Catenated dimer of pea ctDNA. The arrow marks the point of catenation between the two monomer units. The small circular DNA molecules are ØX RF II DNA and the bar indicates 1 μm. This DNA spreading for electron microscopy only contained 3 circular ctDNA molecules per grid square.

of ctDNAs can only be obtained by base sequence analysis of these DNAs. However, the data presented here are as accurate as possible and are quite suitable for most of the studies directed towards obtaining the genetic information of ctDNAs.

Molecular Weights of ctDNAs by Renaturation Kinetics

The sequence complexities of ctDNAs has been analyzed by studying the renaturation rates of these DNAs. The reassociation of the fragmented, denatured ctDNAs follows second order rate kinetics and indicates the presence of only one class of molecules. There are no rapidly renaturing sequences in ctDNAs

TABLE 2. The Frequency of Complex Molecules in ctDNA

DNA	Monomers (%)	Circular dimers (%)	Catenated dimers (%)	Number[a]
Pea (lower band)	94.2	4.1	1.7	520
Pea (middle band)	92.9	3.5	3.6	514
Pea (upper band)	94.4	3.6	2.0	506
Lettuce (lower band)	92.1	5.5	2.4	503
Spinach (lower band)	94.5	3.6	1.9	420
Corn (lower band)	95.8	2.9	1.3	409
Oats (lower band)	95.5	3.0	1.5	200

[a]*Only circular molecules were scored and replicating molecules were considered to be monomers.*

which can be detected by this method (Fig. 7). In this regard, ctDNA behaves like bacterial and viral DNAs and unlike eucaryotic DNAs where a large proportion of chromosomal DNA consists of many families of repeating sequences. The molecular weight of the unique sequences of the ctDNAs from pea, lettuce, spinach, beans, corn, and oats were calculated using the ratio of Cot½(moles/liter/sec) of the ctDNAs to the Cot; of T4 or pea ctDNA (Table 3). The molecular weights of ctDNAs were found to range from 84.8×10^6 to 94.3×10^6. The data again indicate that the ctDNAs from different plants differ in the molecular weights. In general, monocot plants have smaller molecular weight ctDNA than the ctDNA from dicot plants. The exce-lent agreement between the molecular weight of the unique sequences of ctDNAs determined by renaturation kinetics and the molecular weight of the circular ctDNAs determined by electron microscopy suggests that the entire information content of higher plant ctDNA is coded for by the sequence of a circular ctDNA molecule.

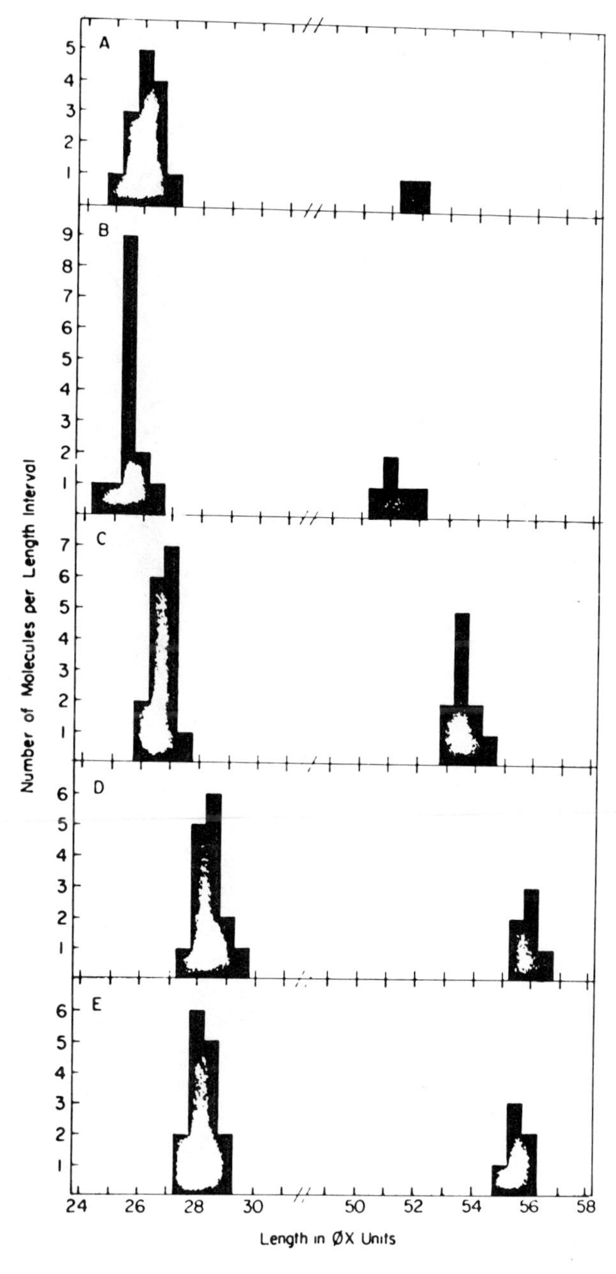

FIGURE 6. Length distribution of the circular ctDNA molecule. (A), oat ctDNA; (B), corn ctDNA; (C), pea ctDNA; (D), lettuce ctDNA; (E), spinach ctDNA. The lengths are expressed in units of ØX length. The large classes of circular DNA molecules are two times the size of the small size class of circular DNA molecules.

TABLE 3

The Molecular Size of ctDNAs

ctDNA	Molecular Size (M)				
	ØX units	$M \times 10^{-6a}$	$M \times 10^{-6a}$	$M \times 10^{-6c}$	$M \times 10^{-6d}$
Pea	26.7 ± 0.1	88.4 ± 0.3	89.1	90.1 ± 2.1	89.1
Lettuce	29.0 ± 0.1	95.8 ± 0.3	96.7 ± 0.5	94.3 ± 2.0	93.1 ± 2.0
Spinach	28.5 ± 0.2	94.3 ± 0.6	95.1 ± 1.0	91.2 ± 2.7	90.1 ± 2.7
Corn	25.6 ± 0.2	84.7 ± 0.6	85.4 ± 1.2	86.6 ± 2.8	85.6 ± 2.9
Oats	25.9 ± 0.1	85.7 ± 0.3	86.4 ± 0.6	84.8 ± 3.0	83.7 ± 3.0

[a] *Calculated using an average M for ØX RF monomer DNA of 3.31×10^6.*

[b] *Calculated assuming an M of 89.1×10^6 for pea ctDNA.*

[c] *Calculated from renaturation rates. The molecular weight of T4 DNA has been taken to be 110×10^6.*

[d] *Calculated from renaturation rates. The molecular weight of pea ctDNA has been taken to be 89.1×10^6.*

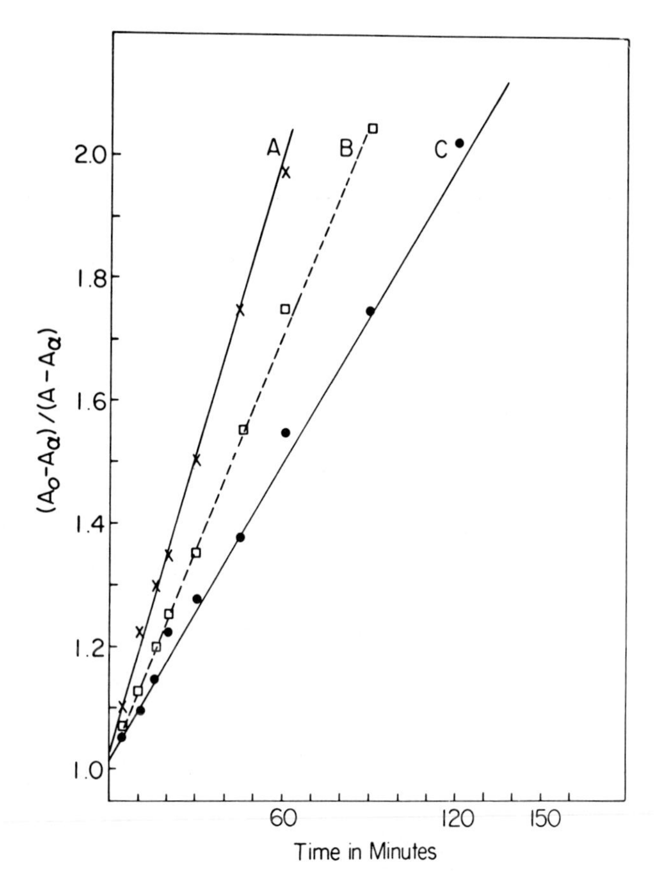

FIGURE 7. Second order rate plot of the renaturation of ctDNA. (A), lettuce ctDNA, 20 μg/ml; (B), bean ctDNA, 15 μg/ml; (C), spinach ctDNA, 10 μg/ml. The DNA was fragmented by passing it through a 27 gauge needle, alkali denatured, and neutralized. The ctDNA was then allowed to renature at T_m-25° in 1 × SSC. A_o, $A_{260\ nm}$ of denatured DNA; $A\alpha$, $A_{260\ nm}$ of native DNA; A, $A_{260\ nm}$ of the sample at a given time.

Molecular Weights of ctDNAs by Sedimentation Studies

The pea Form I ctDNA sediments at about 86S in neutral 3M CsCl (Kolodner *et al.*, 1976), irradiation with 1,000 rads of γ-rays, 50% of this DNA is converted to Form II which sediments at about 58S. Higher doses of γ-rays produce progressively more nicked circular Form II DNA, but no other forms are produced. The sedimentation of Form I DNA in 3M CsCl containing 0.2M NaOH converts all of the ctDNA to the fast moving Form

TABLE 4
The Values of S^{o}_{20,w,Na^+} of ctDNA

Confirguration	Pea S	Spinach S	Lettuce S
Form I	86.1	97.1	94.9
(neutral)	S.D. ± 0.4	S.D. ± 0.6	S.D. ± 0.3
Form II	58.3	60.6	60.8
(neutral)	S.D. ± 0.3	S.D. ± 0.4	S.D. ± 0.1
Form II	55.57	57.7	57.8
(alkaline)	S.D. ± 0.08	S.D. ± 0.5	S.D. ± 0.5
Form IV	240	264	264.1
(alkaline)	S.D. ± 0.7	S.D. ± 3.0	S.D. ± 1.5
Form IV	146.8	---	170
(neutral)	S.D. ± 1.1	---	S.D. ± 10.0

IV, sedimenting at 240S. The sedimentation values of Forms I, II and IV of ctDNAs from pea, spinach, and lettuce are given in Table 4.

The pea Form I ctDNA was purified by preparative sucrose density gradient velocity centrifugation to remove circular and catenated dimers. Different concentrations of purified Form I ctDNA monomers were centrifuged to equilibrium in CsCl, 0.01M EDTA pH 8 (ρ = 1.700 g/cm^3). The values for the apparent dry cesium molecular weight (M_{cs}, app) were determined and extrapolated to zero concentrations ($M°_{cs}$). The resulting $M°_{cs}$ was 118.8 (± 1.0) × 10^6 which is equivalent to a sodium salt molecular weight of 89.1 ± 0.7 × 10^6 using 0.75 as the ratio of the weights of the two DNA salts.

The molecular weight of ctDNA can also be calculated by velocity sedimentation, but the equation $S = 2.7 + 0.01759\ M^{0.445}$ (Gary *et al.*, 1967) derived from theoretical considerations to express the relationship between S and M for open circular molecules does not yield accurately the molecular weights of large circular DNA molecules. We have derived the equation $S = 0.05587\ M^{0.389}$ for the open circular Form II DNA. This equation has been used to calculate the molecular weights of several open circular DNAs whose $S°_{20,w}$ are known. These values have been compared to the values for the molecular

TABLE 5
The Molecular Weights of Assorted Open Circular DNAs

DNA	S°_{20,w,Na^+}	Sedimentation[a] $M \times 10^{-6}$	Electron Microscopy[b] $M \times 10^{-6}$
PM 2	21.2S	6.1	6.0
SV 40	16.7S	3.3	3.49
HeLa Mt-			
monomer	26.1S	10.5	10.2
dimer	33.4S	20.2	20.3
Spinach ct	60.6S	97.2	95.1
Lettuce ct	60.8S	98.2	96.2

[a]*Calculated using equation described in the text.*

[b]*Calculated according to the recommendations of Kolodner and Tewari (1975a).*

weights of these DNAs which were determined by electron microscopy using internal standards (Gary *et al.*, 1967). There is an excellent agreement between the values determined by electron microscopy and those determined from sedimentation coefficients (Table 5).

Denaturation Mapping Studies on Chloroplast DNA

The structure of circular pea chloroplast DNA has been analyzed by denaturation mapping. Denaturation maps were constructed at an average denaturation of 2.5%, 22%, and 44%. At 2.5% average denaturation, 70% of all the circular molecules were partially denatured, containing one to seven denatured regions. A typical molecule from this spreading is shown in Fig. 8. The denaturation map presented in Fig. 9 was produced by starting with the most denatured molecules first and then matching the greatest number of denatured regions. The molecules were also matched with respect to the length of each individual denatured region. The denaturation maps were also produced by using the least denatured molecules first and then arranging them as described above or by randomly picking the molecules. All of the denaturation maps were identical. This map contained six major denatured regions, two minor denatured

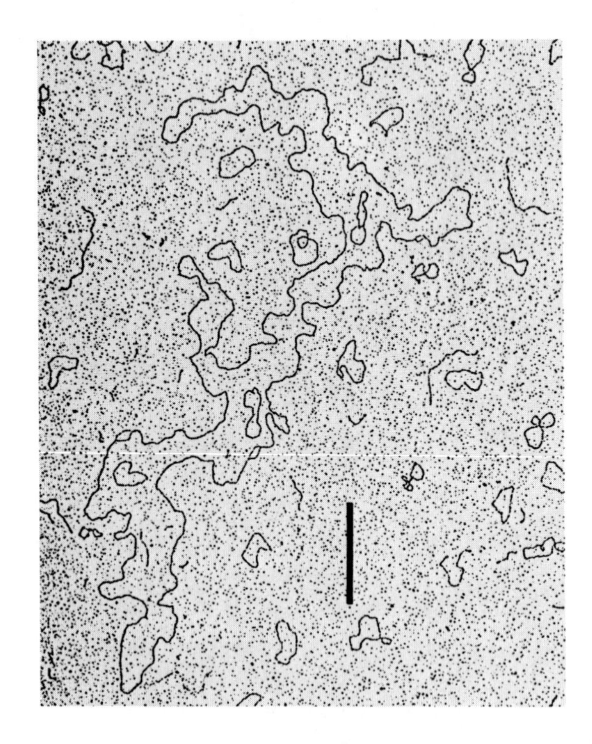

FIGURE 8. A partially denatured pea ctDNA molecule showing three denatured regions. The hypophase contained 45% formamide in 1 × TE (10 mM Tris, 1 mM EDTA) and the spreading solution contained 75% formamide in 10 × TE. The small circular molecules are single-stranded and double-stranded ϕX DNA. The bar indicates 1 μm.

regions, and two places where only one molecule was denatured. Three of the major denatured regions (proceeding from left to right) were found to be an average of 2.3 ϕX units apart. After these three regions, there was a native region corresponding to 4.2 ϕX units. The three other major denatured regions were 1.6 ϕX units apart. The major denatured regions were followed by four regions showing low levels of denaturation. A region corresponding to 34% of the pea ctDNA molecule contained no denatured regions.

At 22% average denaturation, all of the circular molecules were partially denatured and contained 11 to 33 denatured regions. The denaturation map produced from these molecules is presented in Fig. 10. A molecule from this spreading is given in Fig. 11. This map contains 31 distinct denatured regions

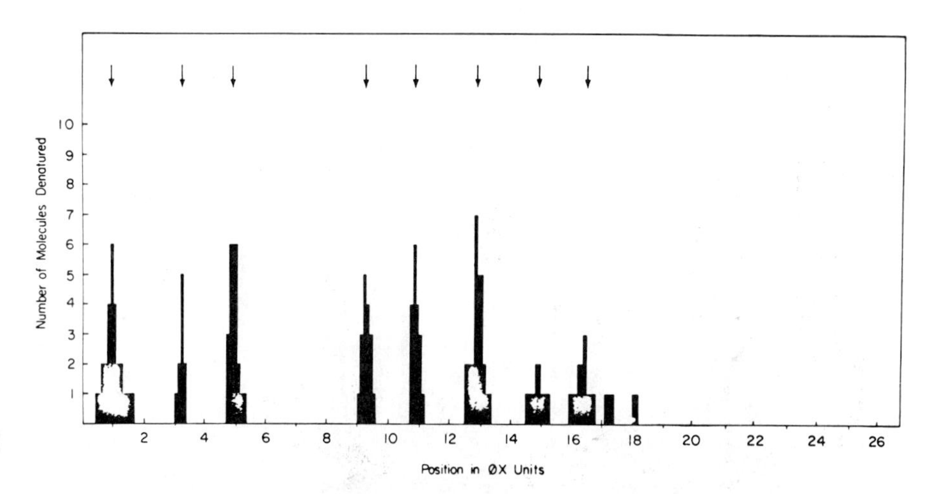

FIGURE 9. The denaturation map of the pea ctDNA that was constructed with molecules from the spreading described in Fig. 8. The average amount of denaturation was 2.5%. The map represents circular DNA molecules and has been linearized for display purposes. The ↓ designates the denatured regions that were found in more than one molecule. The six denatured regions on the left side of the figure are referred to as "major denatured" regions.

as well as six distinct regions showing no denaturation. Three of the undenatured regions were about 0.7 ϕX units long, whereas one undenatured region was 0.4 ϕX units long. There were also two undenatured regions of about 1.2 and 1.5 ϕX units in length. It was possible to locate all of the denatured regions produced at the 2.5% level of denaturation, and they are indicated by the arrows in Fig. 10.

At 44% average denaturation, all of the circular molecules were partially denatured and contained 18 to 43 denatured regions. A denaturation map was constructed using these molecules as described above, and is illustrated in Fig. 12. There were only two distinct undenatured regions of 0.40 ϕX units and 0.70 ϕX units in length. These two undenatured regions matched two undenatured regions in the map of Fig. 10. The five denatured regions between these two undenatured regions were also present in the denaturation map of Fig. 10 and are indicated by the boxes symbols.

The circular dimers of ctDNA have been analyzed by denaturation mapping for evidence of their formation from monomers. In a circular dimer formed by head to tail fusion, the two monomer units will be in a tandem repeat. Therefore, in

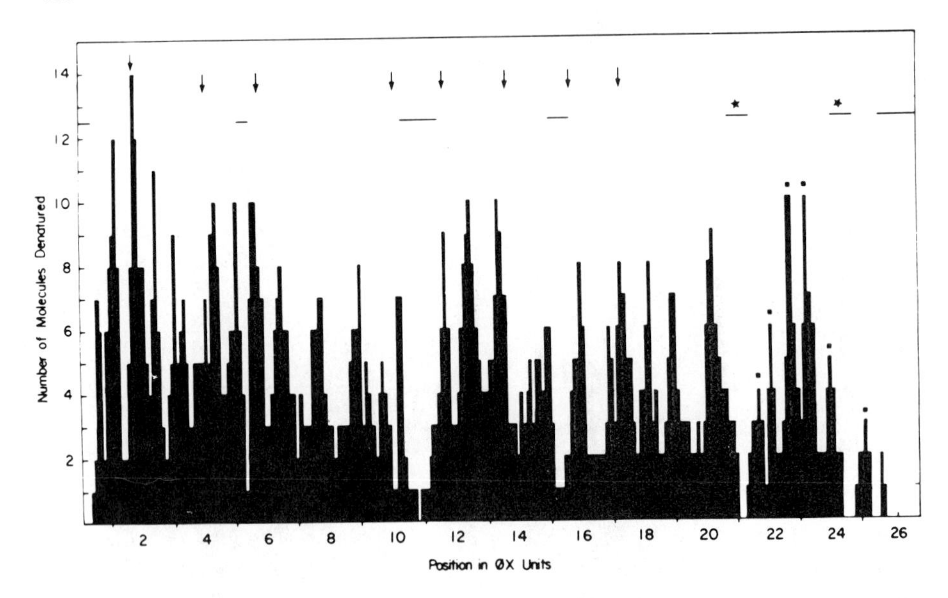

*FIGURE 10. The denaturation map of pea ctDNA constructed with the molecules from the spreading described in Fig. 11. The average amount of denaturation was 22%. The ↓ designates the positions of the denatured regions that were present at the 2.5% level of denaturation (Fig. 9). The -- indicates the positions of regions that consistently remain undenatured. The * marks two undenatured regions that appear in Fig. 12 and the ■ indicates 6 denatured regions that are also present in Fig 12.*

a partially denatured circular dimer, every denatured region will have a corresponding denatured region one monomer length away. In a circular dimer formed by head to head fusion, the two monomer units will be integrated in a reverse repeat. Therefore, in a partially denatured circular dimer, it will be possible to locate two points of symmetry on the molecule. The corresponding denatured regions will then be equidistant on either side of these points. The structure of the denatured circular dimers was examined for the two types of integration described above. There were more matched denatured regions when the data were analyzed by the head to tail test than the head to head test indicating that the circular dimers are integrated in a head to tail fashion.

The extensive use of internal standards to correct for the shortening of single stranded DNA relative to the double stranded DNA has enabled us to compare accurately the denaturation maps of pea ctDNA that were produced at different levels of denaturation. The data have shown that the six most AT-rich

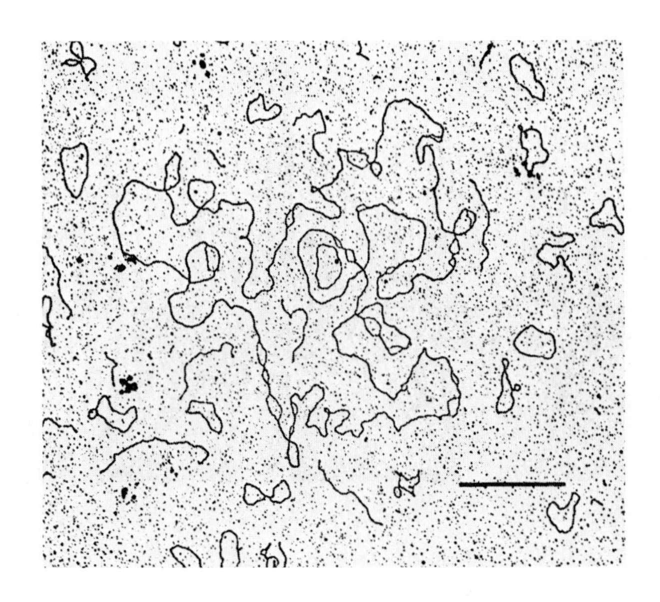

FIGURE 11. A partially denatured pea ctDNA molecule showing 22 denatured regions. The hypophase contained 47% formamide in 1 × TE and the spreading solution contained 77% formamide in 10 × TE. The small circular molecules are single-stranded and double-stranded ØX DNA. The bar indicates 1 μm.

regions in peat ctDNA (Fig. 8), comprising 1 to 2@ of the DNA, are located in one-half of the pea ctDNA molecule. In molecules that are as much as 70% denatured, two regions comprising 1 to 2% of the DNA remain undenatured (Fig. 12). These two relatively GC-rich regions are located on the opposite side of the pea ctDNA molecule in relation to the AT-rich regions. The rest of the pea ctDNA molecule was found to denature uniformly. The heterogeneity observed here is very minor when compared to λ DNA, where one-half of the DNA molecule was found to be almost completely denatured under conditions where the other half of the molecule was just beginning to denature (Inman and Schnos, 1970). The base sequence heterogeneity in pea ctDNA is also less extensive than the heterogeneity found in *Euglena* and *Chlorella* ctDNA, where the ctDNAs can be fractionated into two components using csCL or Ag^+-Cs_2SO_4 density gradients (Stutz and Rawsson, 1970; Vedel *et al.*, 1970). These observations indicate that the base sequences of the ctDNAs from higher plants and algae have undergone major divergence.

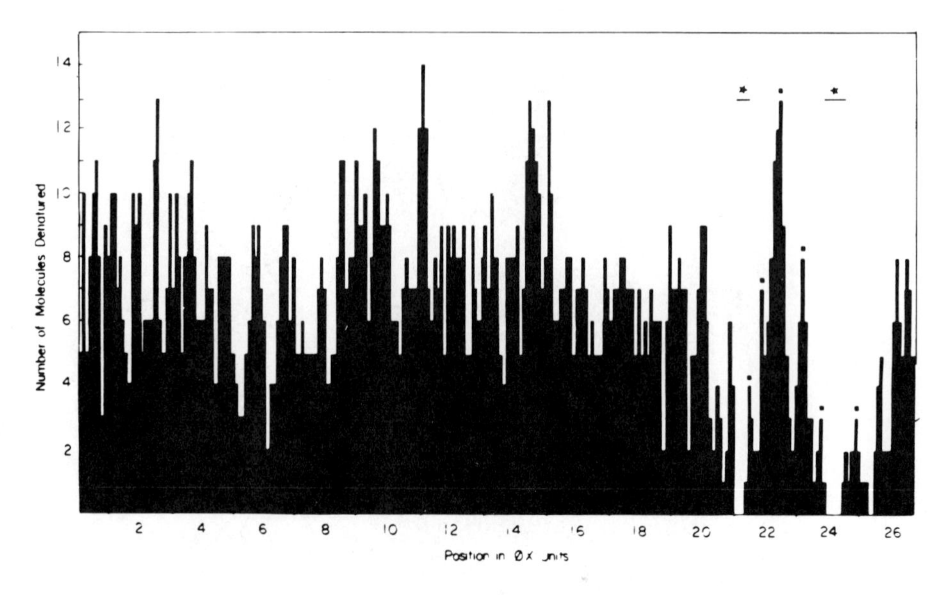

*FIGURE 12. The denaturation map of pea ctDNA constructed from molecules having an average denaturation of 44%. The * indicates two undenatured regions present in the 22% denaturation map (Fig. 10). The ■ indicates six denatured regions that are also present in the denaturation map of Fig. 10.*

The molecular weights of the ctDNAs from pea, lettuce, spinach, corn, and oat plants have been determined by renaturation kinetics and by electron microscopy of circular DNA molecules. In all cases, there has been excellent agreement between the values obtained by the two different methods. These results imply that the total information content of these ctDNAs is coded for by the sequence of one circular DNA molecule. The above denaturation mapping experiments with circular pea ctDNA have tested this point directly. At all levels of denaturation, all of the pea ctDNA molecules that were found were consistent with the denaturation maps that were constructed. Furthermore, large repeating sequences were not detected in the denaturation maps of pea ctDNA. These results show that all of the pea ctDNA molecules are the same, and that the entire information content of pea ctDNA is coded for by the sequence of a circular DNA molecule having a molecular weight of about 90×10^6. It should be pointed out that the denaturation mapping technique cannot detect small deletions, repeats, insertions, or point mutations in a molecule. Similar results have been obtained with lettuce, spinach, and corn ctDNAs.

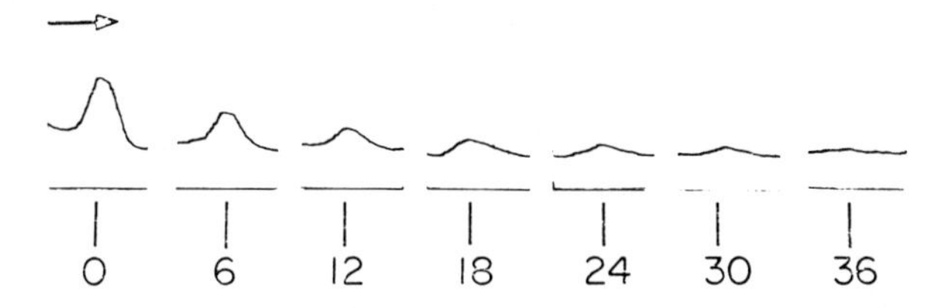

FIGURE 13. The alkali breakdown of pea ctDNA. Pea Form I ctDNA was centrifuged in the alkaline sedimentation solvent at 16,000 rpm. The Form IV peak from scans taken at 6 minute intervals during this expe-iment are shown to illustrate the loss of DNA from the Form IV zone. No other zones were present during this experiment, or in other experiments of this type. The zero time peak presented here was scanned at 4 minutes after layering and the 36 minute peak represents 4% of the applied DNA. The experiment presented here represents a period of 4.2 half-lives. Sedimentation is from left to right.

Covalently Linked Ribonucleotides in Chloroplast DNA

The centrifugation of covalently closed circular pea ctDNA through alkaline csCl has shown that the DNA zone of Form IV DNA decreased with time with a half-life of 10 ± 1 min (Fig. 13). Under similar conditions, ϕX174 RF I and G4 RF I monomers and dimers did not show any detectable loss. Similarly, there was no loss of DNA when closed circular pea ctDNA was centrifuged through neutral sedimentation solvent. Thus, the disappearance of closed circular pea ctDNA from the Form IV zone during sedimentation through the alkaline sedimentation is due to alkali induced single strand breaks which convert rapidly sedimenting Form IV configuration (240S) to the more slowly sedimenting denatured single strand form of pea ctDNA (55.6S). The closed circular ctDNA from spinach and lettuce were also found to undergo degradation with a half-life of 10 ± 1 and 15 ± 1 min., respectively. The kinetics of alkali induced degradation was first order and monophasic. There was no evidence for the existence of two classes of alkali labile Form I ctDNA molecules. The rates of nicking of ctDNAs are 100 to 150 times faster than the rates of nicking of viral and *E. coli* DNAs, which do not contain covalently inserted ribonucleotides (Hill and Fangman, 1973).

The alkali sensitivity of ctDNA indicated the presence of covalently inserted ribonucleotides. This was tested by incubating the Form I DNA with RNase A and RNase T_1. The covalent-

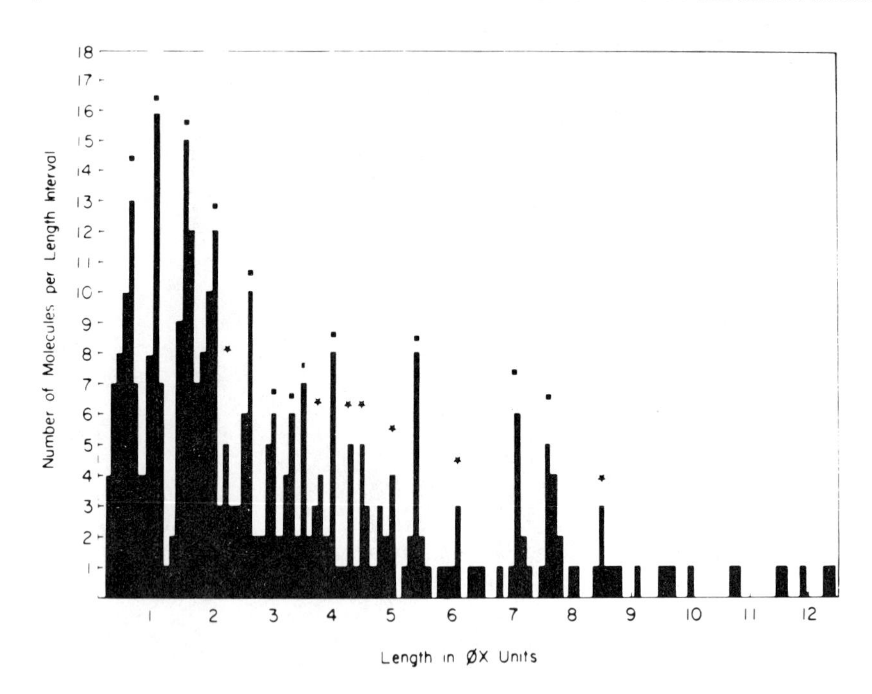

*FIGURE 14. The length distribution of fragments produced by the alkali hydrolysis of pea form I ctDNA. Pea form I ctDNA (10 μg/ml) was incubated at 20°C in 0.2 M NaOH, 0.04 M EDTA for 16 hr (≃ 96 half-lives) and was neutralized with 1.8 M Tris HCl, 0.2 M Tris. This DNA was spread with single-stranded ØX 174 DNA by the formamide technique; the spreading solution contained 50% formamide and they hypophase contained 20% formamide. Fields were selected and photographed randomly, and all of the molecules on a negative were measured. The ■ indicates the size classes that match the fragment lengths predicted by the map that is presented in Fig. 15. The * designates the size classes that match fragment lengths resulting from incomplete digestion of pea Form I ctDNA, as predicted by the map presented in Fig. 15.*

ly closed circular DNA was successively converted to open circular form. Under similar experimental conditions, G4 RF I DNA monomers and dimers were not affected. The preceding experiments indicated the presence of ribonucleotides in the Form I ctDNA. To determine the number of ribonucleotides present in the ctDNA, the rate of alkaline hydrolysis of ctDNA has been compared to the rate of alkaline hydrolysis of *E. coli* [^{32}P]RNA under identical conditions. The kinetic data of hydrolysis of the [^{32}P]RNA were first order, and the RNA had a half-life of 180 min. This half-life represents the rate of breakage of

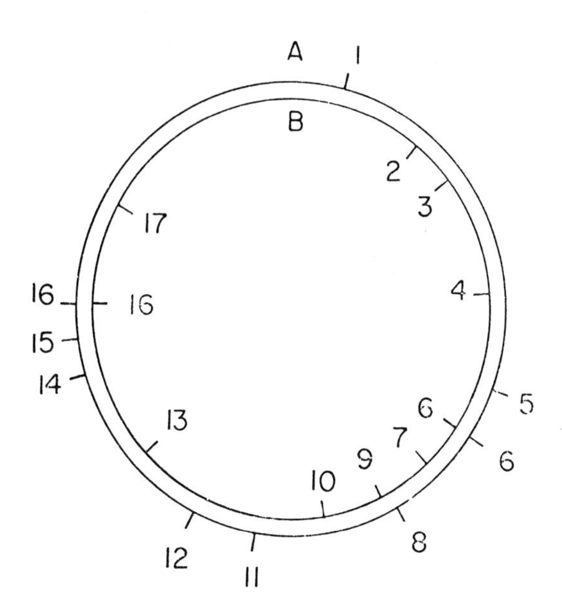

FIGURE 15. The map of the positions of the alkali labile sites in pea ctDNA.

a single RNA-RNA phosphodiester bond. The rate of nicking of pea and spinach ctDNA was 18 times as fast as the rate of breaking a single RNA=RNA phosphodiester bond. Therefore, pea and spinach Form I ctDNA nick at rates that would be expected if they each contained a maximum of 18 ± 2 ribonucleotides. Similarly, the lettuce Form I ctDNA was nicked 12 times as fast as an RNA phosphodiester bond, which corresponds to the presence of 12 ± 2 ribonucleotides in the lettuce ctDNA.

The ctDNAs could contain the ribonucleotides located at one or several sites in either one or both strands of the DNA. To investigate this, pea Form I ctDNA was degraded for 96 half-lives and was mounted for electron microscopy by the formamide technique to visualize single stranded DNA. The length distribution of the fragment sizes produced by this treatment is presented in Fig. 14. The distribution of fragment sizes clearly indicates that the ribonucleotides are not randomly distributed in the ctDNA. The fragment sizes also indicate that the alkali labile sites in pea ctDNA were located in both strands of the DNA since there were no unit length, single stranded circular molecules. To map the position of ribonucleotides, the alkali digested ctDNA fragments were reannealed to produce molecules that generally had single stranded tails and internal duplex sections. This procedure generates a number of overlapping molecules from which a map can be constructed containing the positions of the nicks relative to each other. If the alkali labile sites are at specific sites, this map should be circular with a monomer repeat length of 26.7 ϕX units (the length of pea ctDNA). A map of the relative positions of the alkali labile sites in pea ctDNA is presented in Fig. 15. Nineteen single strand breaks were located at distinct sites. All of the molecules were consistent with this map. It should be noted that at sites 6 and 16, two single strand breaks were mapped on opposite strands at the same position. When two nicks mapping on opposite strands at the same site were found on a hybrid molecule, one end of the molecule appeared to be fully duplex. The repeat length of the map was 26.7 ϕX units. If the map presented in Fig. 15 accurately represents the locations of the alkaline labile sites, it should be possible to make unambiguous strand assignments for each of the nicks. The two nicks that define a single strand tail of a reannealed molecule are located on opposite strands. Using this criteria, it was possible to make a list of the nicks that were located on opposite strands from each other. The nicks at positions 1, 5, 8, 11, 12, 14 and 15 (Fig. 15, Strand A) were located on one strand, while the nicks at positions 2, 3, 4, 7, 9, 19, 13, and 17 (Fig. 15, Strand B) were located on the other strand. This method located nicks on both strands at positions 6 and 16, which agrees with the previous finding that there was one nick located on each strand at these two positions. The significance of the individually inserted ribonucleotides in Form I ctDNA is not understood at this time. The ribonucleotides could arise from a nonstringent DNA polymerase, but in that case they would not be located at specific positions in the ctDNA molecule. It is possible that the ribonucleotides are remnants of RNA primers which are now generally believed to initiate DNA replication. In pea ctDNA, replication is initiated at two sites located on opposite strands of the DNA molecule. Replication is bidirectional and at least 50% of the ctDNA is synthesized bidirectionally. If the ribonucleotides resulted from incomplete excision of the primers that are used for the initiation of pea ctDNA replication, we should expect to observe only two alkali labile sites. If they were remnants of primers for "Okazaki fragments," we would expect to observe on the order of 40 to 50 alkali labile sites. In addition, they would be located more uniformly than the alkali labile sites we have observed. It should be pointed out that in all systems in which RNA primers for DNA replication have been studied, the primer is completely excised in the mature DNA except under abnormal conditions. We consider it possible that these ribonucleotides have some function in ctDNA. These sites could be involved in

TABLE 6
Inverted Repeat Sequence in ctDNAs

ctDNA source	Molecular size $\times 10^{-6}$		
	Single strand region	Double strand region	SS region
Lettuce	12.0	15	53.5
Spinach	11.4	15	52.8
Corn	11	15	43.7

transcription, recombination, or some other process that requires specific recognition sites. It will require further experimentation to test these possibilities.

Inverted Repeat Sequence in ctDNA

The ctDNA from spinach, lettuce, and corn has been found to contain one long stretch of inverted repeat sequences (Tewari *et al.*, 1976). The repeat length of this sequence corresponds to a molecular size of 15×10^6. The size of the repeat length is almost the same in all three ctDNAs (Table 6). Denaturation mapping has shown that repeat sequences occur in the same position in all of the circular molecules of ctDNA. No inverted repeating sequences have been found to be present in pea ctDNA.

INFORMATION CONTENT IN ctDNA

Chloroplasts have been shown to contain ribosomes, tRNAs, and messenger RNAs. What kind of RNA transcripts are formed from ctDNA? The following experimental results describe the genetic information contained in ctDNA.

Ribosomal RNA Genes in ctDNA

Chloroplasts contain ribosomes sedimenting at 70S compared to cytoplasmic ribosomes which sediment at 80S (Fig. 16). Purified chloroplasts essentially contain only 70S ribosomes (Fig. 16). There are many reports of 80S ribosomes being present in chloroplasts, but all those data appear to be because of contaminating endoplasmic reticulum in the chloroplast frac-

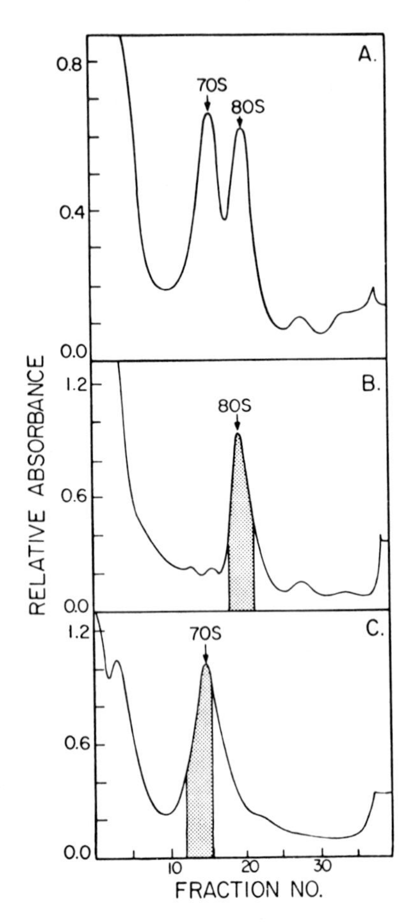

FIGURE 16. Sedimentation of ribosomes in sucrose gradient. Ribosomes were centrifuged in a 15-to-30% linear sucrose gradient at 24,000 rpm for 5 hrs in SW27 rotor. (A), total ribosomes of leaf; (B), purified cytoplasmic ribosomes; (C), purified chloroplast ribosomes.

tion. Purified 70S ribosomes contain 23S, 16S and 5S RNA. Fig. 17 shows that the 70S ribosome preparations used in the following studies do not contain any cytoplasmic 25S and 18S RNA. The data in Fig. 17 also show that the purified 23S and 16S RNAs are not contaminated with each other.

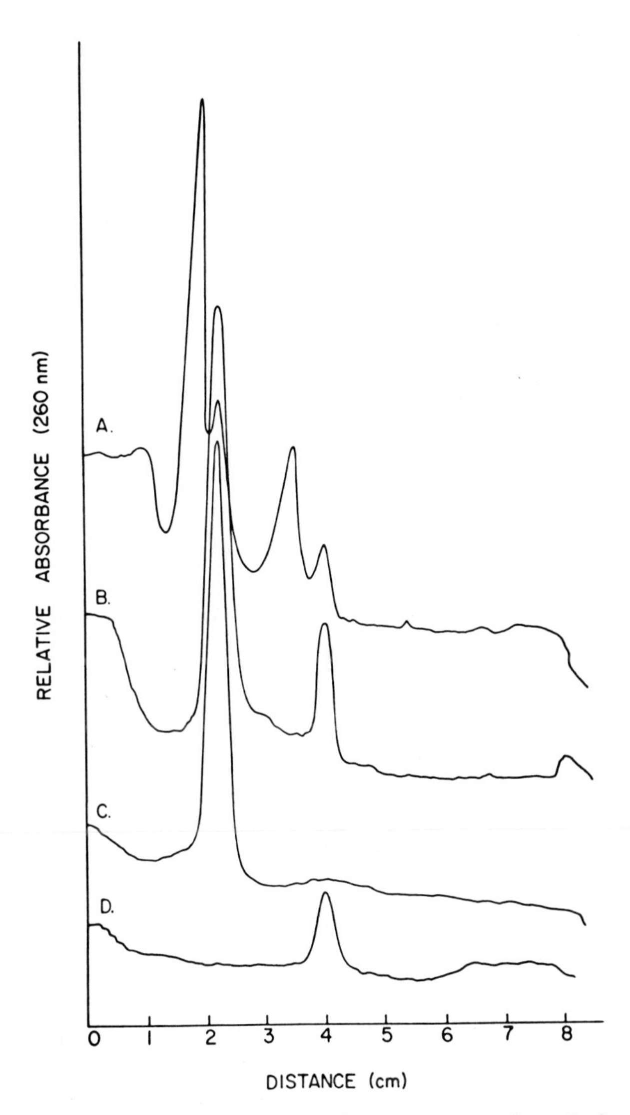

FIGURE 17. Acrylamide gel electrophoresis of fractionated and unfractionated plant rRNAs. (A) pea total leaf rRNA; (B) purified pea ctrRNA; (C) pea 23S rRNA; (D) pea 16S rRNA. Electrophoresis was carried out in MgB-buffer (9) for 5 hrs at 5 ma/tube using 2.6% acrylamide gels (5% bis-acrylamide).

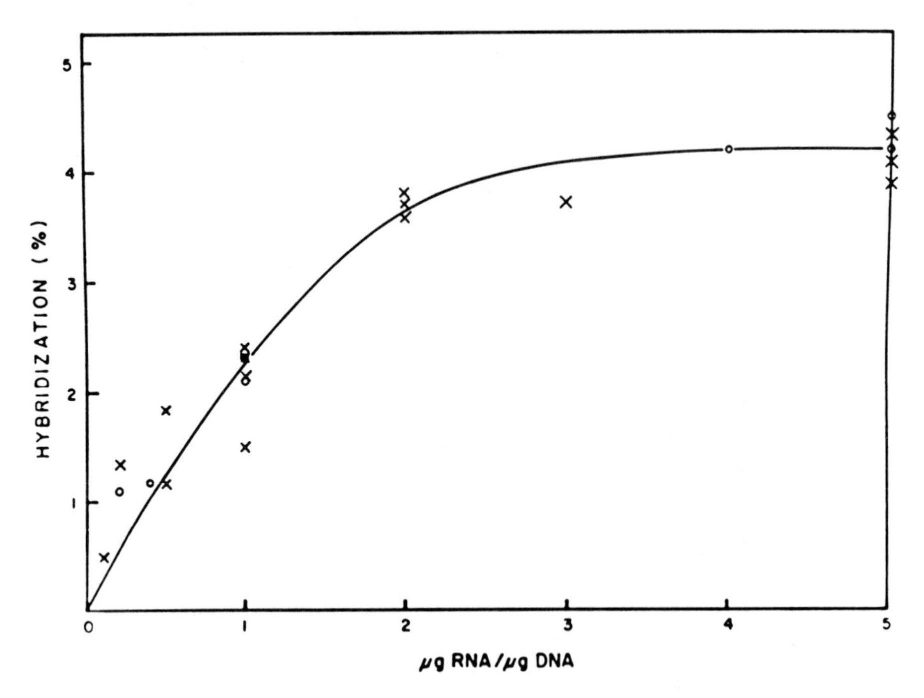

FIGURE 18. Saturation hybridization between pea ctDNA and increasing amounts of ^{32}P-labeled pea ctrRNA. The specific activities of the ^{32}P-labeled pea ctrRNA preparations ranged from 9,800 to 14,400 counts/min per µg. The percentage hybridizations were calculated on the basis of the amount of double stranded ctDNA on the filter. (X) 2.5 µg ctDNA on the filter; (O) 1.0 µg ctDNA on the filter.

Number of rRNA Genes in Pea ctDNA

Purified *in vivo* labeled pea ctrRNA was hybrodized with pea ctDNA. From the saturation-hybridization curve of Fig. 18, the maximum hybridization of 4.2% was obtained. The shape of the saturation curve showed that there were no heterogeneous populations in the ctrRNA preparations. 4.07% ± 0.35% of the ctDNA was found to be complementary to rRNA in different experiments involving varying concentrations of RNA and DNA. DNA-rRNA hybridization was also carried out in solution and the hybrids fractionated by hydroxyapatite chromatography. The maximum hybridization obtained was about 4.0% at the end of 3 h. In this experiment, the size of the fragmented DNA was about 2.8×10^6 daltons and self-annealing in 3 h at this size and concentration should be about 10%. Therefore, the amount of hybridization obtained with solution hybridization was very close to that obtained by filter hybridization. The specificity of ctDNA and ctrRNA has been confirmed by competition experiments. In addition, the thermal stability characteristics of the DNA-rRNA hybrids indicate the absence of mismatched bases.

The saturation hybridization experiments have also been carried out with purified 23S and 16S RNA. About 2.5% of the ctDNA is complementary to 23S RNA and about 1.45% of the ctDNA is complementary to 16S RNA. There was no cross-competition between the two rRNA species. The molecular size of pea ctDNA has been found to be able 90×10^6 and about 4% of its base sequences hybridize with ctrRNA. Assuming a molecular weight of 1.7×10^6 for ctrRNA, the hybridization data indicates that there are two genes for ctrRNA in pea ctDNA. The hybridization data also show that the genes for 23S and 16S RNA are non-overlapping.

To further characterize the rRNA genes in the ctDNA, ctDNA was centrifuged in a csCl gradient and each fraction was analyzed by hybridization. The plot of the counts hybridized to the various fractions of ctDNA showed that the hybridization peak for the ctrRNA is slightly to the heavier density side of the bulk of ctDNA. When the values of percent hybridization are analyzed against the fractions, the DNA on the heavier side of the gradient was found to hybridize 2-3 times more than the main-band DNA. A clear picture emerges when similar experiments are carried out with fragmented ctDNA of 3×10^6 daltons (Fig. 19). Here, the absorbance pattern of ctDNA showed a slightly broader band compared to the previous experiment. CtrDNA clearly banded at a higher density than the main-band; the percent hybridization being 4-5 times more in the heavy region than in the main-band.

Conservation of Chloroplast rRNA Genes in Higher Plants

To find out the base sequences of ctrRNA genes in higher plants, the radioactively labeled pea ctrRNA has been hybridized with ctDNA from spinach, bean, lettuce, and corn. The saturation hybridization data are shown in Fig. 20. At saturation, 4.2% of the pea ctDNA is found to hybridize with pea ctrRNA (homologous system). When spinach ctDNA is substituted for pea ctDNA, about 4% of the spinach ctDNA is found to be complementary to labeled pea ctrRNA (heterologous). The hybridization between labeled pea ctrRNA and ctDNAs is about the same, irrespective of the source of ctDNA.

From the level of hybridization and the molecular sizes of ctDNA, it appears that there are two gene equivalents of ctrRNA in the ctDNAs of higher plants. In order to find out

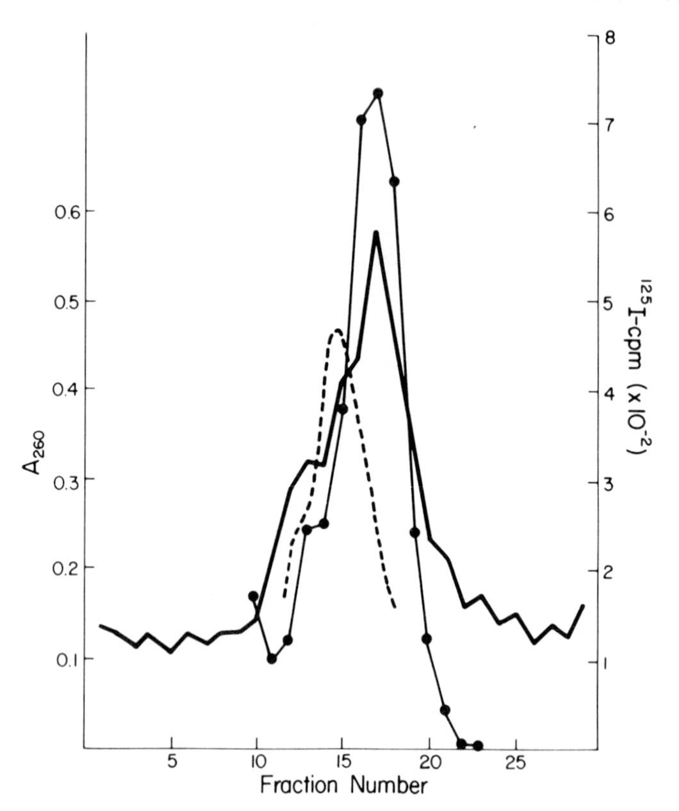

FIGURE 19. Hybridization of labeled pea ctrRNA and pea cttRNAs with the various fractions of pea ctDNA banded in a CsCl gradient of 1.710 g/cm³ in a Spinco fixed angle 50 rotor. ——, A_{260} nm, ----, hybridization with ctrRNA, ●——●, hybridization with cttRNAs.

the specificity of heterologous hybridizations, competition experiments were carried out involving the labeled pea ctrRNA and unlabeled ctrRNA from spinach, bean, lettuce, pea, and corn. The data given in Fig. 21 indicate that the competition was of the same order whether the homologous or heterologous rRNA was used in the reaction forming heterologus DNA-rRNA hybrids. The homologous and heterologous DNA-rRNA hybrids have been analyzed by thermal denaturation. All heterologous hybrids showed a sharp Tm at about 20°C above the hybridization temperature. The thermal stability profiles do not reveal significant heterogeneity in the hybrids. The maximum difference between the lowest and the highest Tm could result from no more than 5% differences in the base sequences of RNAs.

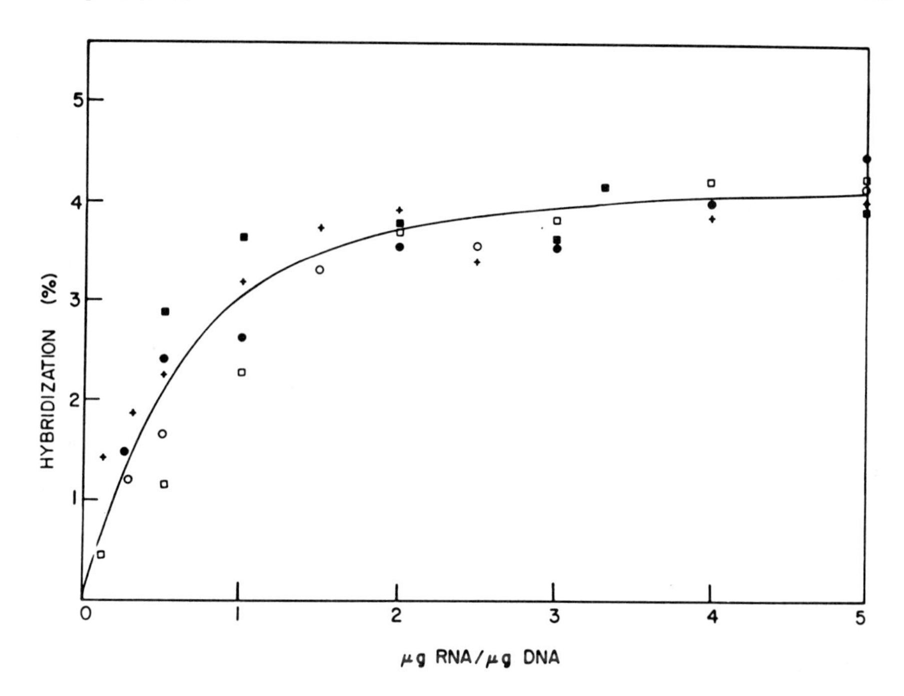

FIGURE 20. Hybridization of ^{32}P-labeled pea ctrRNA with ctDNA from different plants. Each filter contained 1.0 μg ctDNA which were prepared as described (9). (□), pea; (O), bean; (●), lettuce; (+), spinach; (■), corn. The specific activities (cpm/μg) of pea ctrRNA preparations used in the experiments with pea, spinach, bean, lettuce, and corn ctDNAs were 13, 560, 11, 400, 11,400, 10,800, and 7,540 respectively. The percent hybridization on the ordinate is based upon the weight of double stranded ctDNA.

The competition and thermal stability studies show that the nucleotide sequences of the rRNA genes have been exceptionally invariant during the evolution and divergence of higher plants.

tRNAs Genes in Chloroplast DNA

Total tRNAs from chloroplasts have been found to hybridize with 1.23% ± 0.35% of the ctDNA (Fig. 22). This level of hybridization accounts for 40 ± 10 tRNAs genes in ctDNA. The sequence specificity of ctDNA-tRNA hybrid has been confirmed by competition hybridization and thermal stability studies on DNA-tRNA hybrids. Competition experiments have shown that: 1) unlabeled ct-tRNA competes with labeled tRNA at the theo-

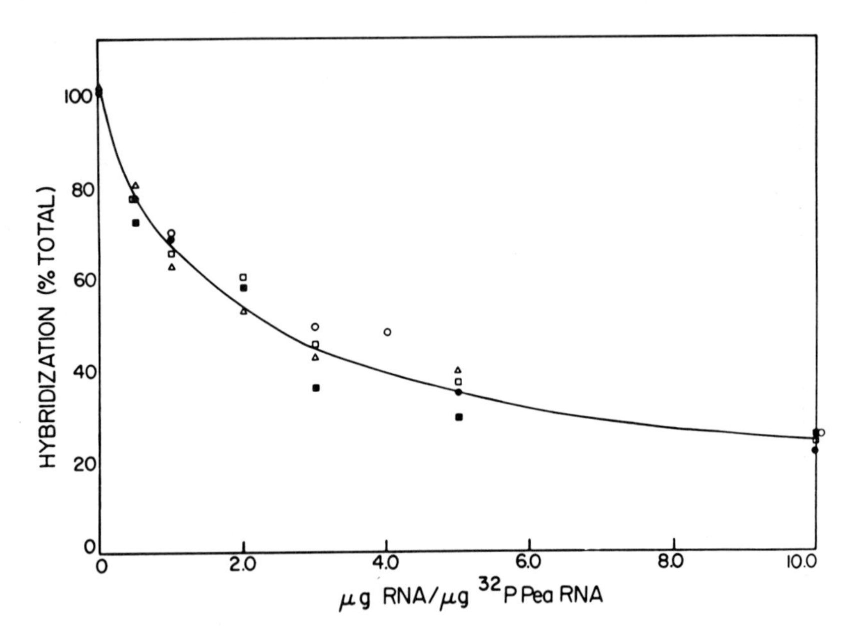

FIGURE 21. Competition hybridization experiments. CtDNAs from different plants were hybridized with ^{32}P-labeled pea ctrRNA in the presence of increasing amounts of cold ctrRNA from different plants. All the data reported are an average of 3 experiments. (O), hybridization with 5.0 μg bean ctDNA in the presence of unlabeled ctrRNA from lettuce. (△), hybridization with 1.0 μg corn ctDNA in the presence of increasing amounts of ctrRNA from spinach. (●), hybridization of 1.5 μg pea ctDNA in the presence of increasing amounts of ctrRNA from oats. (■), hybridization of 1.5 μg pea ctDNA with 3.75 μg ^{32}P-labeled pea ctrRNA in the presence of increasing amounts of ctrRNA from corn. (□), hybridization of 2.5 μg pea ctDNA with ^{32}P-labeled pea ctrRNA in the presence of increasing amounts of rRNA from peas.

retically expected level, 2) pea cytoplasmic tRNAs do not compete significantly with ^{125}I-cttRNA, and 3) the bacterial, yeast, and calf liver tRNAs do not compete with ^{125}I-cttRNA (Fig. 23). These competition experiments clearly indicate that the base sequences of cttRNA genes are unique. Further analysis of the ctDNA by fractionation in a CsCl density gradient and hybridization of the various fractions with tRNAs and rRNA has shown that the rRNA and tRNAs genes are not closely localized in the DNA molecule (Fig. 19). What is the nature of the 30-40 tRNA genes in ctDNA? Are they composed of repeating units of one or two individual aminoacyl tRNAs or are

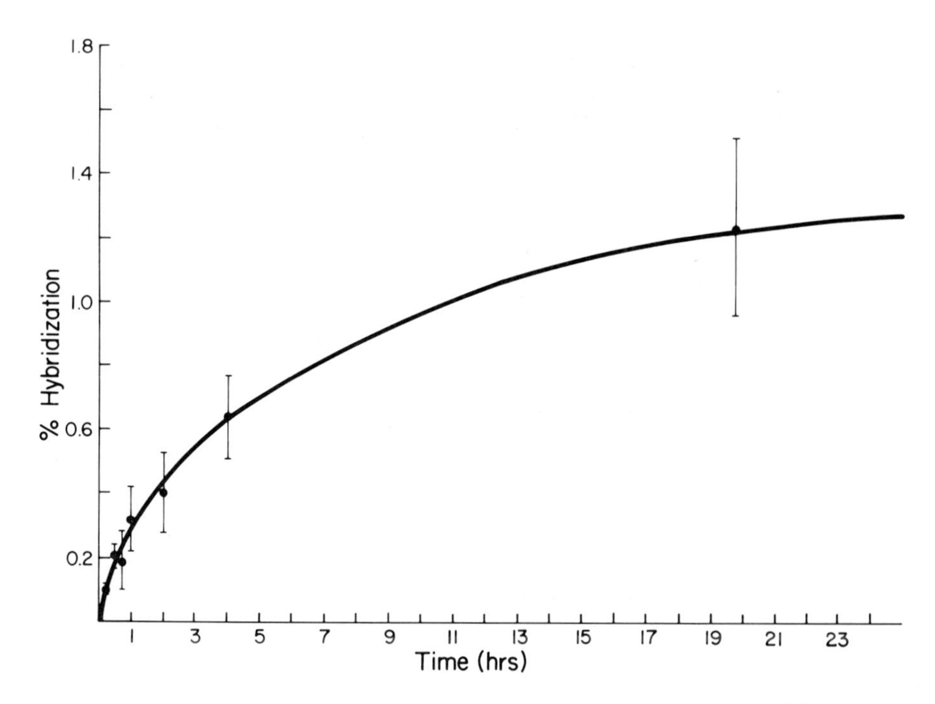

FIGURE 22. Hybridization of ctDNA with in vitro ^{125}*I-labeled tRNA. 1 μg of ctDNA and 2 μg of labeled tRNAs were used in the experiment.*

there copies of tRNA for each of the 20 different amino acids? To answer this question, we have aminoacylated ct-tRNAs with ^{3}H-labeled amino acids. The individually aminoacylted tRNA was hybridized with ctDNA. A saturation hybridization curve for tyr-tRNA is shown in Fig. 24. The data obtained from such experiments are given in Table 7. The hybridization data indicate that 17 aminoacyl tRNA genes are present in ctDNA. Three amino acids, cysteine, glutamine and glutamic acid, were found to be poorly acylated under our experimental conditions. Thus, it is difficult to conclude that these three aminoacyl tRNAs are not coded by ctDNA.

Complete Transcription of ctDNA

It is apparent from the above data that only about 5×10^6 daltons of ctDNA is utilized in the coding of rRNA and tRNAs. The remaining 45×10^6 daltons of ctDNA is available for the formation of mRNA transcripts. Are all of the base sequences of ctDNA transcribed in chloroplasts? The saturation hybridi-

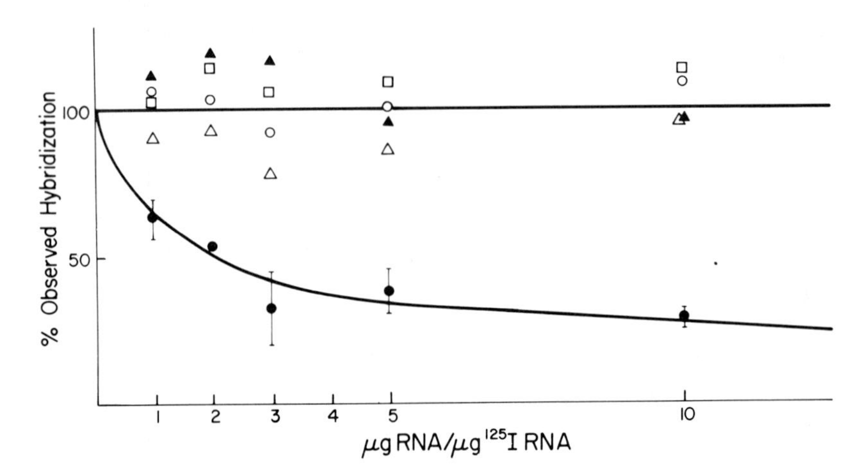

FIGURE 23. Competition hybridization experiments. Pea ctDNA was hybridized with 2 μg of ^{125}/cttRNA in the presence of increasing amounts of pea cttRNAs (●—●), pea cyt. tRNAs (△—△), [E. coli *tRNAs* (○—○)], *yeast tRNAs* (□—□), *and calf liver tRNAs* (▲—▲).

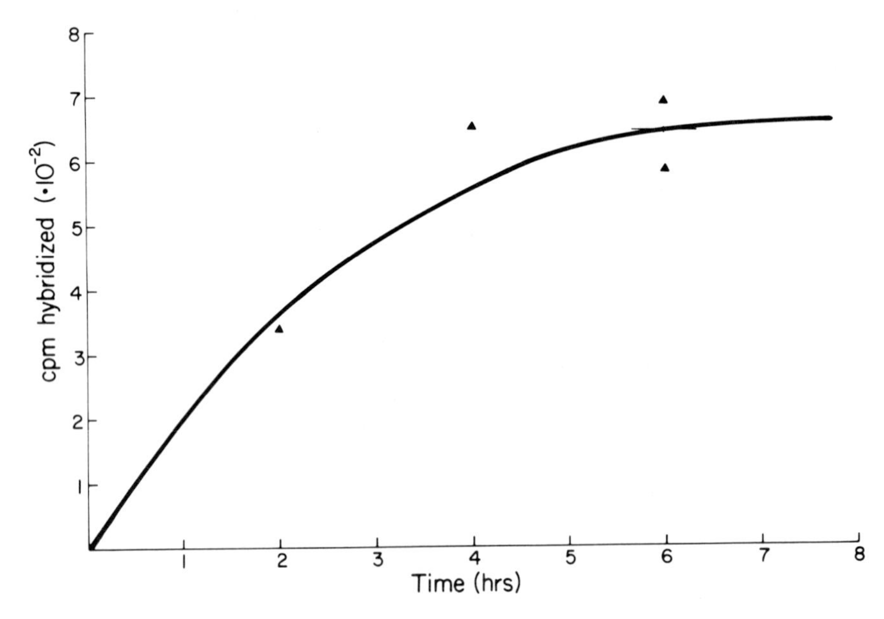

FIGURE 24. Saturation hybridization of ctDNA with increasing concentrations of ^{3}H-tyrosyl tRNA.

TABLE 7

Hybridization of Aminoacyl tRNAs with ctDNA

Amino acid	Specific activity of aminoacyl tRNA (counts per min/ g total tRNA)	Amount of ctDNA on filter	Number of counts bound to ctDNA
Alanine	102	10	12
Valine	444	10	30
Leucine	2,100	5	50
Isoleucine	1,184	10	134
Phenylalamine	2,200	5	143
Tryptophan	120	20	12
Methionine	340	10	24
Glycine	12	10	14
Serine	592	10	24
Threonine	300	10	12
Cysteine	255	10	5
Tyrosine	2,453	10	665
Asparagine	1,595	10	66
Aspartic acid	95	10	161
Lysine	503	10	106
Arginine	659	10	65
Histidine	132	10	21

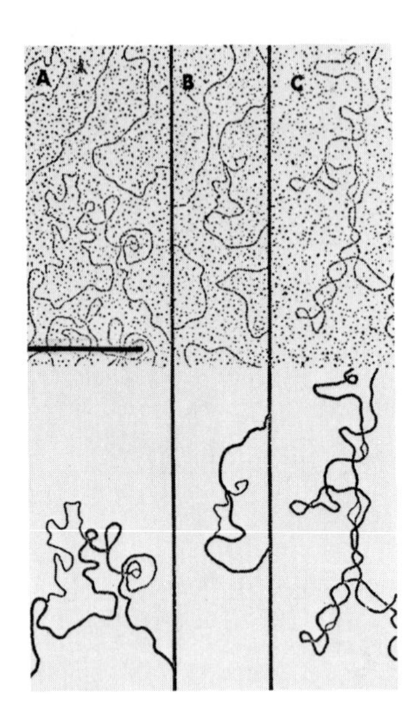

FIGURE 25. High magnification electron micrographs of ctDNA molecules. a) A molecule containing one displacement loop (D-loop) and one denaturation loop (Den-loop) from the pea ctDNA. b) A branch migrating D-loop from corn ctDNA. c) A pea ctDNA molecule containing three Den-loops. The thin and thick lines in the drawings represent the single-stranded regions, respectively. The bar indicates 1 μm.

zation with the *in vivo* labeled RNA and ctDNA has shown that about 50 to 60% of the base sequences of ctDNA are transcribed. The *in vivo* labeled RNA was fractionated on a poly-U sepharose column or on poly-U cellulose filters. About 0.5% of the total RNA was found to bind with the poly U in the presence of 25% formamide, 0.7M NaCl, 50mM Tris, 10mM EDTA, pH 7.5. The bound RNA (presumably containing poly A) was eluted from the column by passing 90% formamide, 10mM EDTA, 10mM KPO_4, pH 7.5 through the column. Non-poly A and poly A containing RNAs were found to hybridize with 45-50% and 15-20% of the ctDNA, respectively. These experiments indicate the presence of both poly A and non-poly A containing RNA transcripts of ctDNA in chloroplasts. Current efforts are directed towards further characterization of these mRNA transcripts.

REPLICATION OF CHLOROPLAST DNA *IN VIVO*

Displacement Loops in ctDNA

The closed circular pea ctDNA molecules have been found to contain two displacement (D) loops which are located at two adjacent sites (Fig. 25). The smallest size of D-loops was 820 base pairs and the average distance between the outside edges of the two D-loops was 7.15 kilobase pairs. The inner distance between the two D-loops was highly variable, indicating that the two D-loops expand towards each other. Small "Cairns type" replicative forked structures ranging in size from 7.2 to 10.6 kilobases have been observed in closed circular pea ctDNA. These Ciarns structures map at the position of the two D-loops and probably result when the two displacing strands expand past each other on opposite strands. This confirms that the two displacing strands are located on the opposite parental strands of the pea ctDNA molecule. The two D-loops in corn ctDNA are also located at two adjacent sites. The size of the corn ctDNA D-loop corresponds to 860 base pairs and the outer distance between the two D-loops was 7.06 kilobase pairs.

Cairn's Replicative Intermediates

Cairn's type of replicative forked structures were found in the DNA from lower, middle, and upper bands (Figs. 2; 26). In pea ctDNA, three percent of the circular molecules in the lower band were Cairns' structures and the extent of their replication ranged from 5.2% to 8.2%. The Cairns replicative intermediates made up an average of 5.7% of the circular DNA in the middle band and the extent of their replication ranged from 7% to 50%. The Cairns structure consisted of 2.9% of the circular ctDNA molecules in the upper band and the extent of their replication ranged from 38% to 87%. The finding of Cairns replicative intermediates in the lower and middle bands of the CsCl-propidium diiodide density gradients, and the correlation between higher banding position and larger amount of replication again suggests that replication takes place on a covalently closed circular template and is accompanied by nicking and closing cycles. In corn ctDNA, Cairns replicative intermediates made up 4.5% of the total circular ctDNA molecules and the extent of their replication ranged from 9.4% to 70%.

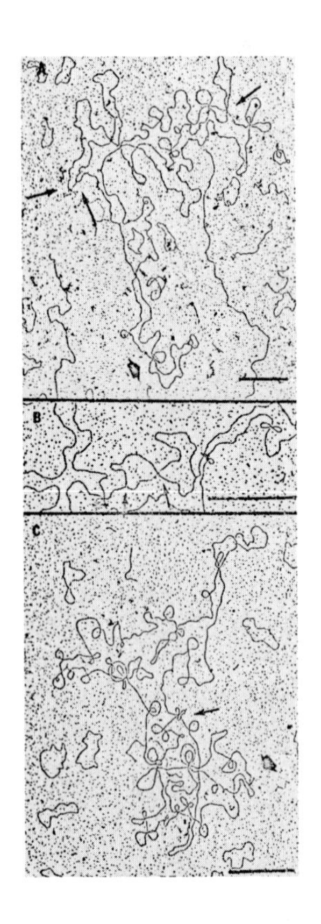

FIGURE 26. Cairns replicative intermediates. (A) A molecule that is 37% replicated and contains a single-strand region (2 small arrows) at one growing fork and a denatured region (large arrow) in the unreplicated portion of the molecule. (B) A portion of a molecule that is 5.2% replicated and contains a single strand region (2 small arrows) at one growing fork. (C) A Cairns replicative forked molecule showing branch migration at one of the replicative forks (large arrow). The molecules were mounted for electron microscopy by the formamide technique; the spreading solution contained 50% formamide. Single-stranded and double-stranded ØX DNA molecules were used as internal standards. The bars indicate 1 μm.

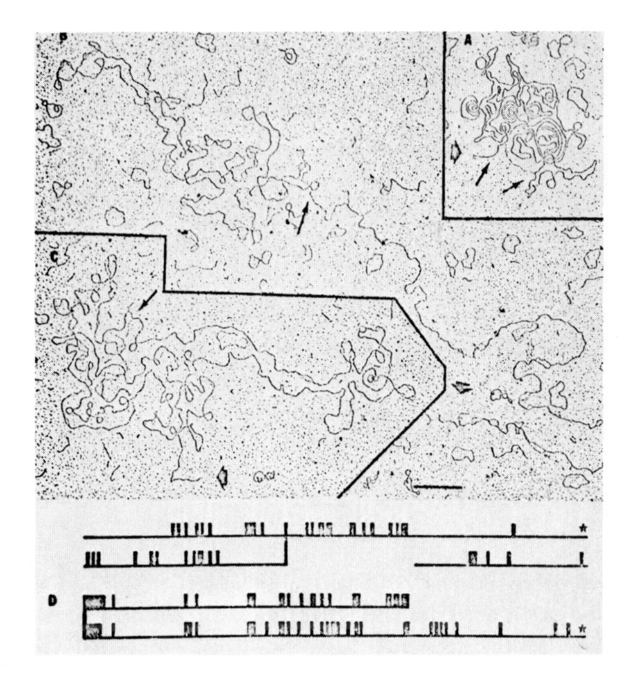

*FIGURE 27. Rolling circle molecules. (A) A molecule with a tail that is 19% of length of the attached circle. There is a single-strand region at the growing fork that is 3,500 bases long (indicated by arrows). (B, C) Partially denatured molecules with tails that illustrate the two different denatured patterns at the tip of the tails. (D) Graphical representation of molecules B and C. The boxes indicate the denatured regions, the * indicates the circular part of the molecule that has been linearized. The unmarked free end is the tip of each tail. The bars indicate 1 μm.*

Rolling Circle Replicative Intermediates

In the pea ctDNA preparations circular molecules with an attached double stranded tail accounted for 4.9% of the circular molecules in the upper band, and 2.8% of the circular molecules in the lower band (Fig. 27). There were no circular molecules with tails in the lower band of pea ctDNA. The lengths of tails in the pea ctDNA ranged from 1.5% to 124% of the length of the attached monomer length circular molecule. In corn ctDNA, 11% of the total circular molecules had tails which ranged in length from 2% to 140% of the attached monomer length circular molecule. The finding of tails that were longer than

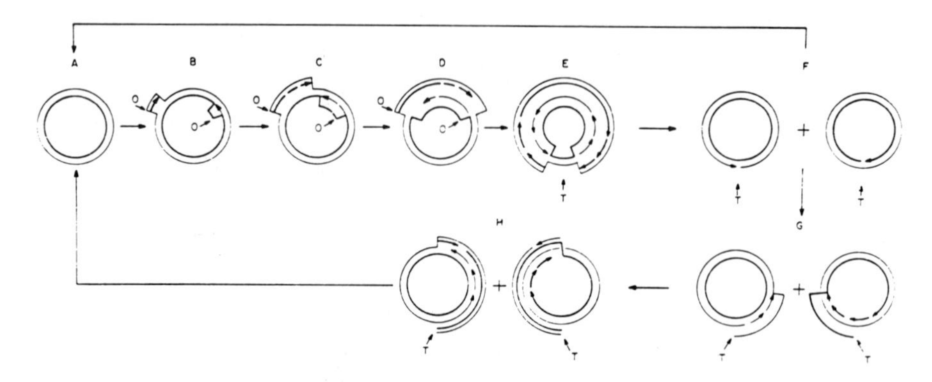

FIGURE 28. A model for the replication of ctDNA. (a), closed circular parental molecule; (b), D-loop containing molecule; (c), expanded D-loop containing molecule; (d, e), Cairns type of replicative intermediate; (f), nicked progeny molecules; (g, h), rolling circles. The thin and thick lines mark the opposite strands of a molecule. The lines with the arrows are the daughter strands. The "O" indicates the positions of the two origins of D-loop synthesis which are 5.2% of pea ctDNA apart. The "T" indicates the terminus of the Cairns round of replication which is 180° opposed to the origins of D-loop synthesis.

the attached monomer length circular molecule eliminated the possibility that the tails arose by breakage of a Cairns forked structure at a replicative fork.

A Model For the Replication of Chloroplast DNA

Based upon the above results, a model for the replication of pea and corn ctDNA is presented in Fig. 28. In this model, ctDNA replication is initiated by the formation of two displacement loops whose displacing strands are complementary to the opposite parental strands of ctDNA (Fig. 28b). The two displacing strands expand toward each other and initiate the formation of Cairns replicative forked structures (Fig. 28c, d). The small Cairns forked structures (Fig. 28e) expand bidirectionally until termination takes place at a site that is 180° around the circular ctDNA molecule from the initiation site. Separation of the daughter molecules takes place yielding two circular molecules that each have a single strand break or small gap at the same site located in opposite daughter strands (Fig. 28f). In ctDNA, the nicked circles could be sealed to close the circles or the 3'OH of each nicked progeny molecule could be extended by a DNA polymerase molecule. This would displace a single stranded tail (Fig. 28g) from the molecule and this tail could be filled in by discontinuous duplex synthesis to yield a molecule with a double stranded tail (Fig. 28h). The tails might then be converted to circular molecules by an intrastrand recombination event. If both progeny from a Cairns round of replication initiated rolling circle synthesis, two types of rolling circles would be formed. In each case, the tip of the tail would map at the same site but the sequence of the two types of tails would extend in the opposite direction from this site. Denaturation maps with pea ctDNA rolling circle confirm that this is the case.

DNA Synthesis *In Vitro*

The above experiments on replication have been carried out with *in vivo* replicative intermediates. In order to further understand the replication process, we must resolve and reconstitute each function outside the cell, and understand its regulation inside the cell. As a first step we have studied DNA synthesizing activity in the isolated chloroplasts. Sucrose density gradient purified pea chloroplasts were found to carry out a very active DNA synthesis (DS) in the presence of dNTP utilizing endogenous ctDNA. The lysis of chloroplasts with triton, deoxycholate, and EDTA failed to solubilize any significant amount of DS activity from chloroplasts. The DS activity was completely solubilized when chloroplasts were treated with 0.5% Lubrol.

The solubilized fraction was adjusted to 0.3M KCl and passed through a DEAE-cellulose column equilibrated with the same buffer. The effluent from the column contained all the activity of the isolated chloroplasts and this activity was totally dependent upon the externally added DNA. This DEAE-cellulose column fraction was able to synthesize DNA to the extent of 1 to 2% of the template DNA and was active against both double-stranded and single-stranded DNA. Using single-stranded ϕX174 and G4 DNA as a template, the *in vitro* synthesized product was found to be very close to the molecular size of the template. This fraction was allowed to synthesize DNA in the presence of ctDNA by substituting bromodeoxyuridine triphosphate for thymidine triphosphate in the DS reaction system. The isolated product had a buoyant density expected of a DNA in which one entire strand contained bromodeoxyuridine instead of thymidine. This DEAE-cellulose column fraction was further fractionated on a DEAE cellulose column with 0 to 0.3M linear KCl gradient. The data are shown in Fig. 29. DS activity was found in three different peaks. The first two peaks

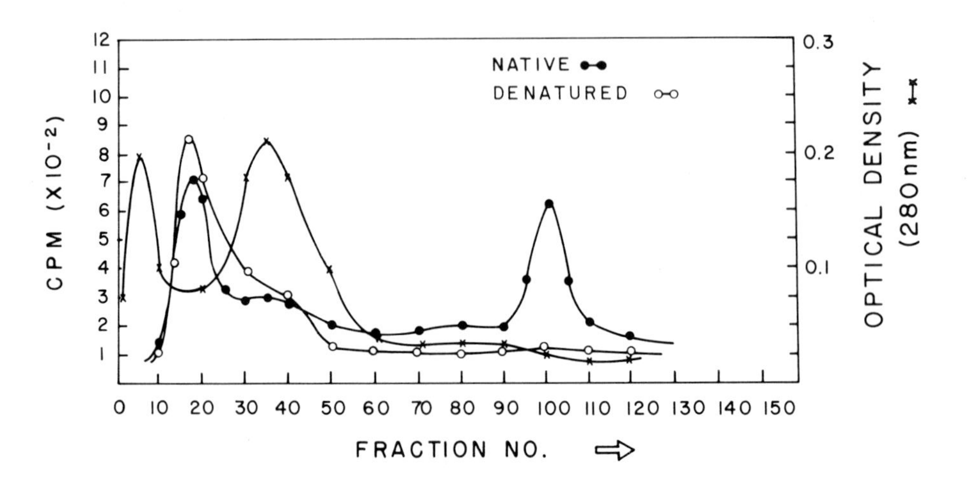

FIGURE 29. The elution profiles of DNA synthesizing activities and protein from a DEAE-cellulose column. The elution was carried out with a linear gradient of 0 to 0.3M KCl.

eluted at approximately 0.07 and 0.13M KCl. These two peaks were found to be active in the presence of both double-stranded and single-stranded DNA. The third peak eluted at 0.25M KCl and was active only when double-stranded DNA was used as a template. The three different peaks were each further purified on a phosphocellulose column. The activity from peak I eluted at 0.3M KCl from the phosphocellulose column. The activity found in peak II was found to elute without any salt. The activity corresponding to peak III eluted with 0.4M KCl.

Experiments are currently in progress to completely purify the three DS activities. Using these purified proteins and the native ctDNA, we propose to study replication *in vitro*. It is clearly recognized by the authors that the complete replication of ctDNA might require proteins other than these three fractions. The knowledge of the *in vivo* replicative intermediates will help us to identify all the proteins necessary for the replication of ctDNA.

ACKNOWLEDGMENT

The work presented here reflects the contribution of many of my graduate and undergraduate students. I would like to especially thank Dr. J. R. Thomas, Dr. R. Kolodner, Dr. M. White, Mr. N. Chu, Mr. W. Dobkin, Mr. T. Sumnicht, and Mr. R. McKown. The assistance of Ms. J. Keithe with the electron microscopy is gratefully acknowledged. I would also like to thank Ms. Ann Rule for help in the preparation of this manuscript. This work was supported by N.S.F.grant PCM 76-10186.

REFERENCES

Barrel, B. G., Air, G. M., and Hutchison, C. A. (1976). Overlapping genes in bacteriophage ϕX174. *Nature 264*, 34-41.

Bayen, M., and Rode, A. (1973). Heterogeneity and complexity of *Chlorella* chloroplastic DNA. *Eur. J. Biochem. 39*, 413-420.

Gray, H., Bloomfield, V. A., and Hearst, J. E. (1967). Sedimentation coefficients of linear and cyclic wormlike coils with excluded-volume effects. *J. Chem. Phys. 46*, 1493-1498.

Hill, W. E., and Fangman, W. L. (1973). Scission of *Escherichia coli* deoxyribonucleic acid in alkali. *Biochemistry 12*, 1772-1774.

Inman, R. B., and Schnös, M. (1970). Partial denaturation of thymine- and 5-bromouracil-containing λ DNA in alkali. *J. Mol. Biol. 49*, 93-98.

Kolodner, R., and Tewari, K. K. (1972a). Molecular size and conformation of chloroplast deoxyribonucleic acid from pea leaves. *J. Biol. Chem. 247*, 6355-6364.

Kolodner, R., and Tewari, K. K. (1972b). Genome sizes of chloroplasts and mitochondrial DNA's in higher plants. In *Proceedings of the 30th Annual Meeting of the Electron Microscopic Society of America* (Arceneaux, C. J., ed.), pp. 190-191. Claitor's Publishing Division, Baton Rouge, La.

Kolodner, R., and Tewari, K. K. (1975a). The molecular size and conformation of the chloroplast DNA from higher plants. *Biochim. Biophys. Acta 402*, 372-390.

Kolodner, R., and Tewari, K. K. (1975b). Denaturation mapping studies on the circular chloroplast deoxyribonucleic acid from pea leaves. *J. Biol. Chem. 250*, 4888-4895.

Kolodner, R., and Tewari, K. K. (1975c). Presence of displacement loops in the covalently closed circular chloroplast deoxyribonucleic acid from higher plants. *J. Biol. Chem. 250*, 8840-8847.

Kolodner, R., and Tewari, K. K. (1975d). Chloroplast DNA from higher plants replicates by both the Cairns and rolling circle mechanism. *Nature 256*, 708-711.

Kolodner, R., Tewari, K. K., and Warner, R. C. (1976). Physical studies on the size and structure of the covalently closed circular chloroplast DNA from higher plants. *Biochim. Biophys. Acta 447*, 144-155.

Kolodner, R., Warner, E. C., and Tewari, K. K. (1975). The presence of covalently linked ribonucleotides in the closed circular deoxyribonucleic acid from higher plants. *J. Biol. Chem. 250*, 7020-7026.

Meeker, R., and Tewari, K. K. In preparation.

Stutz, E., and Rawson, J. R. (1970). Separation and characterization of *Euglena gracilis* chloroplast single-strand DNA. *Biochim. Biophys. Acta 209*, 16-23.

Stutz, E., and Vandrey, J. (1971). Ribosomal DNA satellite of *Euglena gracilis* chloroplast DNA. *FEBS Letters 17*, 277-280.

Tewari, K. K., and Wildman, S. G. (1976). DNA polymerase in isolated tobacco chloroplasts and nature of the polymerized product. *Proc. Nat. Acad. Sci. U.S. 58*, 689-696.

Tewari, K. K., and Wildman, S. G. (1968). Function of chloroplast DNA. I. Hybridization studies involving nuclear and chloroplast DNA with RNA from cytoplasmic (80S) and chloroplast (70S) ribosomes. *Proc. Nat. Acad. Sci. U.S. 59*, 569-576.

Tewari, K. K., and Wildman, S. G. (1970). Information content in the chloroplast DNA. *Soc. Exp. Biol. 24*, 147-179.

Tewari, K. K. (1971). Genetic autonomy of extranuclear organelles. *Ann. Rev. Plant Physiol. 22*, 141-168.

Tewari, K. K., Kolodner, R., Chu, N. M., and Meeker, R. (1977). In *N.A.T.O. Advanced Study Institute on Nucleic Acids and Protein Synthesis in Plants* (Bogorad, L., and Weil, J. H., eds.). Plenum Press, N. Y.

Tewari, K. K., Kolodner, R. D., and Dobkin, W. (1976). Replication of circular chloroplast DNA. In *Genetics and Biogenesis of Chloroplasts and Mitochondria* (Bucher, T., Neupert, W., Sebald, W., and Werner, S., eds.), pp. 379-386, North-Holland Publishing Co., N. Y.

Thomas, J. R., and Tewari, K. K. (1974a). Ribosomal-RNA genes in the chloroplast DNA of pea leaves. *Biochim. Biophys. Acta 361*, 73-83.

Thomas, J. R., and Tewari, K. K. (1974b). Conservation of 70S ribosomal RNA genes in the chloroplast DNAs of higher plants. *Proc. Nat. Acad. Sci. U.S. 71*, 3147-3151.
Vedel, F., Quetier, F., Bayen, M., Rode, A., and Dalmon, J. (1970). Intramolecular heterogeneity of mitochondrial and chloroplastic DNA. *Biochim. Biophys. Res. Commun. 46*, 972-978.
Wells, R., and Sager, R. (1971). Denaturation and the renaturation kinetics of chloroplast DNA from *Chlamydomonas reinhardi*. *J. Bol. Biol. 58*, 611-622.

SECTION II
Molecular Aspects of Transcription and Translation in Plants

PLANT RIBOSOMAL RNA GENES - A DYNAMIC SYSTEM?

J. Ingle

Botany Department
University of Edinburgh
Edinburgh EH9 35H, Scotland

One fascinating aspect of genome structure relates to the question of whether the genome is static or dynamic. Is the genome constant in all cells of an individual? That is, does the DNA undergo differential replication whereby certain parts of the genome are under- or over-replicated? There are several experimental approaches to this problem of differential replication. Evidence suggesting its occurrence may be obtained from determining the amount of DNA in individual nuclei. Most tissues contain nuclei within a polyploid series, 2C, 4C, 8C, 16C,...2^nC. If the DNA is fully replicated at each stage, then the peak values for DNA contents will fall within a doubling series. If the nuclear DNA values, determined by Feulgen staining and microspectrophotometry, are distributed in a non-doubling series this may indicate a differential replication of the genome. Fox (1970) has reported in locusts a tissue specific, non-doubling of DNA during the development of polyploid cells. Visual evidence of differential replication can be obtained from cytological studies involving heterochromatin and euchromatin. Cultured orchid protocorm tissue contained two populations of nuclei, one the

ISBN 012-601950-9

result of regular endopolyploidy of both the eu- and heterochromatin, while the other contained a much larger content of heterochromatin (Nagl, 1972).

The most satisfactory way to look for differential replication is to see whether particular sequences are preferentially under- or over-replicated. This requires that the particular sequences can be assayed, and there are few sequences which may be quantitatively determined. One such sequence is satellite DNA, defined as a minor component in a neutral CsCl equilibrium gradient. Satellite DNAs are present in many dicot species. Where they are clearly resolved from the main band, such as in citrus or melon (Fig. 1), the amount of the satellite DNA may be readily determined quantitatively.

Satellite DNA appears to be under-replicated during fruit development in melon and cucumber (Pearson *et al.*, 1974). The results, summarized in Table 1, show that with all the melons examined the fruit tissue contained a smaller percentage of satellite DNA than the corresponding seed. With the range of material analysed, the difference between fruit and seed was highly significant ($p<0.1\%$). The difference was not limited to the relative distribution of main band and satellite DNA's; the buoyant densities of both main-band and satellite peaks were greater in the fruit than in the seed. The difference was again highly significant ($p<0.1\%$). Cucumber DNA was similar in that the percentage of satellite DNA was lower in the fruit than in the seed. The density of both satellite components was significantly ($p<0.1\%$) greater in the fruit. These results indicate the occurrence of gross differential replication during fruit development.

The under-replication of the satellite DNA may be correlated with under-replication of heterochromatin. Whereas meristematic nuclei from root-tips and seed of cucumber contained approximately 25% heterochromatin (Fig. 2c), the bulk of the highly

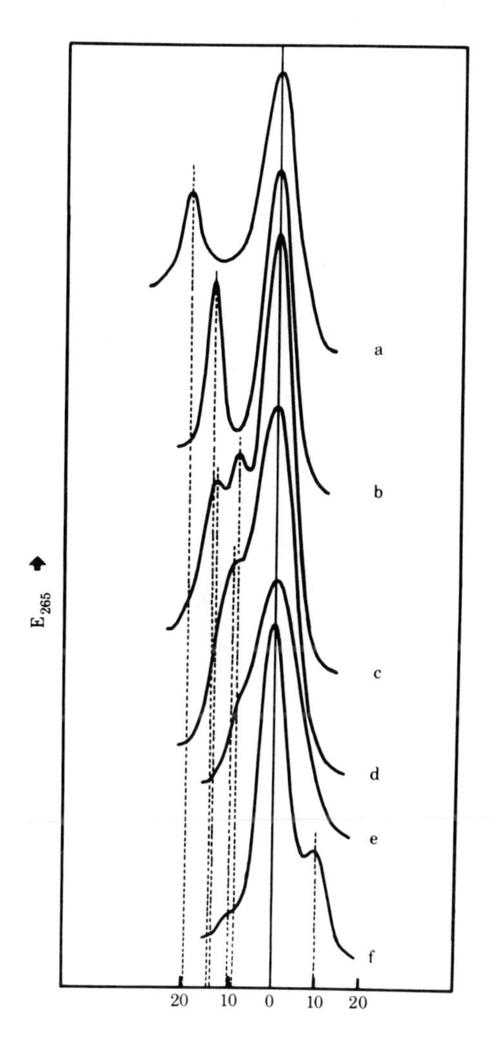

Fig. 1. Plant satellite DNAs. Total DNA prepared from leaf tissue was analysed by neutral CsCl equilibrium centrifugation, at 44,000 rpm, 25° C for 20 hr. The buoyant densities were calculated relative to the internal make DNA of Micrococcus lysodeikticus (1.731 g. cm^{-3}) a) Fortunella DNA, 1.712 and 1.693 g. cm^{-3}. b) Melon DNA, 1.706 and 1.692 g. cm^{-3}. c) Cucumber DNA, 1.706, 1.702 and 1.694 g. cm^{-3}. d) Runner bean DNA, 1.702 and 1.693 g. cm^{-3}. e) Turnip DNA, 1.704 and 1.696 g. cm^{-3}. f) Flax DNA, 1.699 and 1.689 g. cm^{-3}. From Ingle et al. *(1973).*

TABLE 1.

Distribution and buoyant densities of main-band and satellite DNAs from seed and fruit

Variety	Satellite DNA (% of total)		Buoyant density ($g\ cm^{-3}$)			
			Satellite		Mainband	
	Seed	Fruit	Seed	Fruit	Seed	Fruit
Melon						
Hero of Lockinge	26.1	17.1	1.7058	1.7085	1.6924	1.6938
Honeydew (e)	29.2	21.8	1.7062	1.7076	1.6919	1.6933
(f)	24.7	18.4	1.7056	1.7079	1.6922	1.6936
(g)	23.4	18.1	1.7056	1.7066	1.6922	1.6926
(h)	25.5	17.3	1.7063	1.7082	1.6928	1.6939
Ogen (a)	30.6	18.2	1.7056	1.7088	1.6927	1.6937
(c)	24.9	19.1	1.7063	1.7067	1.6929	1.6930
Musk (a)	24.2	16.3	1.7057	1.7070	1.6923	1.6927
(b)	29.5	17.1	1.7056	1.7071	1.6926	1.6940
Spanish Winter (c)	27.7	18.5	1.7063	1.7075	1.6917	1.6931
(e)	24.3	16.7	1.7068	1.7076	1.6931	1.6932
(g)	30.3	15.2	1.7064	1.7079	1.6923	1.6928
(h)	24.8	20.9	1.7057	1.7070	1.6918	1.6928
(j)	34.1	23.9	1.7056	1.7069	1.6921	1.6925
Cucumber						
Kariha* sat. 1	41	34	1.7011	1.7034	1.6939	1.6943
sat. 2			1.7060	1.7084		

* Data are the means from seven samples of cucumber seed and fruit DNA. Due to lack of resolution of the two satellites in cucumber DNA the combined satellites are expressed as a percentage of total DNA. Results from Pearson *et al.* (1974).

polyploid nuclei in the fruit contained only 5% heterochromatin (Fig. 2d). A minor proportion of the fruit nuclei, however, contained 37% heterochromatin, indicating over-replication of the heterochromatin in these particular nuclei (Fig. 2e). The differential replication of satellite DNA sequences may therefore be readily investigated. However, there are limitations to using satellite sequences, since such sequences are restricted to certain plant species, and furthermore, these sequences have no known function.

Another type of sequence which may be readily assayed are those coding for stable RNAs, particularly those coding for the ribosomal RNAs (rRNA). The amount of sequences coding for rRNA, [*i.e.*, the rRNA genes or ribosomal RNA coding DNA (rDNA),] may be determined reasonably accurately and easily by molecular hybridization with radioactive rRNA. These measurements may be made for any plant species, and more importantly, a function may be assigned to these sequences in that the rRNAs are part of the ribosomes, which are involved in protein synthesis. I would like to consider the rRNA genes in the context of differential replication, and also consider whether such differential replication may be related to phenotypic expression.

Justification for this interest in differential replication of rRNA genes comes from studies with animal cells. The classic example of gene amplification is that of rRNA genes during oogenesis in *Xenopus*. The somatic *Xenopus* cell contains about 0.3% rDNA, equivalent to about a thousand copies of the rRNA gene. In the oocyte the rDNA may constitute as much as 70% of the total DNA, accounting for approximately a million copies of the gene. The more subtle phenomenon of gene compensation occurs in *Drosophila*. The rRNA genes are clustered at the nucleolar organizing region (NOR) of the sex chromosome, which contains 250 copies when present with another X or Y chromosome. When present by itself, or present with a X-chromosome containing

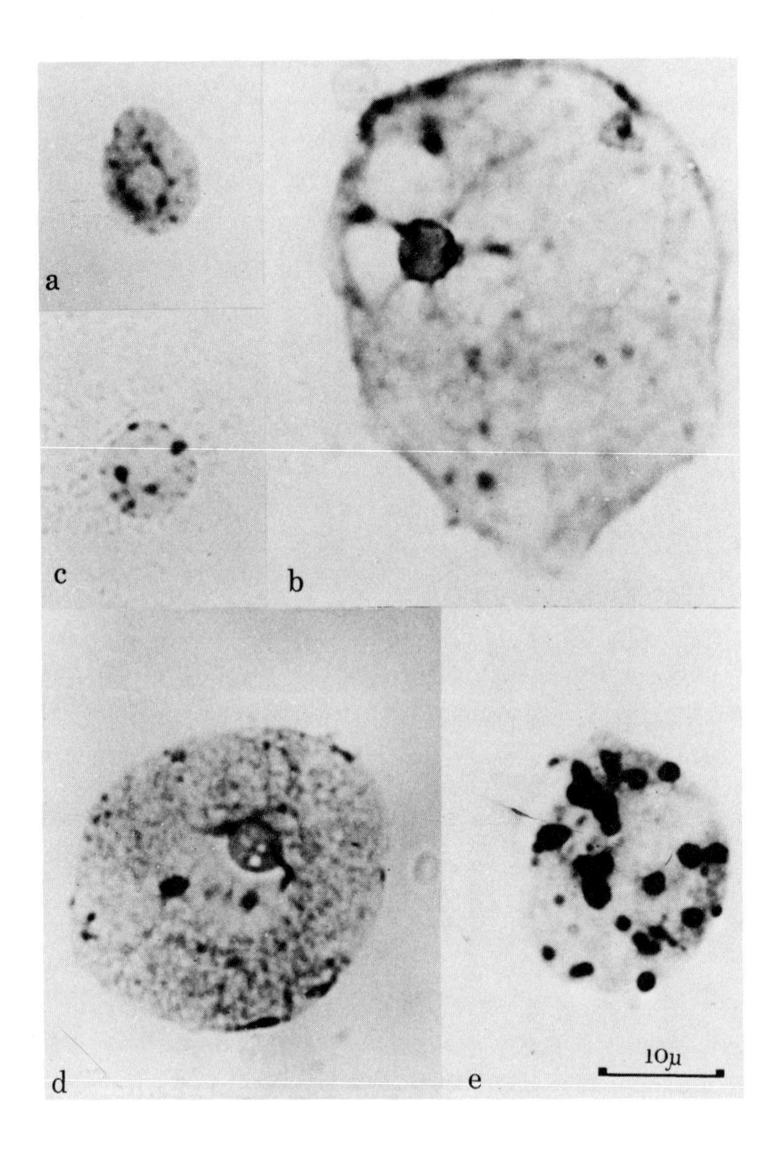

Fig. 2. Nuclei from cucumber and melon tissues. Feulgen stained nuclei from melon root tip, 4C (a); melon fruit, 32C (b); cucumber root tip, 4C (c); cucumber fruit, euchromatin 16C, heterochromatin 4C (d); euchromatin 16C, heterochromatin 32C (e). From Pearson et al. *(1974).*

an NOR deletion, the sex chromosome contains approximately 400 copies of the rRNA gene (Tartof, 1971). If this X-chromosome is then transferred back into an individual where it is associated with a normal X-chromosome, the gene redundancy is reduced back to the 250 level. A related phenomenon, called gene magnification, occurs in the 'bobbed' mutant of *Drosophila*. The 'bobbed' mutation is at the NOR, and results in a reduced number of rRNA genes. Reduction of the number of genes below 130 copies results in reduced vigor of the phenotype, the 'bobbed' character. The 'bobbed' NOR accumulates rRNA genes when maintained within a 'bobbed' nucleus. The number of genes in such a magnified 'bobbed' locus is reduced when associated with a wild type locus (Henderson and Ritossa, 1970). Such examples clearly illustrate the dynamic state of rRNA genes in animal cells.

The number of rRNA genes varies markedly in different plant species, from 1,250 to 31,900 (Table 2). Large differences are also noted among different varieties of a single species. The magnitude of the gene redundancy in the plant species compared to *Drosophila* (500) and *Xenopus* (1,000) and the large variation among different species, suggests that there is perhaps an excess of rRNA genes, and that they may not all be needed for normal growth. The total growth of a plant species certainly does not appear to be limited by rRNA gene redundancy, as attested by a plant of Jerusalem Artichoke (9 feet in height) containing 1,600 copies of the gene, and the hyacinth (only 9 inches in height) with 17,000 copies. Such observations themselves suggest that the efficiency of rRNA gene utilization varies considerably between different species.

A direct analysis of the amount of rRNA gene product in species containing a wide variation of gene redundancy confirms this indirect conclusion. Measurements were made on the meristematic region of the root-tip of species containing few genes, an intermediate number of genes, and a high number of rRNA genes

TABLE 2
Ribosomal RNA genes in plant species

Species	Genes per telophase nucleus
Citrus sinensis (orange)	1,250
Thallictrum aquilegiifolium	1,400
Helianthus tuberosus (artichoke) [102]	1,580
Passiflora antioquiensis (passion flower)	1,800
Vicia benghalensis	1,900
Linum usitatissimum (flax)	1,980
Cucumis melo (melon)	2,000
Lagenaria vulgaris	2,100
Nicotiana tabacum [48]	2,200
Nicotiana rustica [48]	2,200
Beta vulgaris (swisschard)	2,300
Taxus baccata (yew)	2,500
Nicotiana glutinosa [24]	3,200
Oenothera fructicosa	3,400
Luffa cylindrica	3,600
Phaseolus coccineus (runner bean)	4,000
Juniperus chinensis pyramidalis	4,100
Aquilegia alpina	4,600
Tulbaghia violaceae [2X]	4,650
Tradescantia paludosa [12]	4,800
Nicotiana sylvestris [24]	4,900
Momordica charantia	5,500
Secale cereale (rye)	5,700
Zea mays (maize)	6,200
Trillium grandiflorum	6,300
Helianthus annuus (sunflower) [34]	6,700
Pseudotsuga douglasii (Douglas fir)	7,200
Cucumis sativus (cucumber gherkin)	7,700
Pisum sativum (pea)	7,800
Brassica rapa (turnip)	8,600
Tradescantia virginiana [24]	8,600
Cucumis sativus (cucumber)	8,800
Cucurbita pepo (pumpkin)	9,800
Cucurbita pepo (marrow)	10,500
Pinus sylvestris (Scots pine)	10,700
Triticum aestivum (wheat)	12,700
Allium cepa (onion)	13,900
Picea albertiana conica	13,900
Tulbaghia violaceae [4X]	16,800
Hyacinthus orientalis (hyacinth) [2X]	16,800
Picea abies (Norway spruce)	19,300
Hyacinthus orientalis [3X]	22,600
Picea sitchensis (Sitka spruce)	24,700
Larix decidua (larch)	26,800
Hyacinth orientalis [4X]	31,900

Results from Ingle *et al.* (1975).

(Table 3). The ratio of total RNA to DNA was calculated from the amounts of DNA and RNA present in the root-tip; the amount of RNA per telophase cell was estimated from this ratio and the amount of DNA per telophase cell (determined from quantitative measurements of Feulgen stained nuclei). Although within this selection of plant species the DNA per cell varied 25-fold and the gene redundancy varied 10-fold, the variation in RNA per cell was much less. Artichoke, with only one-tenth the gene redundancy of hyacinth, accumulated a similar amount of rRNA per cell, suggesting that its genes are utilized 10 times as effectively as those in hyacinth. Such analyses indicate that the rRNA genes are used with varying efficiencies in different plant species, but inherent differences between the various species prevent a more detailed interpretation.

A more complete study on the relationship between gene dosage and gene product was possible on a single species, *Hyacinth orientalis*, which is available as various euploid and aneuploid varieties (Darlington *et al.* 1951) containing a range of rRNA gene redundancies. In these studies we used a diploid, triploid, and tetraploid, together with three aneuploids either lacking one nucleolar organizing chromosome or containing an additional one (Fig. 3).

These varieties allowed us to determine whether the aneuploids lacking one NOR showed compensation in their total number of rRNA genes. The number of rRNA genes in the euploids was proportional to the ploidy and the number of NORs per nucleus (Table 4). In Eros, essentially a triploid but lacking one NOR, the rRNA gene redundancy was certainly not compensated up to the normal triploid level of 26,000 genes. In fact the redundancy was slightly lower than the normal diploid level of 18,000. Similarly in Delft Blue, a tetraploid lacking one NOR, the number of genes was not compensated up to the tetraploid level but was again slightly lower than the triploid value. The presence of

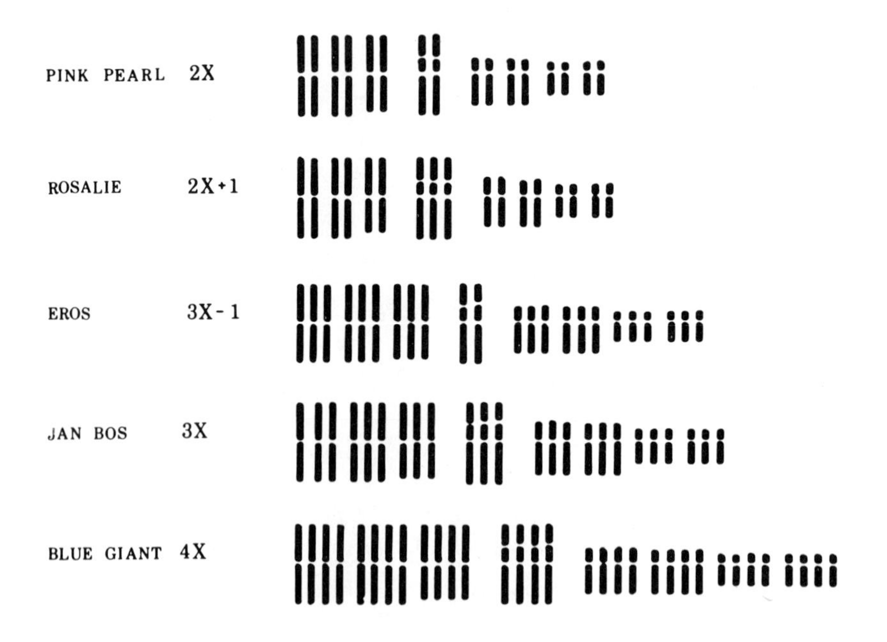

Fig. 3. Chromosome complements of hyacinth varieties. The NOR is shown as a secondary constriction on the long arm of one of the large chromosomes. From Timmis and Ingle (1975b).

an additional NOR in Rosalie resulted in a greater number of rRNA genes, although again not as high as the regular triploid value of 26,000. It was noted that in these and other aneuploids the number of rRNA genes per NOR was significantly lower than that present in the euploids.

Eros and Delft Blue therefore make do with their reduced levels of rRNA gene redundancy. Determination of the amount of rRNA accumulated per cell in these hyacinth aneuploids will therefore show whether the amount of gene product is proportional to the specific gene dosage, or to the total genome. The result of such analyses (Table 5) showed that the amount of RNA

TABLE 3
Relationship between rRNA gene redundancy and rRNA gene product in some higher plants

	rRNA genes/telophase nucleus	DNA/tip (10^{-6}g)	RNA/tip (10^{-6}g)	RNA/DNA	DNA/telophase cell (10^{-12}g)	RNA/telophase cell (10^{-12}g)	RNA/gene (10^{-15}g)
Artichoke	1,580	2.44	9.90	4.0	24.0	97	62.0
Melon	2,000	0.17	4.74	27.9	1.9	53	26.5
Swisschard	2,300	0.10	1.48	14.4	2.5	36	15.6
Sunflower	7,600	0.46	6.27	13.6	10.0	136	17.9
Cucumber	8,800	0.07	4.20	58.2	2.0	116	13.3
Wheat	12,700	1.05	5.02	4.8	30.0	144	11.3
Onion	13,300	4.94	9.30	1.9	32.0	61	4.6
Hyacinth	16,700	6.50	13.40	2.1	49.0	101	6.1

Tips (1-3 mm long) were removed from rapidly growing roots of melon, swisschard, sunflower, cucumber and wheat seedlings, Jerusalem artichoke tubers, and onion and hyacinth bulbs. Total DNA, and total RNA, per root-tip was determined as described by Timmis and Ingle (1975b). DNA per telophase nucleus was determined by comparative microdensitometry of Feulgen stained preparations (Ingle *et al.*, 1975).

TABLE 4
The rRNA gene redundancy in hyacinth varieties

Variety	Chromosome complement	No. of NORs	rRNA genes telophase nucleus	rRNA genes NOR
Euploids				
Pink Pearl	2x	2	18,426	9,213
Jan Bos	3x	3	26,025	8,675
Blue Giant	4x	4	34,680	8,670
Aneuploids				
Eros	3x-1	2	14,000	7,000
Delft Blue	4x-2	3	21,420	7,140
Rosalie	2x+1	3	20,650	6,883

Results from Timmis *et al.* (1972)

accumulated per telophase cell in the euploids was proportional to the genome ploidy and to the number of NORs, resulting in a constant value for the amount of RNA accumulated per NOR. In the aneuploids the amount of RNA accumulated was again proportional to the nominal genome, such that RNA per haploid genome remained constant. However, when expressed as RNA per NOR, significant differences were seen. Rosalie, with an additional dose of genes, accumulated less RNA per NOR, whereas Eros and Delft Blue, aneuploids lacking one NOR, accumulated more RNA per NOR than the euploids. These results show that the amount of gene product is very strictly regulated, and that it is regulated at the total genome level rather than at the specific gene dosage level. The lack of compensation seen in these hyacinth varieties is perhaps not surprising, since this species probably has an excess of rRNA genes. The triploid, containing

24,000 copies appears to be able to lose 8,000 copies and yet still make the normal triploid amount of rRNA.

TABLE 5
The relationship between gene product and gene dosage in euploid and aneuploid varieties of hyacinth

Variety	Chromosome complement	No. of NORs	RNA (10^{-12}g) per telophase cell	per haploid genome	per NOR
Euploids					
Pink pearl	2x	2	75.5	37.7	37.7
Jan Bos	3x	3	115.0	38.3	38.3
Blue Giant	4x	4	146.5	36.6	36.6
Aneuploids					
Eros	3x-1	2	117.5	39.2	58.7
Delft Blue	4x-2	3	156.0	39.0	52.0
Rosalie	2x+1	3	72.7	36.3	24.2

Results from Timmis and Ingle (1975b).

If plants have an excess of rRNA genes, then the phenomenon of compensation may be exceptional, and certainly this type of investigation should concentrate on species containing the lower rRNA gene redundancies. Although gene amplification (Ingle and Sinclair, 1972) and gene compensation do not appear to be prevalent phenomena in plants, there are examples of differential replication of rRNA genes in various plant systems. I would like to summarize these situations and particularly see whether the rRNA gene redundancy may be related to phenotypic expression.

Polytene Cells

The development of the polytene chromosomes in salivary gland cells of *Drosophila* involves an under-replication in the total amount of DNA, under-replication in the amount of heterochromatin, under-replication in satellite DNA, and under-replication of rRNA genes (Pearson *et al.*, 1974). Polytene chromosomes are also present in the suspensor cells of developing embryos of *Phaseolus*. Microdensitometric measurements of Feuglen-stained nuclei indicated a normal doubling series up to 8,192 C (Brady, 1973). This suggests that no gross differential replication occurs during the development of polytene cells, although under-replication of certain sequences may be compensated for by over-replication of others. *Phaseolus* DNA contains a satellite, and there is some controversy as to whether this is normally replicated during the production of the polytene cells. The satellite in *Phaseolus coccineus* is only a broad shoulder on the dense side of the main band. Analyses of DNA prepared from young embryos and from the young suspensors indicated no significant difference in the density or amount of satellite (Ingle and Timmis, 1974). In certain preparations of DNA from *Phaseolus coccineus* however, two satellite components are resolved, and the resolution of the two satellites was thought to be limited to DNA preparations of slightly higher molecular weight. Similar results have been obtained by Lima-de-Faria *et al.* (1975), with the difference being that the two satellite components were resolved from suspensor cell DNA (Fig. 4). The interpretation was in terms of a new satellite being limited to the suspensor tissue. These results stress the need for caution in the interpretation of Model E ultracentrifuge pictures. Lima-de-Faria *et al.* (1975) also determined the number of rRNA genes in various tissues of *Phaseolus*. The results suggested a small degree of under-replication of rRNA genes in the suspensor DNA, although the large differences in hybridization between root and shoot tissue DNAs leave some doubt. The production of polytene cell in *Phaseolus* appears to be considerably different from that in *Drosophila* in that there is evidence for only minimal differential replication of the DNA during this developmental process.

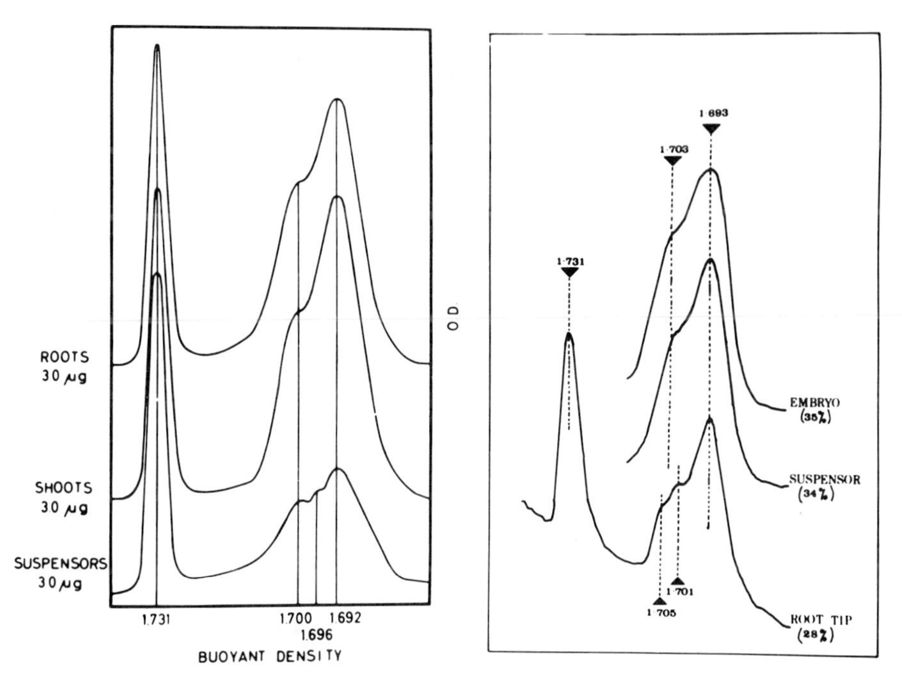

Fig. 4. Satellite DNA in Phaseolus *tissues. DNAs were prepared and analysed as described in Ingle and Timmis (1974) (right hand side) and Lima-de-Faria* et al. *(1975) (left hand side).*

Polyploidy Cells

Many plant tissues contain polyploid cells. The development of such cells in cucumber and melon fruits involve the under-replication of satellite DNA and under-replication of heterochromatin (Pearson *et al.*, 1974). It was therefore of interest to determine whether rRNA genes were also under-replicated. The hybridization experiments indicated a slightly higher level of hybridization with fruit DNA than with the seed DNA (Table 6). However, if the hybridization is expressed as a percentage of the main band DNA, on the assumption that the rRNA genes are not associated with satellite sequences (Ingle *et al.*, 1975), then the hybridization is identical between the seed and the fruit. The rRNA genes therefore appear to be fully replicated during the production of fruit polyploid cells under conditions where satellite sequences are under-replicated.

TABLE 6

The rRNA gene redundancy in melon tissues

	Seed	Fruit
Percentage satellite DNA	25	18
Percentage hybridization		
1.3 x 10^6 rRNA (2 μg/ml, 3 hr.)	0.19	0.22
1.3 + 0.7 x 10^6 rRNAs (2 + 1μg/ml, 3 hr)	0.29	0.32
1.3 x 10^6 rRNA (5μg/ml, 1 hr)	0.16	0.20

Results from Ingle *et al.* (1975)

The cotyledons of certain seeds, such as peas and beans, also contain polyploid cells. Feulgen measurements of the cotyledon nuclei showed that the polyploid cells fell in a regular doubling series, suggesting normal replication of the DNA. The determination of the number of rRNA genes in cotyledons and root-tip tissue in peas and beans also suggested normal replication of this particular sequence (Ingle, unpublished data).

However, a much fuller investigation has been undertaken by Cullis and Davies (1975). They determined the rRNA gene redundancy in 11 varieties of pea (Table 7). In ten of these varieties the gene redundancy was similar in both whole seedling DNA and in cotyledon DNA. However, in the variety JI 813, the gene redundancy was significantly lower in the whole seedling DNA, but was at the normal level in the cotyledons. This suggests that in variety JI 813 the number of rRNA genes (5,000) is sufficient for most stages of development, but not for the fast rate of development of the cotyledons during embryogenesis, when the number of genes is increased to the normal number of 8,000. These results suggest a relationship between differential replication and development, and further indicate the caution which is required in extrapolating results from a single variety of a species.

TABLE 7

The rRNA gene redundancy in pea varieties

Variety	Seed wt. (mg)	Percentage rDNA		
		whole seedling	root tip	cotyledon
JI 181	83	.144	.151	.145
JI 552	129	.183	-	.172
JI 483	142	.170	-	-
JI 537	149	.163	-	-
JI 430	198	.148	.155	.153
JI 644	221	.136	-	-
JI 313	307	.160	-	-
JI 634	336	.145	-	-
JI 373	396	.170	-	-
JI 314	421	.136	-	-
JI 813	476	.099	.095	.172

Results from Cullis and Davies (1975)

Linum usitatissimum

Flax represents an intriguing system in that some varieties can be induced by certain fertilizer treatments to form large or small genotrophs, which are then inheritied in subsequent generations under normal growth conditions (Durrant, 1962). The situation is summarized in Fig. 5. The initial plastic genotroph may be induced to the large form (L_1) by treatment with high nitrogen, or to the small form (S_1) by treatment with high phosphate. The seed from this L_1 or S_1 generation will then produce large or small genotrophs in subsequent generations, and these genotrophs are stable provided that the initial germination of the seed takes place in a warm greenhouse.

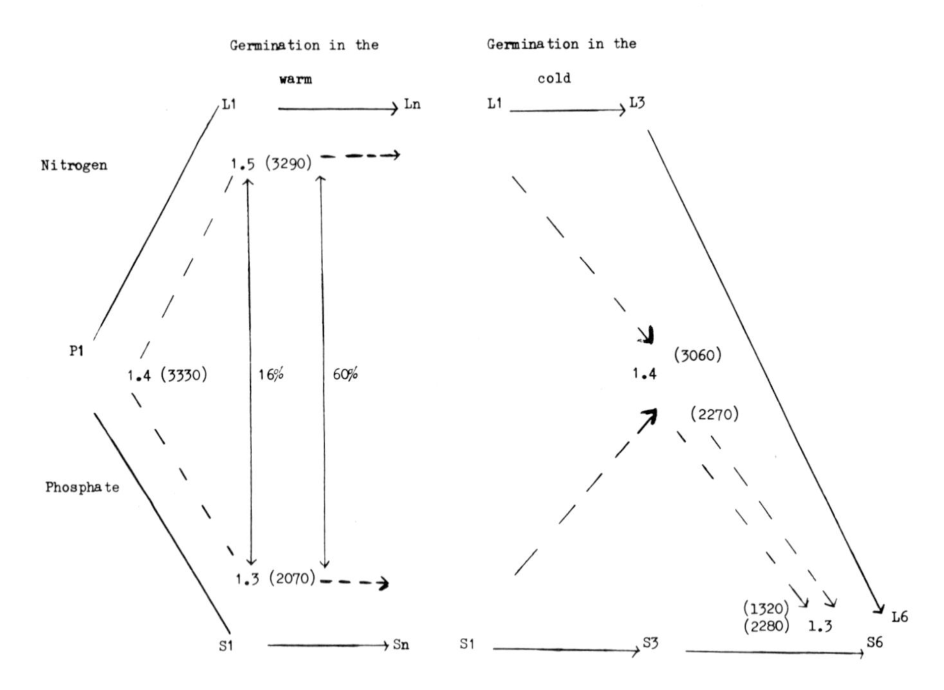

Fig. 5. The induction, stability and reversion of L and S genotrophs in flax.

Associated with this production of the large and small forms is a change in the size of the genome. The genome in the large genotroph is increased relative to the plastic genotroph whereas that of the small genotroph is decreased. The genomes of L_1 and S_1 vary by about 16% in their amount of DNA. This difference in DNA content is inherited stably provided that germination occurs in a warm greenhouse. Analysis of the DNAs from the large and small genotrophs suggested that the additional 16% DNA in the large genotroph was representative of the entire genome in that it was not made up of sequences of any particular base composition, of any particular complexity, or any particular redundancy (Timmis and Ingle, 1974). Only in respect to the redundancy of rRNA genes was the large genome significantly different from that of the small. The large genome contained approximately 70% more rRNA genes than the small (Timmis and Ingle, 1973). The redundancy of genes in the small genotroph (2,000) is low (Table 2); the presence of more genes in the large genotroph may therefore benefit the plant and hence result in more growth. The large genotroph could be related to the higher redundancy of rRNA genes. Fortunately, the flax system was flexible enough to allow one to test this hypothesis. Although the genotrophs are stable when plants are germinated in a warm greenhouse, germination in the cold produces some rather interesting reversions. After three generations of germination in the cold, the phenotypes are still large and small; however, the genome size has reverted back to the initial plastic value of 1.4×10^{-12}g. Although the 16% difference in genome size has been lost, the difference in rRNA gene redundancy is still maintained in this third generation, which again could account for the maintenance of the large and small phenotypes (Timmis and Ingle, 1975a). After six generations of cold germination there is a striking reversion of the phenotype such that the large genotroph has now reverted to a small phenotype.

Furthermore, the genome size has further decreased down to about the normal, small-genome size. By the sixth generation, the rRNA gene redundancy also has reverted with the L_6 containing fewer genes than the S_6 (Cullis, 1976). This is again consistent with the suggestion that the large genotroph may be the result of the higher rRNA gene redundancy. These results clearly illustrate the dynamic state of rRNA genes within this species. They also suggest that the rRNA gene redundancy may limit the growth and hence the phenotypic expression.

Brassica oleraceae

Our interest in *Brassica oleraceae* arose from the tremendous range of phenotypic expression (Table 8) exhibited by the species.

TABLE 8

Phenotypes of *Brassica oleraceae*

Subspecies	Common name
capitata	cabbage
gemmifera	brussels sprout
botrytis	cauliflower
italica	sprouting broccoli
gongylodes	kohl-rabi
acephala	marrowstem kale
fimbriata	curly kale
fruiticosa	thousand headed kale

Assuming a common ancestral form we were interested to see whether changes occurred at the genome level, *i. e.* to see if certain base sequences were favored. One approach is to study rRNA genes. *Brassica* contains a low number of rRNA genes, about 2,000 copies, a level at which rRNA gene redundancy may influence the phenotype. The preliminary results shown in Table 9 indicate that there is considerable intra-species variation in the number of rRNA genes, and that part of this variation appears to be developmental. Studies are underway to determine the developmental variation, the varietal variation within a sub-species, and inter sub-species variation.

TABLE 9

The rRNA gene redundancy in *Brassica oleraceae*

	rRNA genes/telophase nucleus	
Plant	Young leaves	Mature tissue
Cabbage, Danish keeping	-	2,640 (head)
Savoy	1,750	3,080 (head)
Brussels sprout	-	1,280 (sprout)
Cauliflower	1,260	2,450 (curd)
Kohl rabi	2,020	1,880 (stem)
Thousand headed kale	-	2,670 (leaves)

Cytoplasmic rRNA Genes and Chloroplast rRNA Genes

To this point we have only considered the nuclear rRNA genes. However, the situation is slightly more complicated than this. Cytoplasmic rRNA hyridizes to about 0.2% of Swisschard nuclear DNA. Chloroplast rRNA also hybridizes to 0.2% of the nuclear DNA. This result may indicate either that there are two different types of genes present in the DNA, or that there is only one type of gene with cross hybridization between this gene and the two kinds of rRNA. Competition experiments suggest that there are two types of genes, but that there is also cross hybridization between these two types of genes and the cytoplasmic and chloroplast rRNAs. The results from such experiments are rather unsatisfactory and difficult to interpret. A physical separation of the two types of genes in the nucleus is required. Although the rRNA genes may be partially fractionated from the bulk of the DNA by equilibrium centrifugation in CsCl, hybridization profiles with cytoplasmic and chloroplast rRNAs are coincident. Even when the rRNA genes are completely resolved from the bulk of the DNA in a Hg^{++}/Cs_2SO_4 gradient there is still no resolution between cytoplasmic and chloroplast rRNA hybridizations. Recently, however, it has been possible to separate these two types of genes in "restricted" DNA. Restriction enzymes are endonucleases which recognize specific sites of 5 or 6 base pairs in the DNA and cleave at that site. The DNA segments so produced can then be separated according to size on agarose gels. Restriction of Swisschard chloroplast DNA with Eco RII enzyme produced a large number of restriction segments. Digestion of the total DNA or nuclear DNA produced only trace bands which, in fact, corresponded to the chloroplast DNA fragments. The lack of restriction of the nuclear DNA may be due in part to the complexity of the DNA, to the high degree of methylation of the cytosine, or to the choice of restriction enzyme. The gel fractionation of the restricted DNA may be transferred to a millipore filter (Southern, 1975) and hybridized with rRNA to detect fragments of the restricted DNA containing the rRNA genes. Chloroplast 1.1×10^6 rRNA hybridized to a series of small fragments in the chloroplast DNA preparation (Fig. 6). Similar patterns of hybridization were present in the total DNA and to a lesser extent in the nuclear DNAs. The low level of hybridization to the top fragment in the nuclear DNA preparation (Stemmed arrow, Fig. 6) indicates that hybridization to the nuclear DNA is not due simply to the presence of contaminating chloroplast DNA.

The integration of chloroplast sequences into the nuclear genome would be expected to alter the restriction pattern to a

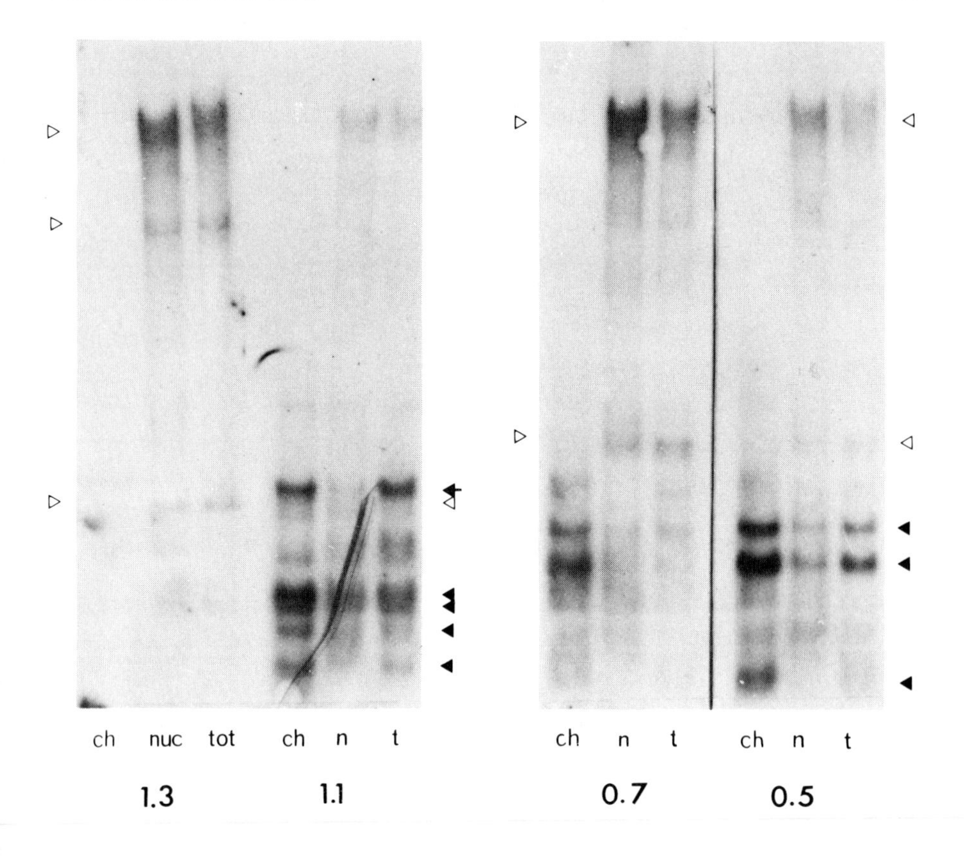

Fig. 6. Cytoplasmic and chloroplast rRNA hybridization to chloroplast, nuclear and total DNA restriction fragments. Chloroplast, nuclear and total DNAs prepared from Swisschard were digested with Eco RII, separated by electrophoresis on 1.5% agarose, transferred to millipore filter and hybridized with cytoplasmic 1.3 and 0.7 x 10^6 rRNAs and chloroplast 1.1 and 0.56 x 10^6 rRNAs. ◀ indicates chloroplast rRNA specific regions; ◁ cytoplasmic rRNA specific regions.

limited extent. The cytoplasmic 1.3 x 10^6 rRNA hybridized poorly to the chloroplast DNA. The bulk of the hybridization to nuclear and total DNA was with the large non-restricted DNA at the top of the gel, although three faint restriction bands also were visible. The chloroplast 0.56 x 10^6 rRNA hybridized to two clear bands in the chloroplast DNA and two similar bands were again present in the total and nuclear DNA preparations. Again the cytoplasmic 0.7 x 10^6 rRNA hybridizied primarily to the non-restricted RNA at the top of the gel, although at least one minor restriction band was visible. This analysis indicates that within the nuclear DNA, chloroplast rRNA hybridizes to restricted fragments, which are probably non-methylated, whereas the cytoplasmic rRNA hybridizes predominantly to non-restricted fragments (possibly methylated sequences). This analysis also indicates the level of cross hybridization between the two rRNAs and the two types of genes. This physical separation of the cytoplasmic rRNA and chloroplast rRNA genes will now allow investigation of the dynamic state of these two types of genes in nuclear DNA.

In conclusion, I think that the type of evidence presented here indicates that the plant genome is dynamic, and specifically that the rRNA genes are dynamic. The normal high redundancy of rRNA genes in many plant species tends to complicate the picture. However, in suitable systems the level of rRNA gene redundancy appears to be correlated with, and possibly influences, phenotypic expression.

REFERENCES

Brady, T. (1973). Feulgen cytophotometric determination of the DNA content of the embryo proper and suspensor cells of *Phaseolus coccineus*. *Cell Differentiation 2*, 65-75.

Cullis, C. A. (1976). Environmentally induced changes in rRNA cistron number in flax. *Heredity 36*, 73-79.

Cullis, C. A. and Davies, D. R. (1975). rRNA amounts in *Pisum sativum*. *Genetics 81*, 485-492.

Darlington, C. D., Hair, J. B., and Hurcombe, R. (1951). The history of the garden hyacinths. *Heredity 5*, 233-252.

Durrant, A. (1962). The environmental induction of heritable change in *Linum*. *Heredity 17*, 27-61.

Fox, D. P. (1970). A non-doubling DNA series in somatic tissues of the locusts *Schistocerca gregaria* (Forskål) and *Locusta migratoria* (Linn.). *Chromosoma 29*, 446-461.

Henderson, A., and Ritossa, F. M. (1970). On the inheritance of rDNA in magnified *bobbed* loci in *D. melanogaster*. *Genetics 66*, 463-473.

Ingle, J, and Sinclair, J. (1972). rRNA genes and plant development. *Nature 235*, 30-32.

Ingle, J., Pearson, G. G., and Sinclair, J. (1973). Species distribution and properties of nuclear satellite DNA in higher plants. *Nature, New Biology 242*, 193-197

Ingle, J., and Timmis, J. N. (1974). A role for differential replication of DNA in development. In *Modification of the information content of plant cells*. (Markham, R., Davies, D. R., Hopwood, D. A., and Horne, R. W., eds.) pp. 37-52, North Holland-American Elsevier.

Ingle, J., Timmis, J. N., and Sinclair, J. (1975). The relationship between satellite DNA, rRNA gene redundancy, and genome size in plants. *Plant Physiol. 55*, 496-501.

Lima-de-Faria, A., Pero, R., Avanzi, S., Durante, M., Stahle, U., D'Amato, F., and Granström, H. (1975). Relation between rRNA genes and the DNA satellites of *Phaseolus coccineus*. *Hereditas 79*, 5-20.

Nagl, W. (1972). Evidence of DNA amplification in the orchid *Cymbidium in vitro*. *Cytobios. 5*, 145-154.

Pearson, G. G., Timmis, J. N., And Ingle, J. (1974). The differential replication of DNA during plant development. *Chromosoma 45*, 281-294.

Southern, E. M. (1975). Detection of specific sequences among DNA fragments separated by gel electrophoresis. *J. Mol. Biol. 98*, 503-517.

Tartof, K. D. (1971). Increasing the multiplicity of rRNA genes in *Drosophila melanogaster*. *Science 171*, 294-297.

Timmis, J. N., and Ingle, J. (1973). Environmentally induced changes in rRNA gene redundancy. *Nature, New Biology 244*, 235-236.

Timmis, J. N., and Ingle, J. (1974). The nature of the variable DNA associated with environmental induction in flax. *Heredity*. *33*, 339-346.

Timmis, J. N., and Ingle, J. (1975a). The status of rRNA genes during nuclear DNA reversion in flax. *Biochem. Gen.* *13*, 629-634.

Timmis, J. N., and Ingle, J. (1975b). Quantitative regulation of gene activity in plants. *Plant Physiol.* *56*, 255-258.

Timmis, J. N., Sinclair, J., and Ingle, J. (1972). rRNA genes in euploids and aneuploids of hyacinth. *Cell Different.* *1*, 335-339.

SELECTIVE TRANSCRIPTION AND PROCESSING IN THE REGULATION OF PLANT GROWTH

Rusty J. Mans, Charles O. Gardner, Jr., and Trevor J. Walter

Department of Biochemistry and Molecular Biology
University of Florida
Gainesville, Florida 32610

Much has been said about the potential of molecular genetics (Watson, 1976) in the treatment of biochemical and genetic diseases of humans; referred to as bioengineering. Perhaps of more immediate relevance to human welfare and certainly freer from moral and societal constraints (Luria, 1973) is the application of the concepts and methodologies of molecular biology to plants; particularly to plants of nutritional (Bonner, 1965) and industrial (McMillen, 1969) consequence to man. What I and others are suggesting is that plant scientists should seriously consider bioengineering of agronomically important crops. Application of the principles of classical genetics and continual selection of propitious phenotypes by plant breeders has fixed many favorable genotypes into agronomically important food, fiber and ornamental crops. Enhanced productivity, disease resistance, cold and drought tolerance, fruit set, seed viability and morphogenic modification for mechanized farming are examples of selected genotypes. The hybrid corn program in the United States, its

ISBN 012-601950-9

extension to Central and South America as well as to Europe, is a stellar example of what the aggressive application of basic genetic information to specfic agronomic problems can accomplish (Wallace and Brown, 1956). I think our appreciation of the physiologic and biochemical mechanisms that underlie the phenotypic expression of favorable genotypes is now adequate to attempt manipulation for agronomic gain. At the least, our appreciation of molecular genetics of higher plants is sophisticated enough to permit correlations between physiologic responses and genetic potential. Fifteen years ago, Barbara McClintock (McClintock, 1961) invited us to examine the apparent relationship between the gene control systems of bacteria, (the experimental tool of the molecular geneticists) and the controlling elements in maize. I thank Dr. McClintock for the invitation.

My purpose is to illustrate some of what we have learned about how corn plants read their genes by applying some of the concepts and methodologies of molecular genetics to this higher plant. I have divided the discussion into three parts. First, I will explain our experimental approach as it relates to current concepts of gene expression. Then I will discuss the isolation and characterization of two enzymes involved in the mechanics of gene expression. Finally, I will discuss one approach to understanding the mechanism of gene expression by following the activity of these two enzymes during kernel maturation.

CENTRAL DOGMA

The "central dogma" of molecular genetics was first enunciated by Watson and Crick in the early 1950's (Watson and Crick, 1953). The first tenet of that dogma would have us assume that all the information on how to build and to operate a cell is encoded in the deoxyribonucleotide sequence of the DNA that is located in the chromosomes and organelles of every cell. That plant DNA is self-replicating was demonstrated by Jack Van't Hof (This volume) in his paper on replication forks in pea roots. The quantitative and qualitative relationships between specific gene products and DNA sequences was illustrated by John Ingle (This volume) in his paper on ribosomal RNA genes in a host of higher plant species. We may wish to reserve judgment as to whether all the plant cell's information is encoded in its own DNA until hearing from the virologists. The second tenet of the "central dogma" held that any expression of information in the DNA required its transcription into an intermediate macro-molecular species called RNA. Transcription of selected sections of DNA at specific times in a cell's life cycle permits selective use of some genetic information at specific times or in response to specific stimuli. The pea chromatin experiments conducted by James Bonner and his colleagues (Bonner *et al.*, 1966) are now classic in demonstrating such selective transcription of plant DNAs. I will discuss this aspect of molecular genetics in more detail. The third tenet of the "central dogma" held that the transcribed information was expressed as a specific protein whose amino acid sequence was translated from the sequence of ribonucleotides in the RNA which served as the message or template for protein synthesis. The *in vitro* translation of a spectrum of messenger RNAs including mammalian hemoglobin RNAs and viral coat protein RNAs by the wheat germ system verifies messenger RNA as an intermediate in genetic expression by plants (Marcus, 1970).

Some of the current notions about the flow of genetic information from the DNA sequences of nuclear genes to the cytoplasmic protein-synthesizing apparatus of eukaryotic cells are diagramatically represented in Fig. 1. Three cell compart-ments are demonstrated by double lines: (1) the nucleolus within the nucleus, (2) the nucleus within the cytoplasm and (3) the cytoplasm. The cell's genome is indicated by the wavy

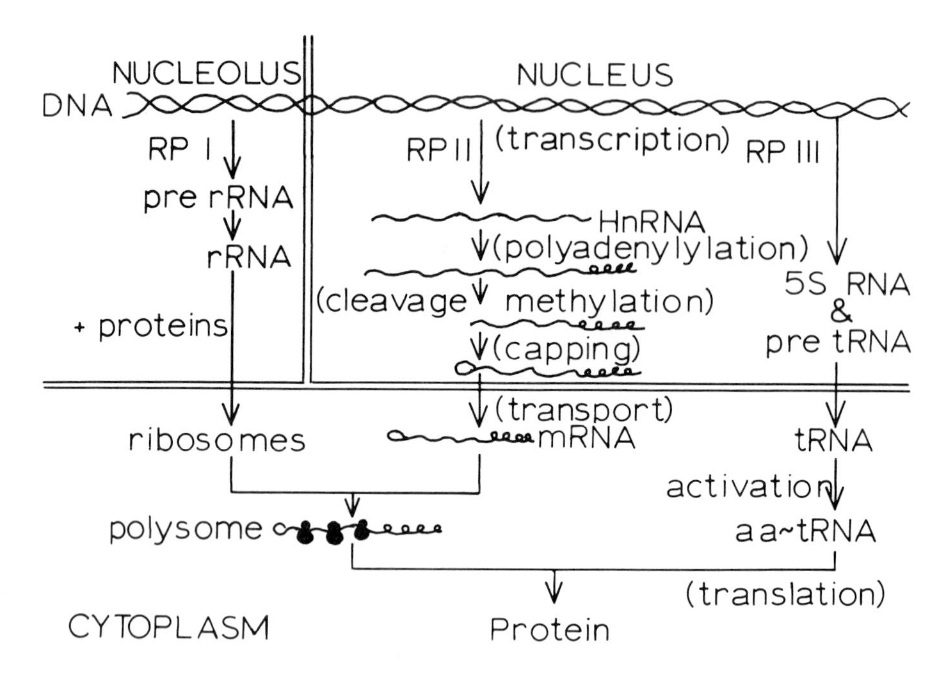

Fig. 1. Diagramatic representation of regulation of genetic expression via transcription and processing in eukaryotic cells.

double line representing a segment of a DNA molecule; it includes ribosomal RNA genes in the nucleolar compartment and structural as well as regulatory genes in the nuclear compartment. I have excluded organellar DNAs for simplification. Three separate nuclear transcription systems are known in eukaryotes (Roeder and Rutter, 1970; Chambon, 1974). The first is associated with the nucleolar organizer region and is called RNA polymerase I or A. It is primarily responsible for transcription of ribosomal precursor RNA which is processed into ribosomal RNAs. The rRNAs associated with ribosomal proteins are transported into the cytoplasm to become functional ribosomes. Two more transcription systems are present in the nucleoplasm: RNA polymerase II or B and RNA polymerase III or C. RNA polymerase III is primarily responsible for transcription of genes for precursor tRNA and 5SRNA (Weinmann and Roeder, 1974). These products are transported to the cytoplasm where they participate in protein synthesis. The third nucleoplasmic transcription system is indicated as RNA polymerase II. The products of transcription by RNA polymerase II are heterogeneous in size and base composition (hence referred to as heterogeneous nuclear RNAs) and may represent transcripts of the structural genes of all the cell's proteins. These transcripts are confined to the nucleus. RNA polymerases I, II and III, although cleanly delineated in Fig. 1, are experimentally difficult to separate as a function of their cellular compartmentalization or by the kinds of genetic information that they transcribe. Fortunately for the molecular biologists, the three polymerase protein complexes can be resolved by DEAE-cellulose chromatography. Furthermore, each polymerase exhibits unique sensitivity to the potent mushroom toxin, α-amanitin. RNA polymerase I is refractory to the toxin; whereas RNA polymerase II is completely inhibited by 10^{-9}M α-amanitin. RNA polymerase III is also inhibited by α-amanitin but at levels in excess of 10^{-5}M. That each of the polymerases may exhibit multiple peaks upon elution from DEAE-cellulose as well as intermediate levels of sensitivity to α-amanitin is known, but the significance is not fully appreciated. Each of these polymerases has been resolved from several plant tissues and the properties of each were summerized in tabular form by Chambon in a recent review article (Chambon, 1974) on eukaryotic polymerases.

A particularly significant aspect of the expression of genetic information in eukaryotic cells is the metabolic processing of RNA polymerase II transcripts. The heterogeneous nuclear RNA molecules undergo some or all of the following processing reactions.

1. Polyadenylylation, *i.e.* the sequential addition of AMP moieties to their 3' hydroxyl termini (Darnell *et al.*, 1971). I will discuss this reaction in plants in some detail.

2. Methylation, *i.e.* 2'-O-methylation of the penultimate nucleotide at the 5' terminus of what eventually becomes the mRNA molecule (Desrosiers *et al.*, 1974).
3. Cleavage to smaller fragments or mononucleotides catalyzed by as yet unidentified nucleolytic activities probably preserving the 3' polyadenylylated terminus.
4. Capping of the 5' terminus of mammalian and viral messenger RNAs with 7 methyl guanosine to yield a m7G(5')ppp(5')N -Np (Furuichi *et al.*, 1975).

These processing reactions seem to be restricted to selected portions of some HnRNA molecules which are then transported to the cytoplasm to serve as messenger RNAs in the assembly of polysomes for the translation of genetic information into proteins. The paper by Abe Marcus (This volume) discusses in detail the association of ribosomal subunits with amino acid-bearing tRNAs and processed messenger RNA under the influence of initiation factors to form functional polysomes as depicted in the lower portion of Fig. 1.

MODEL SYSTEMS IN MAIZE

Over the years we have tried to assemble *in vitro* model systems composed of plant components that will carry out many of the reactions indicated in Fig. 1. Through biochemical dissection of the cellular apparatus combined with *in vitro* reassembly we have begun to understand both the mechanisms and sites of control involved in getting information out of corn DNA and into corn proteins. We have utilized three criteria to decide if a given system warranted further experimental pursuit. First, we insisted that each model system biosynthesize enough macromolecular product to permit isolation and characterization of the product. This necessitated purification of the components of the model system. Second, we insisted that each model system exhibit in a test tube, a manifestation of the physiologic response exhibited by the tissue from which it was isolated. All model systems were derived from the same plant since we thought that any response of the model system to internal regulators or external stimuli would be more likely to occur if all components were taken from plant tissues capable of such a response *in vivo*. Finally, we insisted that each model system be capable of coupling with another; the ultimate objective was to follow the flow of genetic information from genome to protein read-out in an assembled master system.

We chose maize tissue to establish the model systems because: (1) Maize was then and may still be the best genetically characterized agronomic crop (Neuffer *et al.*, 1968), (2) Maize grain is commercially available in genetically stable form and thus is a cheap, constant source of components, (3) Maize kernels and seedlings exhibit a host of physiologic responses to growth regulators and environmental conditions for exploitation in the model systems and (4) Hybrid corn is a major crop in this country and we thought funding of the research program would continue over a protracted period.

This brings me to a discussion of the isolation and characterization of two enzymes involved in gene expression. The enzymes are RNA polymerase II (E.C. 2.7.7.6) and ATP:polynucleotide adenylyltransferase (E.C. 2.7.7.19). The latter enzyme has been variously referred to as poly A polymerase, RNA adenylating enzyme, ATP:polynucleotidylexotransferase and terminal riboadenylate transferase. We refer to it as exotransferase.

MAIZE RNA POLYMERASE II

RNA polymerase from maize was the first soluble, DNA-dependent nucleotide-incorporating enzyme resolved from chromatin by centrifugation of homogenates (Mans and Novelli, 1964). Over the years, Ernest Stout (Stout and Mans, 1967) and Robert Benson (Benson, 1972) applied classical enzymological techniques to the purification of maize RNA polymerase II from the shoots of four-day-old seedlings and worked out the sequence of fractionation procedures that have been widely applied to polymerase isolations from plant tissues. RNA polymerase I (Strain *et al.*, 1971) and a chloroplast polymerase (Bottomley *et al.*, 1971) also have been purified from maize leaf tissue indicating that multiple polymerase transcription systems are present in maize.

Maize RNA polymerase II will polymerize the four nucleoside triphosphates into a DNA-determined polymer by the reaction sequence indicated in Fig. 2. The activity is enriched 3000-fold over that detected in crude extracts, is routinely prepared in 12 hours, is stable at 4°C, is not inactivated by repeated freeze-thaw from liquid nitrogen and catalyzes the polymerization of 15 nmoles of nucleotide per minute per mg protein at 30°C (Mans, 1973). The product has been isolated and chemically characterized

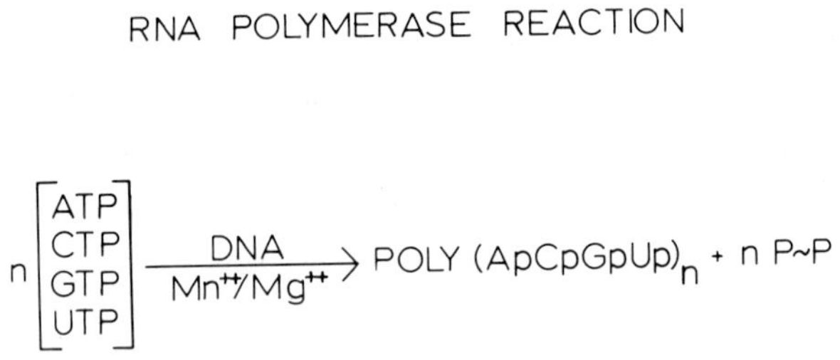

RNA POLYMERASE REACTION

$$n \begin{bmatrix} ATP \\ CTP \\ GTP \\ UTP \end{bmatrix} \xrightarrow[Mn^{++}/Mg^{++}]{DNA} \text{POLY (ApCpGpUp)}_n + n\ P{\sim}P$$

Fig. 2. RNA polymerase reaction.

as DNA-like RNA (Stout and Mans, 1968). The fidelity of base-pairing has been demonstrated by hybridization of isolated product with templating DNA. When offered in the same test tube with calf thymus DNA, maize DNA is the preferred template of the maize enzyme (Brooks and Mans, 1973). The preference of the maize RNA polymerase for its homologous DNA template seems unique among the enzymes purified from eukaryotes (Chambon, 1974). The purified enzyme exhibits three additional characteristics of particular value for investigating the regulation of transcriptive processes in a higher plant: (1) The enzyme will initiate transcription on a circular DNA template (Gardner *et al.*, 1976). (2) The initiation of RNA synthesis seems to be coupled to poly A synthesis (Benson and Mans, 1972). (3) The level of RNA polymerase II activity in dark-grown seedlings is significantly enhanced by brief illumination of the seedlings with white light (Mans and Huff, 1974).

Selective reading of one gene in deference to another, *i.e.*, the transcription of one portion of the templating DNA and not adjacent deoxynucleotide sequences implies the initiation of new RNA chains at internal sites on the templating DNA. One of the difficulties encountered in detecting selective transcription by eukaryotic polymerases has been an inability to measure the initiation of new RNA chains. The presence of single-stranded breaks or nicks in isolated eukaryotic DNAs coupled with the propensity of the polymerase to use these nicks as non-specific initiation points precluded unambiguous detection of specific initiation sites within the DNA sequences. Furthermore, contaminating traces of nuclease activity remaining in purified polymerase preparations can introduce nicks into the most native of DNA preparations again precluding unambiguous detection of RNA initiation at internal DNA sites.

Dr. Charles Gardner in our laboratory has circumvented these difficulties by utilizing the single-stranded, circular DNA of the bacteriophage ØX174 as a template for the purified maize polymerase. The ØX174 DNA was labeled with tritium and the RNA product was synthesized from ^{14}C-labeled ribonucleotides. Reaction mixtures were incubated for an hour and the labeled components resolved on 5 to 20% neutral sucrose gradients as shown in Fig. 3. The bottom panel shows the migration of the circular DNA on a neutral sucrose gradient. The top panel shows the co-migration of the radioactive product with all the

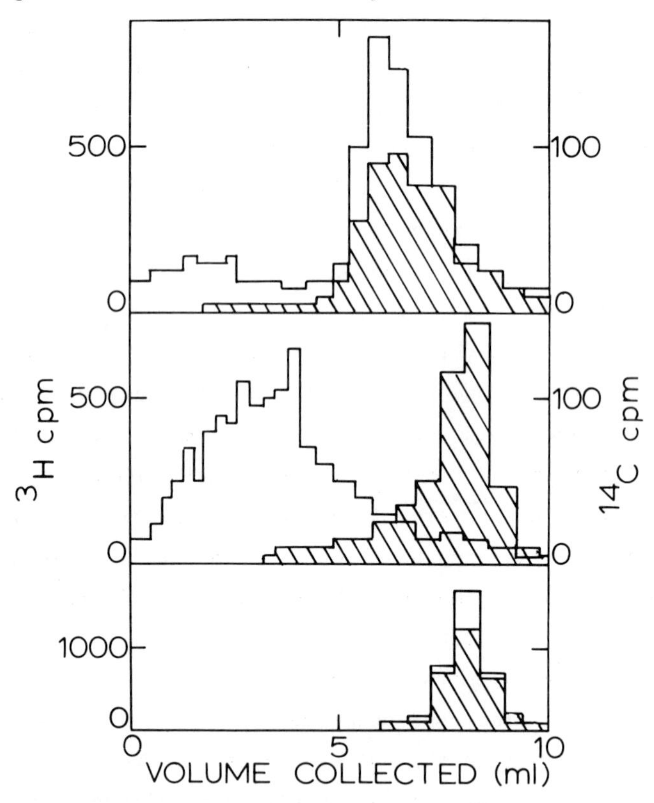

Fig. 3. Sedimentation profiles of products transcribed from ØX174 single-stranded, circular DNA, Top; unheated product. Center; heated and quick-cooled product. Bottom; ØX174 DNA. For experimental details of ^{14}C-labeled product (clear bargraph) preparation with maize RNA polymerase on ^{3}H-labeled ØX174 DNA (shaded bargraph) and analysis on neutral 5 to 20% sucrose density gradients see Gardner et al. *(1976).*

template DNA and their displacement toward the top of the tube. The cosedimentation of the two isotopes indicates that the product accumulated as a DNA-RNA hybrid with a lower sedimentation rate than the single-stranded circular DNA. If the product of the reaction was boiled and quick-cooled to dissociate the putative RNA-DNA hybrid and then centrifuged the template DNA migrated to its original position and the heterodisperse, labeled products sedimented throughout the upper third of the gradient (center panel). A critical test of initiation of RNA chains in the absence of nicks in the template requires that the circular DNA be recovered after the incubation period. Since linear rods and circles sediment at different rates on alkaline sucrose gradients, and since alkali disrupts the hydrogen bonds holding the RNA-DNA hybrids together (as well as hydrolyzes the RNA) the incubated reaction mixture was analyzed on alkaline sucrose gradients. By comparing the profiles in Fig. 4, it is seen that all the input ØX174 circles (80% of the total DNA) were recovered intact after transcription. The smaller peak toward the top of the gradient present in the DNA before and after transcription represents linear strands of DNA that arise as a consequence of alkaline sucrose gradient centrifugation (Achey *et al.*, 1976). Prolonged incubation of the DNA with the maize RNA polymerase does not generate any more rods showing the enzyme is free of DNase knicking activity. Dr. Gardner has repeated this experiment with double-stranded, circular, replicating form of the ØX174 DNA and again found that no ends are required for chain initiation on double-stranded or native DNA (data not shown here). This observation shows that the maize RNA polymerase II, analogous to the calf thymus enzyme (Mandel and Chambon, 1974) can initiate transcription at internal deoxynucleotide sequences on double-stranded circular DNAs.

The second characteristic of the maize polymerase of particular value for studying the regulation of eukaryotic transcription is the apparent coupling of chain initiation with poly A synthesis. Three observations, made initially with polymerase preparations of relatively low specific activity and consistently observed by Dr. Benson (Benson, 1972) with much more active enzyme, suggest that poly A is synthesized by the maize RNA polymerase. Data in Table 1 show the high ApA frequencies routinely observed among the alkaline digests of products made from [α-^{32}P]ATP on native and especially on

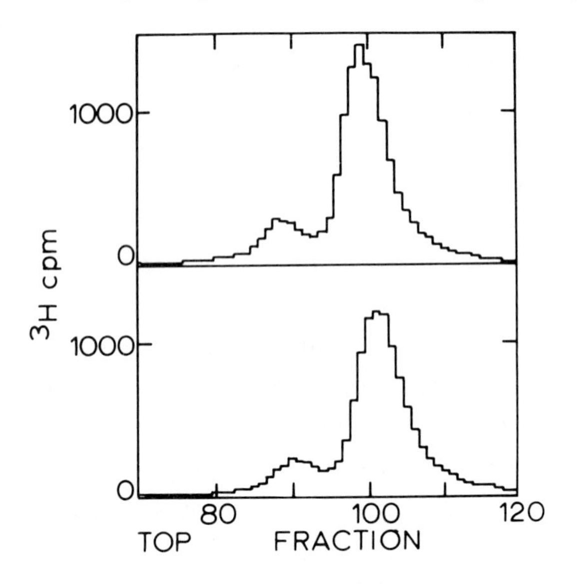

Fig. 4. Sedimentation profiles of ØX174 single-stranded, circular DNA. Top; before transcription. Bottom; after transcription. For experimental details of transcription of ^{3}H-labeled ØX174 and analysis on alkaline 5 to 20% sucrose density gradients see Gardner et al. *(1976).*

TABLE 1

Nearest Neighbor Frequency of Polymerase Products

System	CpA	ApA	GpA	UpA
Native DNA	0.208	0.308	0.228	0.255
Denatured DNA less NTPs	0.009	0.895	0.045	0.050
Denatured DNA	0.440	0.440	0.181	0.133

denatured eukaryotic DNA templates. A significant amount of AMP was incorporated into an acid-insoluble product in the absence of other nucleotides and the nearest neighbor frequency analysis indicated that this product was an A-rich polymer. The second observation was that the acid-insoluble radioactive products synthesized on native DNA templates always showed bimodal size distributions. In panel A of Fig. 5, calf thymus DNA sedimenting near the bottom of the sucrose gradient (shaded portion) when used as template for *in vitro* transcription gave rise to the products seen in panel B. The major product sedimented to the lower half of the gradient and a second portion of the acid-insoluble product remained at the top of the 5 to 20% sucrose gradient. In panel C, the templating DNA was selected from the center of the sucrose gradient and was presumably shorter. The major product transcribed from it was correspondingly shorter, *i.e.* sedimented slower in the sucrose gradient. Again, an additional peak of acid-insoluble product accumulated at the top of the gradient. When each of these four products was subjected to nearest neighbor frequency analysis the large RNA products consistently showed a base composition like that of the templating DNA. On the other hand, the acid-insoluble materials remaining at the top of the 5 to 20%

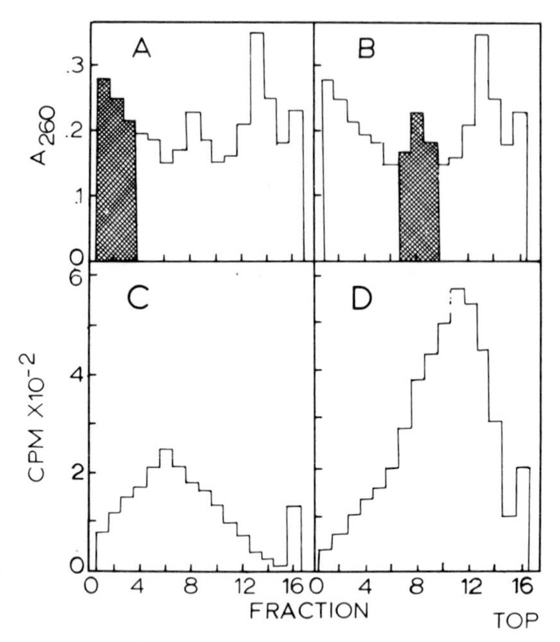

Fig. 5. Sedimentation profiles of products transcribed from sized, calf thymus DNA. Panels A and B; native calf thymus DNA. Panel C; ^{14}C-labeled products transcribed from DNA of shaded portion in panel A. Panel D; ^{14}C-labeled products transcribed from DNA of shaded portion in panel B. For experimental details of product preparation with maize RNA polymerase and analysis on 5 to 20% sucrose density gradients see Stout and Mans (1968).

sucrose gradient consistently showed very high ApA frequencies. It seems that two classes of products were synthesized by the polymerase, larger DNA-like RNAs and smaller but still acid-insoluble AMP-rich polymers. The third observation concerns the nucleotide requirement for AMP incorporation and is reported in Table 2. When denatured calf thymus DNA as template and manganese as the metal cofactor were offered to the purified maize polymerase, almost half of the incorporation of

TABLE 2

RNA Polymerase Reaction with Manganese

System	Labeled Substrate ATP (pmoles)	Labeled Substrate UTP (pmoles)
Native DNA	415	197
Native DNA less NTPs	21	3
Denatured DNA	913	458
Denatured DNA less NTPs	381	22

AMP was independent of the other nucleoside triphosphates. Whereas 95% of the AMP incorporated was dependent upon the presence of CTP, GTP and UTP with native DNA as template. Note that essentially all the UMP incorporated by the maize polymerase was dependent upon the presence of the other nucleotides with either native or denatured templating DNA; as expected for a "respectable" RNA polymerase. The incorporation data taken together with the product analysis data indicate poly A synthesis by the maize polymerase.

Poly A synthesis from ATP on single-stranded DNA in the absence of other nucleotides is most reminiscent of the poly A synthesis catalyzed by bacterial RNA polymerase. The bacterial poly A synthesis was postulated by Chamberlin and Berg (1964) to occur by reiterative copying of short oligo dT sequences in the DNA template. However, as I had indicated earlier, poly A synthesis in eukaryotes is catalyzed by at least one enzyme other than RNA polymerase; *i.e.* the poly A polymerase involved in polyadenylylation of HnRNA. Therefore, it was necessary for us to rule out contamination of the RNA polymerase II preparation with this primer-dependent poly A synthesizing activity. It turns out that the two enzymes are completely resolved by DEAE-cellulose chromatography (Fig. 6). The poly A polymerase from maize was eluted from the column at 90 mM $(NH_4)_2SO_4$ with a relatively shallow salt gradient whereas the maize RNA polymerase required 350 mM $(NH_4)_2SO_4$ for elution from the DEAE cellulose at pH 8 in a steep salt gradient. The elution profile of the ATP incorporating activity independent of NTPs is coincident with the NTP requiring UTP incorporating activity eluated at 350 mM $(NH_4)_2SO_4$. Therefore, the poly A synthesis observed with the RNA polymerase cannot be ascribed to

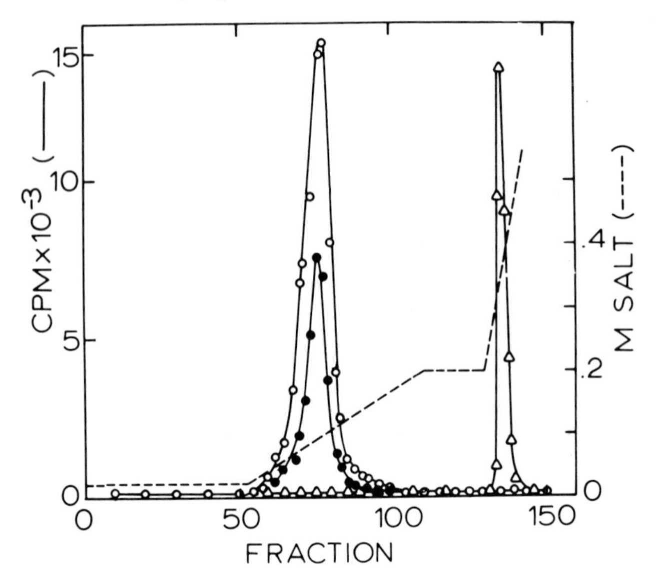

Fig. 6. Elution profile of maize poly A polymerase and RNA polymerase II from DEAE-cellulose. Poly A polymerase was assayed as tRNA (open circles) or single-stranded DNA (closed circles) primer-dependent ^{14}C-AMP incorporation. See Mans and Huff (1975) for experimental details. RNA polymerase II (open triangles) was assayed as DNA-dependent, α-amanitin sensitive, ^{14}C-UMP incorporation. See Mans (1973) for experimental details.

contaminating poly A polymerase. RNA synthesis and poly A synthesis by the RNA polymerase preparation were equally sensitive to 10^{-10} and 10^{-9} M α-amanitin (Table 3). The extreme sensitivity of both reactions to the specific protein-binding inhibitor, α-amanitin and the copurification of the two catalytic activities strongly suggest that the purified polymerase was synthesizing two template dependent products; relatively high molecular weight RNAs and relatively low molecular weight poly A. The kinetics of the accumulation of these two products indicated that the AMP-rich material accumulated early in the reaction. In some very recent experiments utilizing the ØX174 single-stranded DNA as a template, Dr. Gardner has isolated the early transcription products by sucrose density gradient centrifugation and found poly A to be synthesized from ATP in the absence of CTP, GTP and UTP. Initiation of these poly A chains can be detected by the incorporation of [γ-^{32}P]ATP into isolated product. It now seems possible to determine how the poly A chains are initiated on the circular templates and what, if any, relationship poly A synthesis has to initiation of RNA chains by the eukaryotic RNA polymerase.

TABLE 3

α-Amanitin Sensitivity

Drug	Concentration µg/ml	% of Control	
		RNA	Poly A
α-amanitin	0.10	6	7
	1.0	1	2

In addition to high enzymic activity, ability to synthesize DNA-directed products, preference for homologous DNA and ability to initiate chains on circular DNAs, the RNA polymerase II activity from maize seedlings exhibits a physiologic response. Illumination of etiolated seedlings with white light for less than five minutes enhances more than two-fold the level of activity detected in extracts immediately after illumination (Mans and Huff, 1974). Even shorter exposures result in greater increases in soluble polymerase activity. This work I have just described illustrating that we can distinguish initiation of new chains from enhanced elongation rates on circular templates is directly applicable to interpretation of the light response. We can now ask if illumination of the dark-grown seedlings (equivalent to seedling emergence) enhances transcription of the genes already being read by the etiolated seedling or if light induces the reading of different genes *i.e.*, the expression of different information in response to an environmental stimulus.

MAIZE POLY A POLYMERASE

While searching for a sigma-like factor for the maize RNA polymerase several years ago, Trevor Walter and I stumbled upon a poly A synthesizing enzyme (Walter and Mans, 1970). Subsequently, the enzyme was purified from maize seedlings and identified as ATP:polynucleotide adenylyltransferase on poly A polymerase (Mans and Walter, 1971; Mans and Huff, 1975). The enzyme requires a primer bearing a 3' hydroxyl terminus, ATP and a divalent metal for the sequential addition of AMP moieties to the primer to synthesize a covalently attached poly A chain approaching 200 nucleotides (Fig. 7). The enzyme is specific for ATP, will utilize either magnesium or manganese, and adenylylates an array of RNA primers including tRNAs, rRNAs,

POLY A POLYMERASE REACTION

$$\mathrm{RNApN} + \mathrm{nATP} \xrightarrow[\mathrm{Mg^{++}}]{\mathrm{Mn^{++}}} \mathrm{RNApNpApA...pA_n} + \mathrm{nP{\sim}P}$$

Fig. 7. Poly A polymerase reaction.

viral RNAs, homopolymers and histone mRNAs (Thrall *et al.*, 1974) and single-stranded DNA (Mans and Huff, 1975). The physiologic function of the enzyme is assumed to be the polyadenylylation of nuclear transcripts as described earlier (Fig. 1). Polyadenylylation of deoxyoligomers may be unique for the plant enzyme since highly purified enzymes from mammalian sources (Winters and Edmonds, 1973; Tsiapalis *et al.*,1973) and the HeLa enzyme purified in our laboratory (Mans and Stein, 1974) are not DNA-primed. We have asked if there are two poly A polymerases, one specific for RNA and the other for DNA primers. After adsorption in 10 mM $(NH_4)_2SO_4$ the maize poly A polymerase was eluted from DEAE-cellulose at 90 mM $(NH_4)_2SO_4$ with a shallow salt gradient (Fig. 6). The ratio of the tRNA to deoxyoligomer-primed activities was constant across the elution profile suggesting the same enzyme catalyzes both adenylylation reactions (Mans and Huff, 1975).

Further purification of the poly A polymerase by glycerol gradient centrifugation did not resolve the two activities (Fig. 8). More recently, we purified the enzyme further by preparative isoelectric focusing. Both activities exhibit the same profiles with an isoelectric point at pH 6.8. After reduction and denaturation with SDS, the enzyme exhibits one major and one minor protein-staining band after electrophoresis

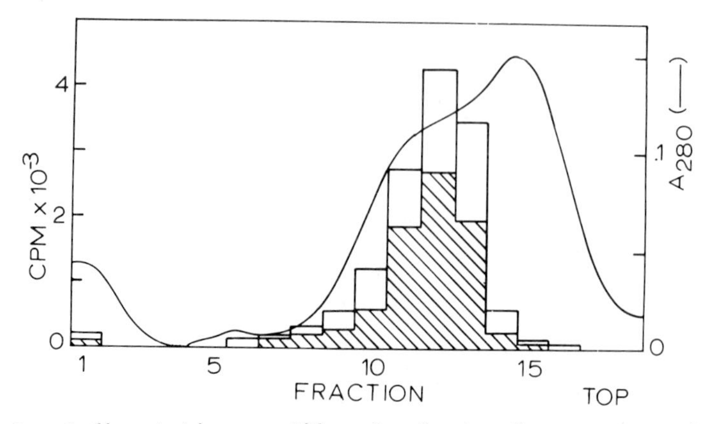

Fig. 8. Sedimentation profile of poly A polymerase. Maize poly A polymerase purified through DEAE-cellulose chromatography was centrifuged on a 5 to 20% glycerol gradient and fractions were assayed as tRNA-primed (clear bargraph) and single-stranded, DNA-primed (cross-hatched bargraph) activities. See Mans and Huff (1975) for experimental details.

on 7.5% polyacrylamide gels. We are becoming more convinced that the one enzyme uses either primer. However, the enzyme may consist of more than one polypeptide. Whether the use of deoxyoligomers as primers is an experimental artifact of the *in vitro* assay or of physiological consequence in maize is not known.

The poly A polymerase is unambiguously distinguished from RNA polymerase II by several experimental parameters (in addition to its complete resolution from RNA polymerase II by DEAE-cellulose column chromatography as seen in Fig. 6). The poly A polymerase requires a polynucleotide with a 3' hydroxyl terminus as a primer whereas the RNA polymerase utilizes a template with no ends. We have demonstrated the covalent attachment of the poly A sequence to the primer molecule in

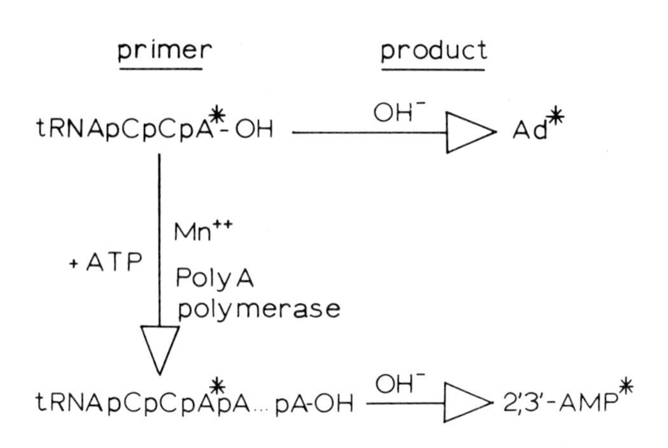

Fig. 9. Mechanism for detecting covalent attachment of poly A to RNA.

several ways. Leucyl tRNA was labeled in its 3' hydroxyl terminus with [8-^{14}C]AMP by ATP:tRNA adenylyltransferase E.C. 2.7.7.20 (Best and Novelli, 1971). As indicated in Fig. 9, the labeled tRNA was incubated briefly with unlabeled ATP and maize poly A polymerase. The product was isolated by Sephadex chromatography and the labeled polynucleotide was hydrolyzed with alkali. The only radioactive moiety resolved by paper electrophoresis was AMP (Mans and Walter, 1971). The terminally labeled tRNA primer yielded only radioactive adenosine when hydrolyzed before incubation with the poly A polymerase. In contrast recall the incorporation of ^{32}P from [γ-^{32}P]ATP into poly A and the complete dissociation of the labeled RNA polymerase product from the ØX174 DNA template on neutral sucrose density gradients (Fig. 3). The covalent attachment of the poly A to a deoxyoligomer by the poly A polymerase was demonstrated electrophoretically. Oligo(dT)$_{10}$ was incubated with [^{14}C]ATP

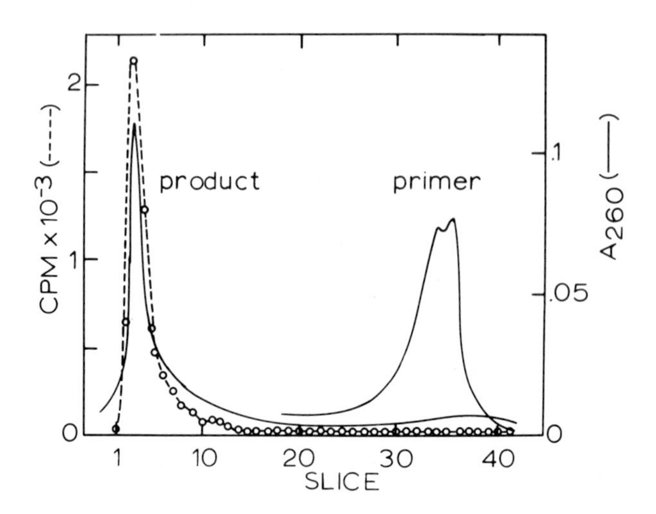

Fig. 10. Gel electrophoretic profile of poly A polymerase product. Reaction mixtures containing oligo(dT)$_{10}$ and ^{14}C-labeled ATP were analyzed on 7.5% polyacrylamide gels with 1% SDS before and after a 3 hour incubation with maize poly A polymerase. See Mans and Huff (1975) for experimental details.

and poly A polymerase for three hours. The product was isolated by gel exclusion, boiled and quick-cooled and subjected to electrophoresis on 7.5% polyacrylamide gels with 1% SDS. Comparison of the profiles (Fig. 10) of the ultraviolet-absorbing primer before and after incubation shows the primer mobility decreased with incubation, *i.e.* the primer was elongated. The acid-insoluble, radioactivity and the ultraviolet absorption profiles of the heated product were coincident after gel electrophoresis, again indicating a heat-stable bond between the primer and the radioactive product (Mans and Huff, 1975). After alkaline hydrolysis the oligo(dT)$_{10}$-primed product retained a single radioactive adenine moiety as expected if the 5' phosphate of the AMP was linked to the 3' hydroxyl terminus of the deoxyoligomer.

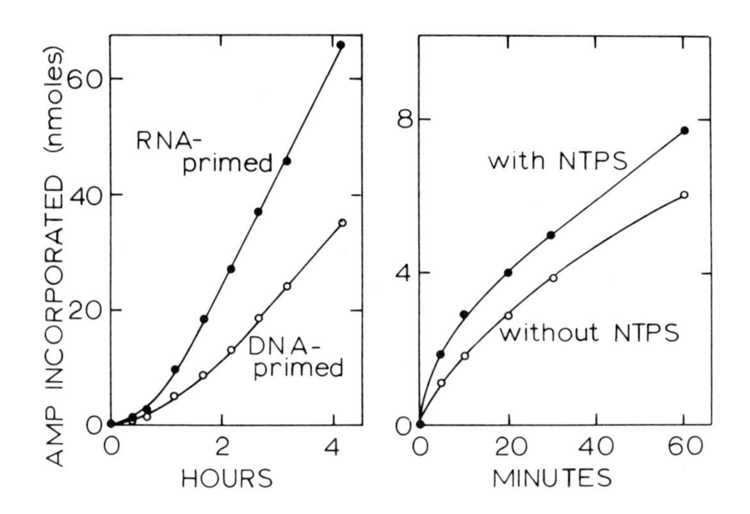

Fig. 11. Rates of product accumulation. Left; incorporation of ^{14}C-AMP into acid-insoluble material by maize poly A polymerase with either tRNA (closed circles) or single-stranded DNA (open circles) as primer. Right; incorporation of ^{14}C-AMP into acid-insoluble material by maize RNA polymerase II on denatured calf thymus DNA in the presence (closed circles) or absence (open circles) of CTP, GTP and UTP.

Several other characteristics distinguish poly A synthesis by the poly A polymerase from that catalyzed by RNA polymerase II. The initial rate of AMP incorporation by the two enzymes differ in that the poly A polymerase exhibits a 20 minute lag with essentially no incorporation on any primer, whereas the polymerase exhibits its most rapid AMP incorporation for the first 10 minutes incubation at 30°C (Fig. 11). The poly A polymerase exhibits strict substrate specificity for ATP whereas the RNA polymerase utilizes all four nucleotide triphosphates. In fact, levels (1 mM) of NTP's that inhibit the exotransferase completely are optimum for the RNA polymerase II reaction. Poly A polymerase is refractory to high levels of α-amanitin whereas the RNA polymerase is completely inhibited at 10^{-9} M. Finally, the molecular weight of the maize poly A polymerase determined by gel exclusion chromatography, SDS gel electrophoresis and glycerol gradient centrifugation is about 65,000 daltons while that of the maize RNA polymerase II is in excess of 450,000 daltons.

We currently view poly A synthesis to be catalyzed by at least two enzymes in maize: (1) a relatively small, α-amanitin insensitive poly A polymerase that covalently links AMP moieties sequentially to a polynucleotide primer and (2) a relatively large, α-amanitin sensitive RNA polymerase that initiates and then reiteratively elongates a poly A chain on a short oligo(dT) sequence in the templating DNA.

PHYSIOLOGIC RELATIONSHIP BETWEEN RNA POLYMERASE II AND A POLY A POLYMERASE IN MAIZE GRAIN

Finally, I want to discuss our recent attempts to understand the mechanisms of gene expression by following the levels of RNA polymerase II and poly A polymerase during maize kernel maturation. We asked if there was a fixed relationship between the level of transcription of structural genes by RNA polymerase II and the processing of these transcripts by the poly A polymerase. If there were a tight coupling of these two activities under several physiologic situations that alter RNA metabolism then we would look for a functionally coupled role for the two enzymes, *e.g.*, polyadenylylation as the termination mechanism for transcription. Such a tight functional coupling could imply a structural relationship between the two enzymes. If, on the other hand, the levels of one enzyme activity changed independently of the other, we would infer a more distant relationship between the two; perhaps no tighter than transcription being required to provide primer for eventual adenylylation by the poly A polymerase. The *in vitro* differential sensitivity of the two enzymes to cordycepintriphosphate suggests that this kind of relationship can exist in the cell (Maale *et al.*, 1975).

Since we could readily distinguish the RNA polymerase II and poly A polymerase activities purified from seedlings, we approached the problem by modifying the fractionation of the two enzymes so that several samples could be processed simultaneously from less than 3 g fresh weight of material within eight hours (Walter and Mans, 1975). Trevor Walter then applied the fractionation technique to developing maize kernels harvested at two-day intervals after pollination. Correlated with an increase in soluble protein and fresh weight, a continuous rise in primer-dependent poly A polymerase activity per kernel occurred up to 25 days after pollination (Fig. 12). A similar increase occurred in α-amanitin sensitive, DNA-dependent RNA polymerase activity.

Within the grain several physiologically and genetically distinct tissues are recognized (Randolph, 1936). Among them is the triploid endosperm which is highly specialized for the synthesis and accumulation of starch and storage proteins and increases in size primarily by cell expansion. The diploid embryo, on the other hand, represents a collection of cells with diverse functions involved in the synthesis of a broad spectrum of plant proteins and increases in size primarily by cell division. The RNA metabolism of each of these tissues necessarily underlies these physiologic differences. The anatomically distinct embryos and endosperms (less glumes and pericarp) were dissected, extracts prepared and both activities assayed in each tissue. The coordinate increase in RNA polymerase II and poly A polymerase activities in both tissues is evident (Fig. 13). Note that the changes in RNA polymerase assayed as either α-amanitin sensitive, DNA-dependent nucleotide-incorporating activity or labeled α-amanitin protein-binding activity were

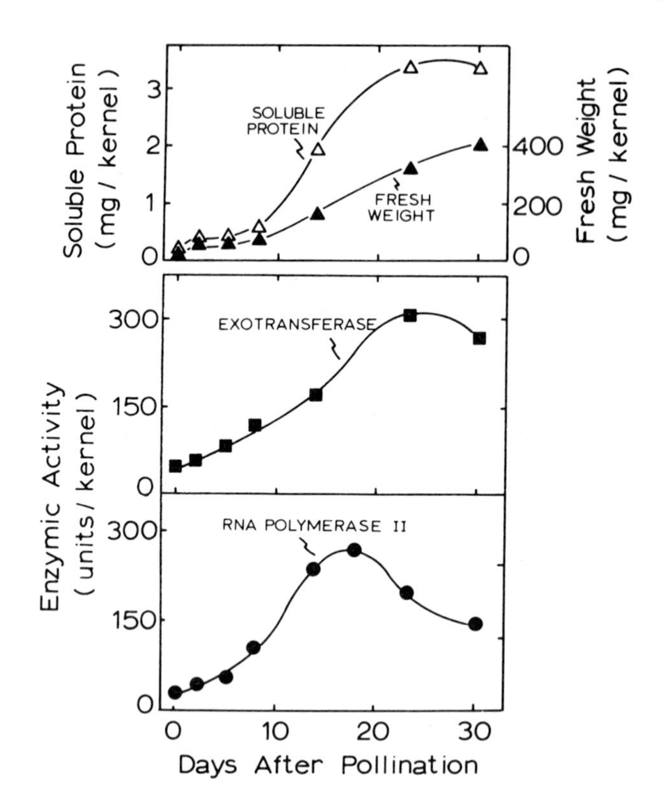

Fig. 12. Changes in RNA polymerase II and poly A polymerase activities in developing maize grain. Top: soluble protein (open triangles) and fresh weight (closed triangles). Center: poly A polymerase. Bottom: RNA polymerase II. For experimental details of preparation and assay see Walter and Mans (1975).

identical. These data strongly suggest that the increase in RNA polymerase activity during early development and the decrease 20 days after pollination reflect changes in the level of enzyme protein. Interestingly, the total elongation rate per tissue of both nucleotide polymerizing enzymes is essentially the same. The differential decay of the two enzymes in the endosperm may reflect differing half-lives of the two enzymes *in vivo* since the poly A polymerase activity is more stable than the RNA polymerase II in partially purified preparations. These data are consistent with but do not prove a tight functional and

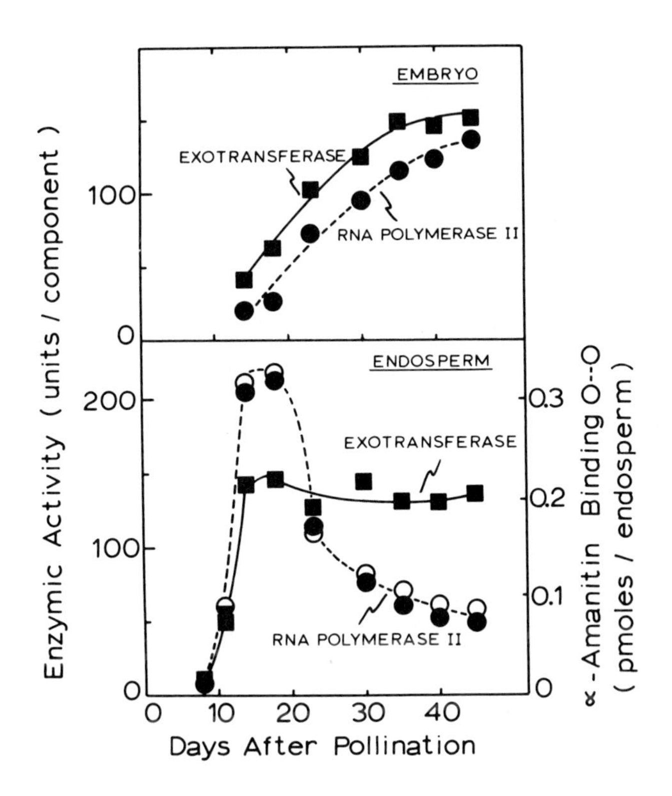

Fig. 13. Comparison of RNA polymerase II and poly A polymerase activities in developing maize embryo and endosperm. Top; embryo poly A polymerase (closed squares) and RNA polymerase II (closed circles). Bottom; endosperm poly A polymerase activity (closed squares) and RNA polymerase II activity (closed circles) and α-amanitin-binding protein (open circles).

perhaps structural relationship between RNA polymerase II catalyzed transcription and poly A polymerase catalyzed polyadenylylation in maize.

SUMMARY

I have attempted to provide an overview of selective transcription and processing as it is currently viewed in higher plants and other eukaryotes. Included were examples of how concepts and methodologies of molecular genetics are applicable in characterizing two enzymes directly involved in the transcription apparatus of maize. I described how we can now recognize the initiation of new RNA chains as a measure of "turning on" the reading of additional genes in the plant genome. I have indicated how we detected a previously unknown reaction, the polyadenylylation of DNA, that may be of physiologic importance in some aspect of plant nucleic acid metabolism. Finally, I have shown the application of experimental know-how gained from model systems to the probing of regulatory processes operational in the expression of genetic information in the grain of an agronomically important plant.

We must now refine our methodology to investigate the mechanism of expression of a specific gene. As I had indicated in my introductory remarks the time is right for attempting to manipulate the biochemical mechanisms that underlie phenotypic expression of plant genotypes favorable to man's existance on this planet.

ACKNOWLEDGMENTS

This investigation was supported in part by the United States Energy Research and Development Administration, Division of Biomedical and Environmental Research, Report No. ORO-3982-44 and by the Institute of Food and Agricultural Sciences of the University of Florida BY-01640.

REFERENCES

Achey, P.M., Billen, D., and Beltranena, H.P. (1971). Single-strand breaks in gamma-irradiated ØX174 DNA induced by exposure to alkali. *Int. J. Radiat. Biol. 20*, 501-504.

Benson, R.H. (1972). Purification of a eukaryotic RNA polymerase II that synthesizes polyadenylic acid. Ph.D. Thesis, University of Florida, Gainesville, Florida.

Benson, R.H. and Mans, R.J. (1972). Polyadenylic acid synthesis by purified eukaryotic RNA polymerase. *Fed. Proc. 31,* 427 abs.

Best, A.N. and Novelli, G.D. (1971). Studies with tRNA adenylyl-(cytidylyl) transferase from *Escherichia coli* B II. Regulation of AMP and CMP incorporation into tRNApCpC and tRNApC. *Arch. Biochem. Biophys. 142,* 539-547.

Bonner, J. (1965). The need for a better understanding of plants. In *Genes to Genus* (F.A. Greer and T.J. Army, eds), pp. 13-21. International Minerals and Chemical Corp., Skokie, Illinois.

Bonner, J., Dahmus, M.E., Fambrough, D., Huang, R.C., Marushige, K., and Tuan, D.Y.H. (1966). The biology of isolated chromatin. *Science 156,* 47-56.

Bottomley, W., Smith, H.J. and Bogorad, L. (1971). RNA polymerases of maize: partial purification and properties of the chloroplast enzyme. *Proc. Nat. Acad. Sci. U.S. 68*, 2412-2416.

Brooks, R.R. and Mans, R.J. (1973). Selection of repeated sequences of homologous and heterologous DNA during *in vitro* transcription by maize RNA Polymerase. *Biochem. Biophys. Res. Commun. 52,* 608-613.

Chamberlin, M. and Berg, P. (1964). Mechanism of RNA polymerase action: characterization of the DNA-dependent synthesis of polyadenylic acid. *J. Mol. Biol. 8,* 708-726

Chambon, P. (1974). Eukaryotic RNA polymerases. In *The Enzymes, 10.* (P.D. Boyer, ed.), pp. 261-331. Academic Press, New York.

Darnell, J.E., Philipson, L., Wall, R. and Adesnik, M. (1971). Polyadenylic acid sequences: role in conversion of nuclear RNA into messenger RNA. *Science 174,* 507-510.

Desrosiers, R., Friderici, K. and Rottman, F. (1974). Identification of methylated nucleosides in messenger RNA from Novikoff hepatoma cells. *Proc. Nat. Acad. Sci. U.S. 71,* 3971-3975.

Furuichi, Y., Morgan, M., Shatkin, A.J., Jelinek, W., Salditt-Georgieff, M. and Darnell, J.E. (1975). Methylated, blocked 5' termini in HeLa cell mRNA. *Proc. Nat. Acad. Sci. U.S. 72,* 1904-1908.

Gardner, C.O., Jr., Achey, P., and Mans, R.J. (1976). Transcription of circular, single-stranded DNA by maize RNA polymerase II. *FEBS Letters 63*, 205-208.

Luria, S.E. (1973). *Life: The Unfinished Experiment.* Charles Scribner's Sons, New York.

Maale, G., Stein, G., and Mans, R. (1975). Effects of cordycepin and cordycepin-triphosphate on polyadenylic and ribonucleic acid-synthesizing enzymes from eukaryotes. *Nature 255,* 80-82.

Mandel, J. and Chambon, P. (1974). Animal DNA-dependent RNA polymerases. Studies on the reaction parameters of transcription *in vitro* of simian virus 40 DNA by mammalian RNA polymerases AI and B. *Eur. J. Biochem. 41,* 367-378.

Mans, R.J. (1973). RNA polymerases in higher plants. In *Methods in Molecular Biology, 4* (A.L. Laskin and J.A. Last, eds.), pp. 98-125. Marcel Dekker, New York.

Mans, R.J. and Huff, N.J. (1974). RNA synthesis induced by brief illumination of etiolated maize seedlings. In *Mechanisms of Regulation of Plant Growth, 12,* (M.M. Cresswell, A.R. Ferguson, R.L. Bieleski, eds.), pp. 325-332. The Royal Society of New Zealand, Wellington, N.Z.

Mans, R.J. and Huff, N.J. (1975). Utilization of ribonucleic acid and deoxyoligomer primers for polyadenylic acid synthesis by adenosine triphosphate: polynucleotidylexotransferase from maize. *J. Biol. Chem. 250,* 3672-3678.

Mans, R.J. and Novelli, G.D. (1964). Ribonucleotide incorporation by a soluble enzyme from maize. *Biochim. Biophys. Acta 91,* 186-188.

Mans, R.J. and Stein, G. (1974). Addition of polyadenylic acid to RNA by ATP: polynucleotidylexotransferase partially purified from HeLa cells. *Life Sciences 14,* 437-445.

Mans, R.J. and Walter, T.J. (1971). Transfer RNA-primed oligoadenylate synthesis in maize seedlings II. Primer, substrate and metal specificities and size of product. *Biochim. Biophys. Acta 247,* 113-121.

Marcus, A. (1970). Tobacco mosiac virus ribonucleic acid-dependent amino acid incorporation in a wheat embryo system *in vitro. J. Biol. Chem. 245,* 962-966.

McClintock, B. (1961). Some parallels between gene control systems in maize and in bacteria. *Amer. Naturalist 95,* 265-277.

McMillen, W. (1969). *The Green Frontier: Stories of Chemurgy.* G.P. Putnam's Sons, New York.

Neuffer, M.G., Jones, L., and Zuber, M.S. (1968). *The Mutants of Maize.* Crop Science Society of America, Madison, Wisconsin.

Randolph, L.F. (1936). Developmental morphology of the caryopsis in maize. *J. Agri. Res. 53,* 881-916.

Roeder, R.G. and Rutter, W.J. (1970). Specific nucleolar and nucleoplasmic RNA polymerases. *Proc. Nat. Acad. Sci. U.S. 65*, 675-682.

Stout, E.R. and Mans, R.J. (1967). Partial purification and properties of RNA polymerase from maize. *Biochim. Biophys. Acta 134*, 327-336.

Stout, E.R. and Mans, R.J. (1968). Template requirement of maize RNA polymerase. *Plant Physiol. 43*, 405-410.

Strain, G.C., Mullinix, K.P. and Bogorad, L. (1971). RNA polymerases of maize: nuclear RNA polymerases. *Proc. Nat. Acad. Sci. U.S.A. 68*, 2647-2651.

Thrall, C.L., Park, W.D., Rashba, H.W., Stein, J.L., Mans, R.J. and Stein, G.S. (1974). *In vitro* synthesis of DNA complementary to polyadenylated histone messenger RNA. *Biochem. Biophys. Res. Commun. 61*, 1443-1449.

Tsiapalis, C., Dorson, J.W., DeSante, D.M. and Bollum, F.J. (1973). Terminal riboadenylate transferase: a polyadenylate polymerase from calf thymus gland. *Biochem. Biophys. Res. Commun. 50*, 737-743.

Wallace, H.A. and Brown, W.L. (1956). *Corn and its Early Fathers.* The Michigan State University Press.

Walter, T.J. and Mans, R.J. (1970). Transfer RNA-primed oligoadenylate synthesis in maize seedlings. I. Requirements of the reaction and nature of the product with crude enzyme. *Biochim. Biophys. Acta 217*, 72-82.

Walter, T.J. and Mans, R.J. (1975). A rapid technique for the estimation of polynucleotide adenylyltransferase and ribonucleic acid polymerase in plant tissues. *Plant Physiol. 56*, 821-825.

Watson, J.D. (1976). *Molecular Biology of the Gene 3rd ed.* W.A. Benjamin, Inc., Menlo Park, California.

Watson, J.D. and Crick, F.H.C. (1953). Genetical implications of the structure of deoxyribonucleic acid. *Nature 171*, 964-969.

Weinmann, R. and Roeder, R.G. (1974). Role of DNA-dependent RNA polymerase III in the transcription of tRNA and 5SRNA genes. *Proc. Nat. Acad. Sci. U.S. 71*, 1790-1794.

Winters, M.A. and Edmonds, M. (1973). A poly(A) polymerase from calf thymus. Characterization of the reaction product and the primer requirement. *J. Biol. Chem. 248*, 4763-4768.

PROTEIN CHAIN INITIATION IN EXTRACTS OF WHEAT GERM

Marcella Giesen, Samarendra N. Seal, Ruth Roman
and Abraham Marcus

The Institute for Cancer Research
The Fox Chase Cancer Center
Philadelphia, Pa. 19111

The study of protein biosynthesis in extracts of wheat embryos began with the observation that this process is one of the first systems "activated" by exposure of embryos to water (Marcus and Feeley, 1964). A major facet of the "activation" process is the attachment of ribosomes to preformed mRNA (Marcus, 1969), and the reaction can, in part, be reproduced *in vitro* (Weeks and Marcus, 1970). In further studies it was found that the ribosome attachment reaction is effective with a number of eucaryotic mRNAs, in particular with plant viral RNAs (Marcus *et al.*, Klein *et al.*, 1972, Davies and Kaesberg, 1974). Utilizing TMV-RNA primarily, we have studied aspects of the initiation reaction in extracts of wheat and the subsequent sections summarize the current status of these studies.

CHARACTERISTICS OF TMV-RNA CATALYZED AMINO ACID INCORPORATION

With isolated ribosomal subunits (Weeks *et al.*, 1972) and an initiation inhibitor, aurintricarboxylic acid (Marcus *et al.*, 1970), it can be shown that the initiation reaction requires

ISBN 012-601950-9

ATP, GTP and mRNA and at least two initiation factors (Marcus *et al.*, 1973). The process occurs on the 40S ribosomal subunit and results in the formation of a 50S complex containing an initiator species of methionyl-tRNA. This complex subsequently combines with a 60S ribosomal subunit forming an 80S ribosome-Met-tRNA initiation complex. Amino acid incoporation into protein requires, in addition, two elongation factors.

RESOLUTION OF THE WHEAT PROTEIN CHAIN INITIATION SYSTEM AND ANALYSIS OF THE FUNCTIONS OF THE RESOLVED FACTORS

One of the more controversial questions with regard to the initiation process is the ordering of the dequence of attachment of the mRNA and the initiator tRNA to the 40S ribosomal subunit. In procaryotic systems, the binding of fMet-tRNA to ribosomes has an almost absolute requirement either for the trinucleotide AUG or for an mRNA containing the AUG codon (Haselkorn and Rothman-Denes, 1973). Furthermore, binding of radioactive mRNA to 30S ribosomal subunits can be carried out in the absence of fMet-tRNA. Thus the procaryotic system seems to function by initially binding the mRNA which then provides an aligning site for the binding of the initiator tRNA. Evidence for the alternative sequence in which fMet-tRNA first binds to the 30S subunit and then directs the attachment of the mRNA was reported by Jay and Kaempfer (1975).

In early studies of Met-tRNA binding to ribosomes with the wheat system, a strong requirement for mRNA was noted (Table 1) suggesting either the simultaneous binding of the mRNA and the initiator tRNA or a sequential process with the mRNA attaching first. Similar results were obtained by Shafritz and Anderson (1970) and Shafritz *et al.* (1972) with a fractionated system obtained from the ribosomal wash of rabbit reticulocytes.

TABLE 1

Requirements for ribosomal binding of methionyl-tRNA

Substrate	Conditions	Met-tRNA bound (pmol)		
		+ TMV-RNA	- TMV-RNA	Δ
Met-tRNA	No factors	0.20		
	C + D	1.44	0.26	1.18
	C alone	0.17		
	D alone	0.20	0.18	
	C + D (ATP omitted)	0.35	0.11	0.24
	C + D (GTP omitted)	0.70	0.11	0.59
Met-tRNA	C + D	1.45	0.18	1.27
Met-tRNA	C + D	0.42	0.22	0.20

The reaction mixture contained in a volume of 0.34 ml: 1.1 mM Met, 30 mM Tris acetate, pH 8, 1.1 mM ATP, 60 μM GTP, 10 μg of TMV RNA, 2.6 mM dithiothreitol, 1.3 mM MgAc , 51 mM KCl, ribosomes (220 μg of RNA), 38 pmol of unresolved [^{14}C]Met-tRNA or 34 pmol of [^{14}C]Met-tRNA or [^{14}C]Met-tRNA (1 pmol = 330 c.p.m.) and initiation factors: C, 170 μg and/or D-final, 125 μg (Seal *et al.*, 1972). After incubating 10 minutes at 20 C the [^{14}C]Met-tRNA retained by nitrocellulose filters (*i.e.* [^{14}C]Met-tRNA bound to ribosomes) was determined.

Studies from a number of other laboratories (Schreier and Staehlin, 1973; Levin *et al.*, 1973 a,b; Dettman and Stanley, 1972; Cashion and Stanley, 1974; Gupta *et al.*, 1973,1975), however, described systems from rabbit reticulocytes and mouse fibroblasts in which a 40S-Met-tRNA complex could be formed in the absence of an mRNA. Furthermore, a factor functioning in all of these systems could bind Met-tRNA in a form adsorbable to nitrocellulose filters. In the partially fractionated wheat system (see Table 1) no such reactions were observed. Subsequent experiments, however, involving the further resolution of the wheat factors, have resulted in very different findings. As seen in Table 2, a factor (C3β) can be isolated that does indeed

TABLE 2

Aminoacyl-tRNA binding to factor C3β

		Aminoacyl-tRNA bound	
		Met-tRNA	Phe-tRNA
Experiment	Conditions	(pmol)	(pmol)
1	complete system	2.24	
	-C β	0.08	
	-GTP	0.61	
	-MgAc	2.02	
	-C β + EF_1	0.04	
2	complete system	2.75	2.14
	+ D	1.34	0.18

The complete system contained in a volume of o.17 ml: 1mM ^{12}C amino acid (corresponding to the radioactive aminoacyl-tRNA used), 30 mM Tris acetate pH 8.0, 2.6 mM dithiothreitol, 60 μM GTP, 70 mM KAc, 5 mM KCl, 1.3 mM $MgAc_2$, 7 mM phosphoenolpyruvate, 5.6 μg pyruvate kinase, 20 μg C3β and 19.6 μg tRNA containing either 3.3 pmol ^3H Met-$tRNA_i^{Met}$ (2300 cpm/pmol) or 14.3 pmol ^{14}C Phe-tRNA (666 cpm/pmol). Where indicated, 90 μg D-final were added or C3β was replaced by 45 μg EF1 (C3γ). Data similar to that presented for EF1 (exp. 1, line 5) were obtained when C3β was replaced either by C3α, A, D-final, or A2, all in amounts as used in Table 3, experiment 1. After 10 minutes incubation at 20 C, aminoacyl-tRNA binding was determined by the nitrocellulose filter assay. The preparation of the various factors was described by Giesen *et al.* (1976).

bind Met-tRNA in a reaction independent of mRNA. The reaction requires GTP and is unaffected by variation of the Mg^{++} concentration from 0.2 to 5 mM.

The resolution leading to the preparation of factor C3β is shown in Fig. 1 and was developed on the basis of a requirement for TMV-RNA-catalyzed transfer of amino acids into protein from radioactive aminoacyl tRNA (see Table 3). A clue to the inability to detect the ribosome and mRNA-independent Met-tRNA

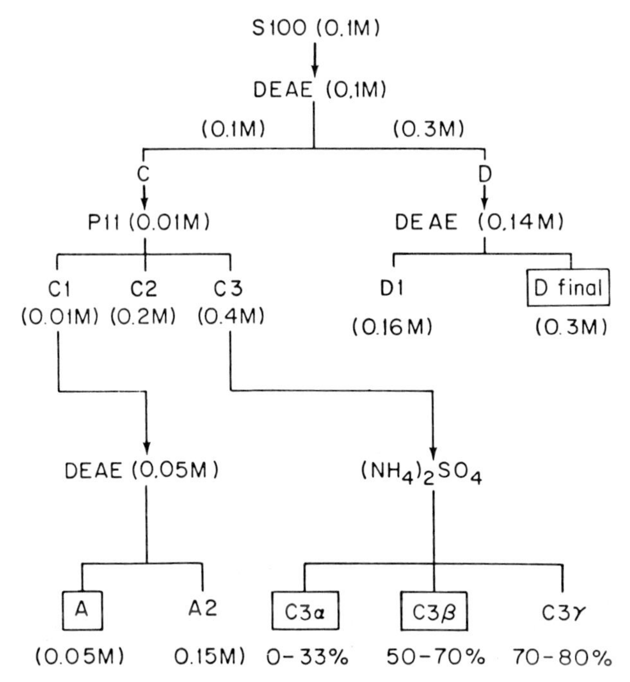

Fig. 1. Schematic representation of the fractionation of wheat germ supernatant. The components surrounded by a box are those required for amino acid acid polymerization (see Table 2).

binding reaction in the cruder fractions is provided by experiment 2 of Table 2. The presence of factor D, a component absolutely required for the overall initiation sequence (see Figs. 2 and 3) results in a considerable decrease in the binding of Met-tRNA. A substantially greater effect of this type in crude fraction C would reduce the Met-tRNA binding to an essentially insignificant level.

A further divergence with the more fractionated preparations was obtained in studies of the formation of a 40S-Met-tRNA complex. As seen in Fig. 2, the reaction is independent of mRNA

TABLE 3

Factor requirements for TMV-RNA-catalyzed amino acid polymerization

Experiment	Conditions	Amino acid incorporation (pmol)
1	complete system	8.6
	-C β	0.4
	-C α	1.9
	-A	2.7
	-D	0.4
2	complete system	8.6
	-A	0.9

The complete system contained in a volume of 0.2 ml: 2.2 mM dithiothreitol, 50 μM GTP, 1 mM ATP, 25 mM Tris-acetate pH 8.0, 3.6 mM $MgAc_2$, 70 mM KAc, 5 mM KCl, 8 mM creatine phosphate, 8 μg creatine phosphokinase, 158 μg $K_{100}M_2$ ribosomes, 5 μg TMV-RNA, 7.2 μg tRNA containing 29 pmol 8-^{14}C-aminoacyl-tRNAs (450 cpm/pmol), and 8 ^{12}C-amino acids (corresponding to the ^{14}C-amino acids; 1mM each). In experiment 1 the following factors were added in the complete system: 20 μg C3β, 20 μg C3α, 27 μg A, 94 μg D-final, 4.5 μg EF1, and 42 μg A2 as a source of EF2. In experiment 2, 600 μg D were added in place of D-final (this level of D contains a saturating level of EF2) and A2 was omitted. Under these conditions, a clearer requirement for fraction A can be demonstrated. After 15 minutes at 30 C, the radioactive material insoluble in hot trichloroacetic acid was determined.

in contrast to the earlier results. The explanation here, however, is clear. The 40S-Met-tRNA complex is fixed with glutaraldehyde which stabilizes an otherwise essentially nondetectable complex. Formation of this complex requires GTP, C3β (the Met-tRNA binding factor), and factor D. The function of factor D may be realted to its ability to lessen the affinity of factor C3β for the Met-tRNA, thereby perhaps allowing the Met-tRNA to bind to an appropriate ribosomal site. Another possible function of factor D might be to insure the specificity

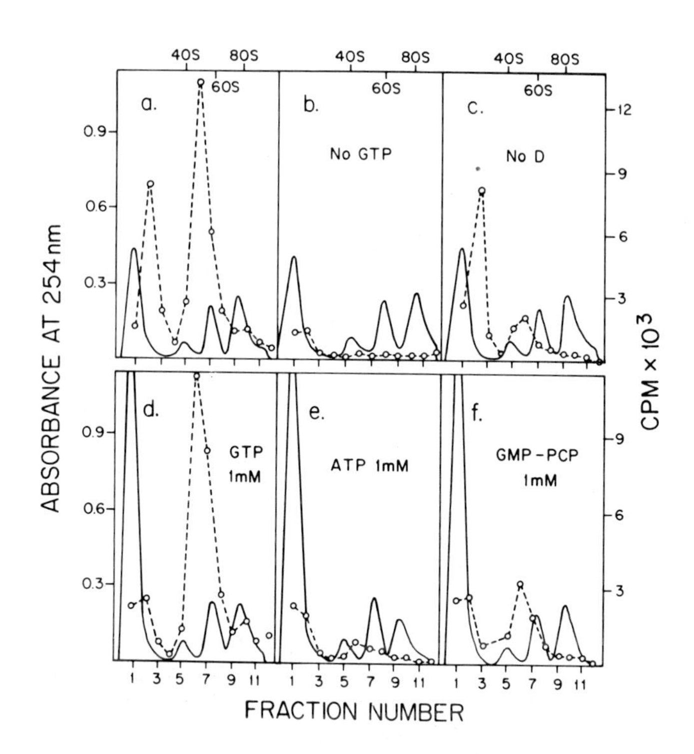

Fig. 2. Requirements for the formation of a 40S·Met-$tRNA_i^{Met}$ complex. In (a) the complete reaction mixture (0.34 ml) contained 1 mM unlabeled methionine, 30 mM Tris acetate pH 8.0, 2.6 mM dithiothreitol, 60 μM GTP, 70 mM KAc, 5 mM KCl, 1.3 mM $MgAc_2$, 7 mM phosphoenolpyruvate, 11.2 μg pyruvate kinase, 39 μg tRNA containing 6.6 pmol [CH_3-^3H]Met-$tRNA_i^{Met}$ (2300 cpm/pmol), 25 μg C3β, 157 μg D final and 115 μg $K_{100}M_2$ ribosomes. In (b) GTP was omitted. In (c) fraction D was replaced by 155 μg bovine serum albumin. In (d)-(f) phosphoenolpyruvate, pyruvate kinase and the low 60 μM GTP were omitted, and 1 mM GTP, ATP, or GMP-P(CH_2)P (guanylyl-5'-methylene diphosphonate) were added as indicated. After incubation for 10 minutes at 20°C, 38 μl of 10% glutaraldehyde in 40 mM triethanolamine were added and the mixtures were kept on ice for 5 minutes. The samples were then centrifuged through a 5-ml 13-21% sucrose gradient in an SW50.1 Spinco rotor for 70-80 minutes at 230,000 x g. Fractions (0.45 ml) were collected and analyzed for bound aminoacyl tRNA by nitrocellulose filtration. The solid line is absorbance and open circles denote radioactivity.

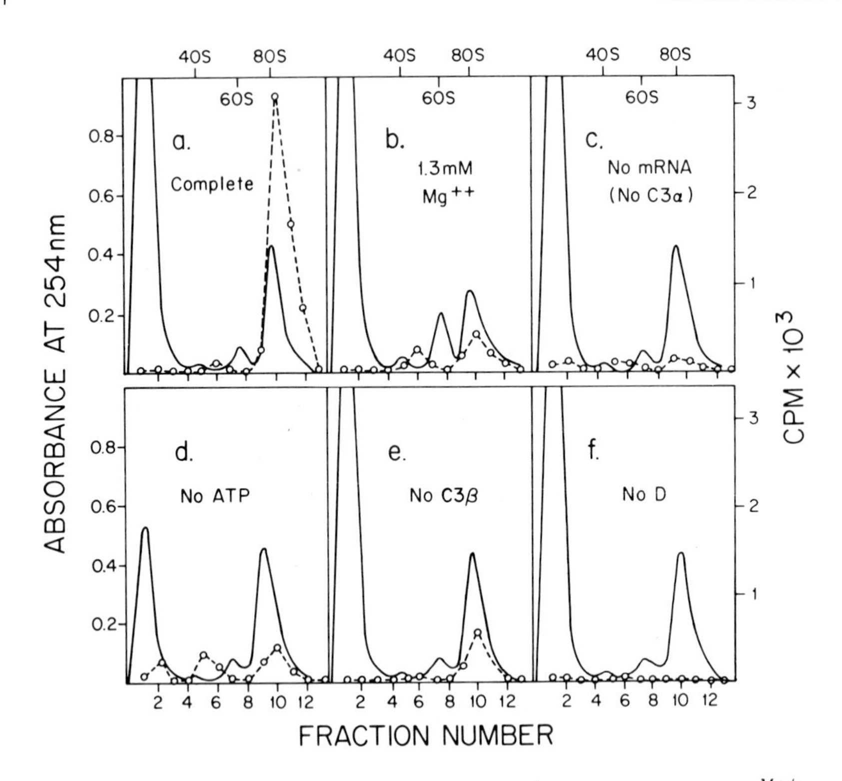

Fig. 3. Requirements for the formation of an 80S·Met-tRNA$_i^{Met}$ complex. The reaction was carried out in two steps. The first incubation contained in 0.34 ml, the complete system as in Fig. 2 except that 52 μg tRNA containing 28.1 pmol [CH_3-^3H]Met-tRNA (2300 cpm/pmol), and 189 μg fraction D-final were used. After 2 minutes at 20°C, 35 μg C3α, 6 μg alfalfa mosaic virus RNA-top a (AMV-RNA), ATP (1 mM), $MgAc_2$ (3.6 mM), and KAc (70 mM) were added to a volume of 0.40 ml, and the incubation was continued for 10 minutes at 20°C. The samples were centrifuged without glutaraldehyde fixation through sucrose gradients and analysed as in Fig. 2. In (b) Mg^{++} was maintained at 1.3 mM, and in (c)-(f) either AMV-RNA, ATP, fraction C3β, or fraction D-final were omitted. In experiments where C3α was omitted a profile similar to that of Fig. 3c was obtained.

of the C3β reaction. As noted in Table 2, experiment 2, the presence of factor D, while only halving the binding of Met-tRNA, almost completely eliminates the binding of Phe-tRNA.

Fig. 3a describes the further function of the resolved fractions in forming an 80S-Met-tRNA$_i^{Met}$ complex. The reaction is carried out in two steps. In the first step, the 40S-Met-tRNA$_i^{Met}$ complex is made at 1.3 mM Mg^{++} with factors C3β and D. In the second step the Mg^{++} concentration is adjusted to 3.6 mM and mRNA, factor C3α, and ATP are added. In contrast to the 40S-Met-tRNA$_i^{Met}$ complex, the 80S-Met-tRNA$_i^{Met}$ complex is stable without glutaraldehyde so that this reagent is not added. Consequently, the 40S-Met-tRNA$_i^{Met}$ complexes that are present in the various control incubations (Fig. 3 b-f) are not seen. Of particular interest are the requirements for mRNA (Fig. 3c) and ATP (Fig. 3d). The latter requirement is consistent with the observations made previously with the unfractionated wheat system, and has also been reported by Schreier and Staehlin (1973) for the reticulocyte system and Kramer *et al.* (1976) for a mixed *Artemia* reticulocyte system. With regard to the function of ATP, it should be noted that a GTP-generating system is added in the initial step of the incubation so that the role of ATP does not appear to be related to the regeneration of GTP. Such a role for ATP, utilizing the nucleoside diphosphate kinase reaction, has recently been suggested (Walton and Gill, 1976).

The requirement for mRNA for the formation of the 80S-Met-tRNA$_i^{Met}$ complex suggests that mRNA participates in this reaction. Figure 4 shows a sucrose gradient analysis of an 80S-reaction with ^{14}C-labeled TMV-RNA. Although there is considerable retention of radioactivity throughout the 40S region, a distinct peak of radioactivity can be seen in the position of the 80S-Met-tRNA$_i^{Met}$ complex. The specificity of the reaction is indicated by the absence of the peak in an incubation in which factor C3α is omitted (Fig. 4b). Two other points relevent to this experiment suggest that the mRNA attachment reaction may be more complex. First, the amount of ^{14}C-radioactivity transferred to the 80S-complex (based on a molecular weight of 2 x 10^6 for TMV-RNA) is only one-fifth that of the Met-tRNA$_i^{Met}$ transferred. Second, it was observed in preliminary experiments that a shift of

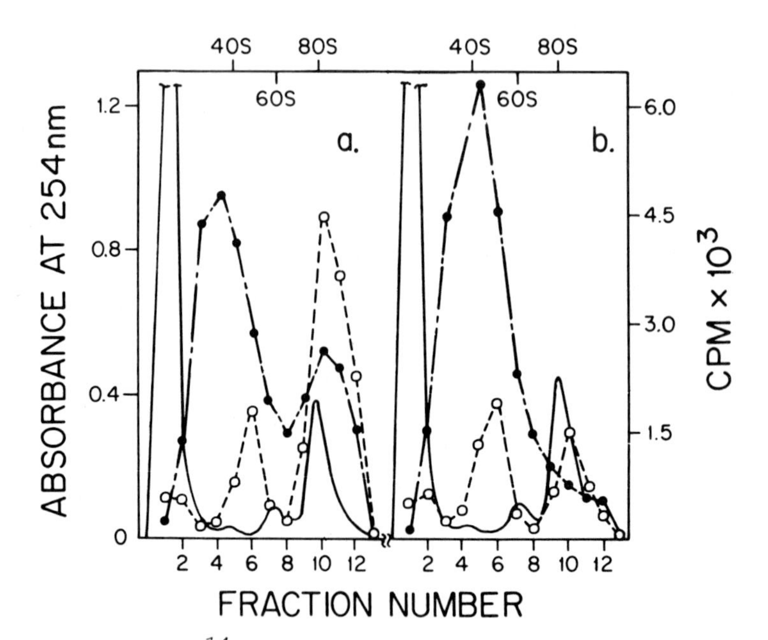

Fig. 4. Binding of ^{14}C-TMV-RNA to an 80S-initiation complex. The incubation conditions were those of Fig. 3 except that the second incubation contained 2 μg ^{14}C-TMV-RNA (4550 cpm/μg) in place of AMV-RNA. This level of viral RNA is below saturation as assayed by Met-tRNA$_i^{Met}$ transfer to the 80S-complex. Higher levels of ^{14}C-TMV-RNA, however, obscured the peak in the region of the 80S complex. In (b) fraction C3α was omitted. The solid line represents absorbance at 254 mμ. The closed circles show ^{14}C-radioactivity from TMV-RNA and the open circles denote ^{3}H-radioactivity from Met-tRNA

^{14}C-TMV-RNA radioactivity similar to that of Fig. 4a could be obtained in the absence of both factor C3β and Met-tRNA$_i^{Met}$. Only ATP and factors C3α and D were required.

In explanation of the stoichiometric balance two points are relevant. It has been reported that TMV-RNA has two initiation sites (Knowland *et al.*, 1975). Furthermore, the mRNA component present finally in the 80S region may only be a fragment of the mRNA that was initially attached. The lack of requirement for C3β and Met-tRNA$_i^{Met}$ for mRNA attachment suggests the possibility that mRNA attachment may be a distinct reaction independent of Met-tRNA$_i^{Met}$ binding and perhaps even precedes it.

Such a scheme would still allow the Met-$tRNA_i^{Met}$ anticodon to align the mRNA codon, but the initial attachment of the mRNA would be relegated to a site on the mRNA other than the initiating triplet. Two possibilities for such binding sites are polypurine-rich stretches preceding the AUG codon (Shine and Dalgarno, 1974) and the $m^7G^{5'}$ppp 5'-termini which have been shown to facilitate the translation of many eucaryotic mRNAs (Furuichi *et al.*, (1975).

ANALYSIS OF THE SEQUENCE OF REACTIONS LEADING TO THE FORMATION OF THE 80S COMPLEX

The studies described above suggest the following sequence of reactions:

(1) 40S + mRNA → 40S-(mRNA)

(2) C3β + Met-$tRNA_i^{Met}$ → C3β-Met-$tRNA_i^{Met}$

(3) 40S-(mRNA) + C3β-Met-$tRNA_i^{Met}$ → 40S-(mRNA)-C3β-Met-$tRNA_i^{Met}$

(4) 40S-(mRNA)-C3β-Met-$tRNA_i^{Met}$ → 40S-mRNA-C3β-Met-$tRNA_i^{Met}$

(5) 40S-mRNA-C3β-Met-$tRNA_i^{Met}$ + 60S → 80S-mRNA-Met-$tRNA_i^{Met}$

It appeared possible to test the role of the C3β-Met-tRNA complex and the 40S-Met-$tRNA_i^{Met}$ complex as intermediates in reactions (3) and (5) respectively by preparing each of these complexes with radioactive Met-$tRNA_i^{Met}$ and then carrying out their respective reactions in the presence of excess unlabeled Met-$tRNA_i^{Met}$. The identical experiments with either no unlabeled Met-$tRNA_i^{Met}$ added or with the unlabeled diluent added prior to the reaction in which the presumed intermediates are made, would serve as controls. When such an experiment was performed with C3β in reaction (3) there was no greater transfer of radioactivity from the preformed C3β-Met-$tRNA_i^{Met}$ than when the radioactive complex was not preformed. This result suggests either that the C3β-Met-$tRNA_i^{Met}$ complex is not an intermediate in the forming of the 40S-complex or that the rate of exchange of Met-$tRNA_i^{Met}$ out of the C3β complex is considerably greater than its rate of transfer to the 40S subunit. When the same type of experiment was done to test the role of the 40S-Met-$tRNA_i^{Met}$ complex in reaction (5), a positive result was obtained.

As seen in Table 4, experiment 1, when the 80S reaction is terminated by glutaraldehyde fixation after three minutes, the added unlabeled Met-$tRNA_i^{Met}$ has little effect and the radioactive Met-$tRNA_i^{Met}$ is transferred essentially to the same extent as in the control lacking the diluent Met-$tRNA_i^{Met}$. These data are consistent with the 40S-Met-$tRNA_i^{Met}$ complex as an intermediate in the formation of an 80S complex. If the 80S reaction is run for 10 minutes (experiment 2) there is considerably greater dilution by the added Met-$tRNA_i^{Met}$. This is due in part to the fact that in the 10 minute incubation about three times as much Met-$tRNA_i^{Met}$ appears in the 80S complex as was present initially in the 40S-complex. Thus at least two-thirds of the 80S-Met-$tRNA_i^{Met}$ complex derives from 40S-Met-$tRNA_i^{Met}$ complexes made with fully diluted Met-$tRNA_i^{Met}$. The longer incubation also allows equilibration of the initially formed 80S-Met-$tRNA_i^{Met}$ with the 40S-Met-$tRNA_i^{Met}$ pool, again resulting in lowered radioactivity in the 80S-complex. The short-term incubation (experiment 1) indicates clearly, however, that the Met-$tRNA_i^{Met}$ present in the 40S complex can be transferred to the 80S complex with little dilution.

THE 5'-7-METHYLGUANOSINE "CAP" AND mRNA TRANSLATION

An important aspect of the initiation process, as noted previously, is the attachment of mRNA to the 40S ribosomal subunit. Both *et al.* (1975) have shown that the attachment of

TABLE 4

Transfer of Met-$tRNA_i^{Met}$ from the 40S complex to an 80S complex

	Components Added		^{3}H-Met-tRNA Bound (pmol)	
	Incubation 1	Incubation 2	40S complex	80S complex
1	2 min 20°C	3 min 20°C		
	^{3}H-Met-tRNA		1.4	0.66
	^{3}H-Met-tRNA + Met-tRNA		0.21	0.13
	^{3}H-Met-tRNA	Met-tRNA	0.99	0.57
2	2 min 20°C	10 min 20°C		
	^{3}H-Met-tRNA		0.49	1.4
	^{3}H-Met-tRNA + Met-tRNA		0.17	0.26
	^{3}H-Met-tRNA	Met-tRNA	0.19	0.37
	^{3}H-Met-tRNA[a]	-	0.56	

The reaction conditions were those of Fig. 3 except that in experiment 1 the second incubation was terminated after 3 minutes. 28 pmol of [CH_3-^{3}H]Met-$tRNA_i^{Met}$ (52 μg of tRNA), and 140 pmol of unlabeled Met-$tRNA_i^{Met}$ (260 μg of tRNA) were added to the particular incubations as indicated, and the reactions were terminated by the addition of glutaraldehyde as in Fig. 2. The data presented are a calculated sum of the nitrocellulose-bound radioactivity in the single fractions corresponding to the specified regions of the gradient.

[a] Reaction terminated at end of first incubation.

reovirus mRNA to ribosomes and the subsequent translation of this mRNA requires the presence of an $m^7G^{5'}$ppp "cap" on the mRNA. As a test of the universality of this conclusion, we studied the translation of satellite tobacco necrosis virus RNA (STNV-RNA). This viral RNA can be translated authentically in the wheat system (Klein *et al.*, 1972, Roman *et al.*, 1976) although its 5' terminus has been reported to be either ppApGpUp... (Lesnaw and Reichmann, 1970) or pppApGpUp... (Horst *et al.*,

1971). If one could ascertain that the translational system was not adding the 5'-"cap", this would indicate that STNV-RNA did not require the "cap" for translation. Table 5 demonstrates exactly this point showing that the addition of S-adenosyl homocysteine (SAH), a methyl transfer inhibitor known to prevent the "capping" reaction (Both *et al.*, 1975) has no effect on the translation of STNV-RNA.

Further evidence supporting the idea that STNV-RNA functions without a 5'-"cap" was obtained in experiments using the inhibitor, 7-methyl-guanosine-5' phosphate, $m^7G^{5'}p$ (Hickey *et al.*, 1976). As shown in Table 6, this inhibitor has no effect on the formation of the 40S-Met-$tRNA_i^{Met}$ complex, but is strongly inhibitory to the mRNA-dependent reaction in which an 80S-Met-$tRNA_i^{Met}$ complex is formed. The effect, however, is strikingly dependent on the specific mRNA added. The reaction with TMV-RNA and AMV-RNA is strongly inhibited and that with STNV-RNA is unaffected. Clearly, the most compatible explanation of these data is that STNV-RNA translation is independent of 5'-"cap".

SUMMARY

Protein chain initiation in extracts of wheat germ proceeds first by the mRNA-independent formation of an unstable 40S ribosome·Met-$tRNA_i^{Met}$ complex that is detected in sucrose gradients only after fixation with glutaraldehyde. Two soluble factors (C3β and D) and GTP are required for this reaction. One of the factors (C3β) binds Met-$tRNA_i^{Met}$ suggesting that a C3β·Met-$tRNA_i^{Met}$ complex may be an intermediate in the formation of the 40S complex. When ATP, mRNA, magnesium acetate (final concentration 3.6 mM), and a third factor (C3α) are added, a stable 80S·Met-$tRNA_i^{Met}$ complex is formed. Radioactive TMV-RNA binds to this complex suggesting that the mRNA is a component.

TABLE 5

^{14}C-Leucine incorporation catalyzed by STNV-RNA in the presence and absence of S-Adenosylhomocysteine (SAH)

Additions	^{14}C-Leucine Incorporated (cpm)
-	1,200
STNV-RNA (2.5 μg)	24,331
" + SAM (4 μM)	26,752
" + SAH (320 μM)	25,250
reovirus RNA (2.5 μg)	8,873
" + SAM (4 μM)	13,944
" + SAH (320 μM)	2,480

A 0.1 ml incubation containing 20 mM HEPES-KOH pH 7.6, 1 mM ATP, 20 μM GTP, 8 mM creatine phosphate, 40 μg/ml creatine phosphokinase, 2.5 mM dithiothreitol, 2 mM $MgAc_2$, 80 μM spermine, 20 mM KCl, 120 mM KAc, 30 μM 19 amino acids-leucine, 0.625 μCi/ml ^{14}C-L-leucine, 15 μg/ml wheat tRNA, 25 μl wheat germ S23, and viral RNA as indicated, was incubated for 60 minutes at 25°C and the hot TCA-insoluble radioactivity was determined. Wheat germ S23 was prepared by grinding 5 g wheat germ (General Mills Inc.) first with 5 g of sand and then with 25 ml 1 mM $MgAc_2$, 2 mM $CaCl_2$, 90 mM KCl using a chilled mortar and pestle. After centrifuging at 23,000 x g for 10 min (pellet discarded), adjusting to 20 mM Tris acetate pH 7.6 and 2 mM $MgAc_2$ and recentrifuging, the supernatant was passed through a column of Sephadex G-25 and eluted with 1 mM Tris acetate pH 7.6, 50 mM KCl, 3 mM $MgAc_2$, 4 mM β-mercaptoethanol. The turbid fraction was collected, centrifuged at 23,000 x g for 10 min (pellet discarded), and stored in small aliquots at -70°C. Reovirus RNA having a 5'-terminus of 75% ppGp and 25% GpppG was a gift of M. Morgan and A. Shatkin of the Roche Institute.

TABLE 6

Effect of $m^7G^{5'}p$ on the formation of 40S-Met-$tRNA_i^{Met}$ and 80S-Met-$tRNA_i^{Met}$ complexes

Experiment	Addition (0.25 mM)	Met-$tRNA_i^{Met}$ bound (cpm)	Inhibition (%)
1) 40S reaction	-	1701	
	$m^7G^{5'}p$	1640	3.6
2) 80S reaction	-	1706	
(TMV-RNA)	$m^7G^{5'}p$	342	80.0
	$m^7G^{2(3')}p$	1468	14.0
	$G^{5'}p$	1397	18.1
3) 80S reaction	-	1953	
(STNV-RNA)	$m^7G^{5'}p$	2147	
4) 80S reaction	-	5297	
(AMV-RNA)	$m^7G^{5'}p$	1363	74.3

The reaction systems were those of Figs. 2 and 3, with the viral RNAs added at 6 µg per 0.34 ml total volume. The data are a calculated sum of the nitrocellulose bound radioactivity; in experiment 1 for the region of the 40S complex and in experiments 2-4 for the region of the 80S complex. The data in experiments 2-4 are corrected for a control in which mRNA was omitted.

Evidence that the 5'-"cap" of the mRNA is a participant in the 80S-forming reaction is obtained in experiments with $m^7G^{5'}p$. The reaction catalyzed by either TMV-RNA or AMV-RNA is inhibited by the "cap" analogue, whereas when STNV-RNA (an RNA lacking the 5' "cap") is used, the analog is without effect. Preformed 40S·Met-$tRNA_i^{Met}$ transfers Met-$tRNA_i^{Met}$ to the 80S complex in preference to free Met-tRNA suggesting that the 40S complex is an intermediate in the 80S reaction.

Analyses of the factor requirements for TMV-RNA-catalyzed amino acid polymerization show that in addition to the three factors already described and the two elongation factors there is a requirement for a fourth factor suggesting that the initiation process may be more complex than thus far indicated.

ACKNOWLEDGMENTS

This work was supported by U.S.P.H.S. grants GM15122, CA-06927 and RR-05539 from the National Institutes of Health, and by an appropriation from the Commonwealth of Pennsylvania. M.S. was a recipient of postdoctoral fellowships from NATO and the Deutsche Forschungsgemeinschaft. R.R. was a recipient of a postdoctoral fellowship from the Universidad Nacional Autonama de Mexico.

REFERENCES

Both, G.W., Banerjee, A.K., and Shatkin, A.J. (1975). Methylation -dependent translation of viral messenger RNAs *in vitro*. *Proc. Nat. Acad. Sci. U.S. 72*, 1189-1193.

Both, G.W., Furuichi, Y., Muthukrishnan, S., and Shatkin, A.J. (1975). Ribosome binding to reovirus mRNA in protein synthesis requires 5' terminal 7-methylguanosine. *Cell 6*, 185-189.

Cashion, L.M., and Stanley, W.M. Jr. (1974). Two eukaryotic initiation factors (IF-I and IF-II) of protein synthesis that are required to form an initiation complex with rabbit reticulocyte ribosomes. *Proc. Nat. Acad. Sci. U.S. 71*, 436-440.

Davies, J.W., and Kaesberg, P. (1974). Translation of virus RNA: protein synthesis directed by several virus RNAs in a cell-free extract from wheat germ. *J. Gen. Virol. 25*, 11-20.

Dettman, G.L., and Stanley, W.M. Jr. (1972). Recognition of eukaryotic initiator tRNA by an initiation factor and the transfer of the methionine moiety into peptide linkage. *Biochim. Biophys. Acta 287*, 124-133.

Furuichi, Y., Morgan, M., Shatkin, A.J., Jelinek, W., Dalditt-Georgiev, M., and Darnell, J.E. (1975). Methylated, bolcked 5' termini in HeLa cell mRNA. *Proc. Nat. Acad. Sci. U.S. 72*, 1904-1908.

Giesen, M., Roman, R. Seal, S.M., and Marcus, A. (1976). Formation of an 80S methionyl-tRNA initiation complex with soluble factors from wheat germ. *J. Biol. Chem. 251*, 6075-6081.

Gupta, N.K., Chatterjee, B., Chen, Y.C., and Majumder, A. (1975). Protein synthesis in rabbit reticulocytes. *J. Biol. Chem. 250*, 853-862.

Gupta, N.K., Woodley, C.L., Chen, Y.C., and Bose, K.K. (1973). Protein synthesis in rabbit reticulocytes. *J. Biol. Chem. 248*, 4500-4511.

Haselkorn, R., and Rothman-Denes, L.B. (1973). Protein synthesis. *Ann. Rev. Biochem. 42*, 397-438.

Hickey, E.D., Weber, L.A., and Baglioni, C. (1976). Inhibition of initiation of protein synthesis by 7-methylguanesine-5'-monophosphate. *Proc. Nat. Acad. Sci. U.S. 73*, 19-23.

Horst, J., Frankel-Conrat, H., and Mandeles, S. (1971). Terminal heterogeneity at both ends of the satellite tobacco necrosis virus ribonucleic acid. *Biochem. 10*, 4748-4752.

Jay, G., and Kaempfer, R. (1975). Initiation of protein synthesis. *J. Biol. Chem. 250*, 5742-5748.

Klein, W.H., Nolan, C., Lazar, J.M., and Clark, J.M. Jr. (1972). Translation of satellite tobacco necrosis virus ribonucleic acid. I. Characterization of *in vitro* procaryotic and eukaryotic translation products. *Biochem. 11*, 2009-2014.

Knowland, J., Hunter, T., Hunt, T., and Zimmern, D. (1975). In In vitro *transcription and translation of viral genomes* (Haenni, A.L., and Beandt, G., eds.), pp. 211-216. Inserm, Paris.

Kramer, G., Konecki, D., Amadevilla, J.M., and Hardesty, B. (1976). Communications ATP requirement for binding of ^{125}I-labeled globin mRNA to *Artemia salina* ribosomes. *Arch. Biochem. 174,* 355-358.

Levin, D.H., Kyner, D., and Acs, G. (1973). Protein synthesis initiation in eukaryotes. *J. Biol. Chem. 248,* 6416-6425.

Levin, D.H., Kyner, D., and Acs, G. (1973). Protein initiation in eurkaryotes: formation and function of a ternary complex composed of a partially purified ribosomal factor, methionyl transfer RNA, and guanosine triphosphate. *Proc. Nat. Acad. Sci. U.S. 70,* 41-45.

Lesnaw, J.A., and Reichmann, M. (1970). Identity of the 5'-terminal RNA nucleotide sequence of the satellite tobacco necrosis virus and its helper virus: possible role of the 5'-terminus in the recognition by virus-specific RNA replicase. *Proc. Nat. Acad. Sci. U.S. 66,* 140-145.

Marcus, A., Bewley, J. D., and Weeks, D. P. (1970). Aurintricarboxylic acid and initiation factors of wheat embryo. *Science 167*, 1735-1736.

Marcus, A., and Feeley, J. (1964). Activation of protein synthesis in the imbibition phase of seed germination. *Proc. Nat. Acad. Sci. U.S. 51*, 1075-1079.

Marcus, A., Luginbill, B., and Feeley, J. (1968). Polysome formation with tobacco mosaic virus RNA. *Proc. Nat. Acad. Sci. U.S. 59,* 1243-1250.

Marcus, A., Weeks, D.P., and Seal, S.N. (1973). Protein chain initiation in wheat embryo. In *Nitrogen Metabolism in Plants* (Smellie, R.M.S. and Goodwin, T.W., eds.), *Biochem. Soc. Symp. 38,* 97-109.

Roman, R., Brooker, J.D., Seal, S.N., and Marcus, A. (1976). Inhibition of the transition of a 40S ribosome-Met-$tRNA_i^{Met}$ complex to an 80S ribosome-Met-$tRNA_i^{Met}$ complex by 7-methylguanosine-5'-phosphate. *Nature 260,* 359-360.

Schreier, M.H., and Staehlin, T. (1973). Functional characterization of five initiation factors for mammalian protein synthesis. (Bautz, E.K.F., Karlson, P., and Kersten, H., eds.), pp. 335-349, *Proc.* 24th Mosbacher Colloq. Springer Verlag, New York.

Seal, S.N., Bewley, J.D., and Marcus, A. (1972). Protein chain initiation in wheat embryo. *J. Biol. Chem. 247,* 2592-2597.

Shafritz, D.A., and Anderson, W.F. (1970). Factor dependent binding of methionyl-tRNAs to reticulocyte ribosomes. *Nature 227,* 918-920.

Shafritz, D.A., Prichard, P.M., Gilbert, J.M., Merrick, W.C., and Anderson, W.F. (1972). Separation of reticulocyte initiation factor M activity into two components. *Proc. Nat. Acad. Sci. U.S. 69*, 983-987.

Shine, J., and Dalgarno, L. (1974). The 3'-terminal sequence of *Escherichia coli* 16S ribosomal RNA: complementarity to nonsense triplets and ribosome binding sites. *Proc. Nat. Acad. Sci. U.S. 71*, 1342-1346.

Walton, G.N., and Gill, G.N. (1976). Regulation of ternary [met-tRNA ·GTP·eukaryotic initiation factor 2] protein synthesis initiation complex formation by the adenylate energy charge. *Biochim. Biophys. Acta 418*, 195-203.

Weeks, D.P., and Marcus, A. (1970). Preformed messenger of quiescent wheat embryos. *Biochim. Biophys. Acta 232*, 671-684.

Weeks, D.P., Verma, D.P.S., Seal, S.N., and Marcus, A. (1972). Role of ribosomal subunits in eukaryotic protein chain initiation. *Nature 236*, 167-168.

HORMONAL CONTROL OF ENZYME FORMATION IN BARLEY ALEURONE LAYERS

Tuan-Hua David Ho*

Department of Biology
Washington University
St. Louis, Missouri 63130

One of the major events during seed germination is to supply nutrients to the growing embryo. For most dicotyledenous seeds the cotyledons are the source of nutrients. For the monocotyledenous seeds, such as cereal grains, the endosperm is the nutrient warehouse. The mobilization of the endosperm reserves is controlled by a hormone called gibberellin (GA) which is formed in the embryo (Fig. 1). Gibberellin then diffuses from the embryo into the endosperm and eventually to the aleurone surrounding the endosperm. The aleurone is composed of homogeneous, non-dividing, protein-rich cells. In response to GA these cells synthesize several hydrolases, including α-amylase and protease. These enzymes are secreted into the endosperm where they hydrolyze the starch and protein reserves into simple sugars and amino acids. The sugars and amino acids are then absorbed by the scutellum and transported to the embryo to support the heterotrophic growth of whole seedlings.

Isolated aleurone also responds to exogenous GA, such as gibberellic acid (GA_3), by synthesizing and secreting α-amylase and protease after a 4 to 8 hour lag period (Fig. 2) (Yomo and Varner, 1971). The rate of production of these emzymes reaches a maximum after 12 hours of GA_3 treatment. At this time, secretion of these enzymes begins. α-Amylase becomes the predominant protein (~40% of total proteins) synthesized in the presence of GA_3 (Varner and Ho, 1976). Therefore, we have been using α-amylase as a biochemical marker to study the mode of action of GA_3. Our strategy has been to describe the most prominent effect of the hormone, *i.e.* the increase of α-amylase

**Present address: Department of Botany, University of Illinois, Urbana, Illinois 61801.*

ISBN 012-601950-9

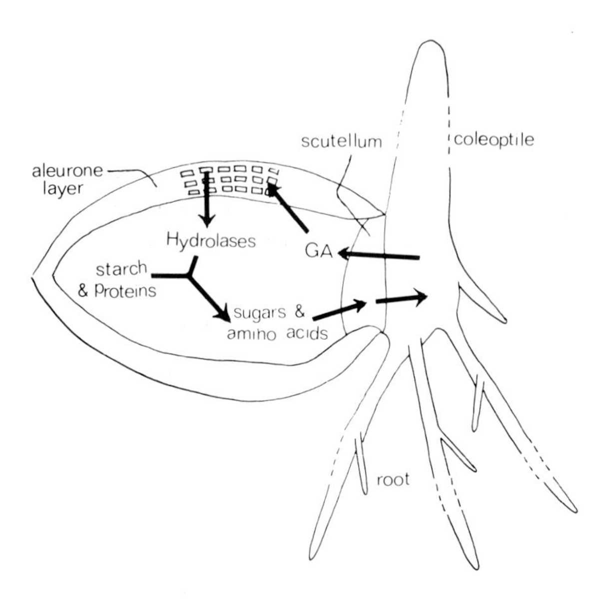

FIGURE 1. A cross section diagram of a barley seedling showing the relationship between the production of GA, the synthesis and secretion of hydrolases, and the mobilization of endosperm nutrients.

activity, and then use the information obtained to work toward the cause of this effect until the primary effect of the hormone is discovered.

DE NOVO SYNTHESIS OF α-AMYLASE MOLECULE

Initially we asked whether the increase in α-amylase activity was due to *de novo* synthesis of the enzyme molecule. This was answered by the incorporation of radioactive amino acids into α-amylase followed by purification of the enzyme (Varner and Chandra, 1964). Fingerprint analysis of the purified α-amylase showed that most tryptic peptides contained radioactivity indicating that the whole α-amylase molecule was synthesized *de novo* (Fig. 3). Furthermore, incubating barley aleurone layers with $H_2{}^{18}O$ and GA_3, Filner and Varner (1967) observed an increased density of α-amylase resulting from the incorporation of ^{18}O into free amino acids during the breakdown of storage protein, and the utilization of the ^{18}O-amino acids for the synthesis of α-amylase (Fig. 4). From the extent of the density shift, they concluded that

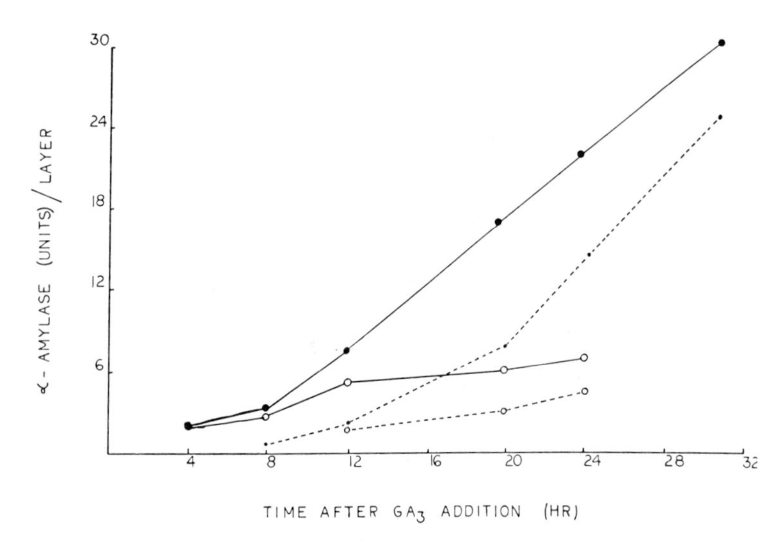

FIGURE 2. The time course of the GA_3 enhanced α-amylase formation in barley aleurone layers. The increase of α-amylase activity reflects the accumulation of α-amylase. ●——● *total α-amylase activity, including enzyme in the aleurone layers and enzyme in the medium, in the presence of 1 μM GA_3.* ●-----● *α-amylase activity in the medium with 1 μM GA_3.* ○——○ *total α-amylase activity without GA_3.* ○----○ *α-amylase activity in the medium without GA_3.*

essentially all the α-amylase molecules were synthesized *de novo* after the addition of GA_3. However, since GA_3 enhanced α-amylase production had a lag period of 4 to 8 hours and most of the α-amylase was formed after 12 hours of GA_3 treatment, one could suggest that a precursor of α-amylase was synthesized *de novo* during the lag period while the activity increase was due to the activation of this precursor. To test this possibility, we incubated aleurone layers with ^{13}C-amino acids either before or after 12 hours of treatment with GA_3 (Ho, 1975). As shown in Table 1, the density shift was much larger when ^{13}C-amino acids were added after 12 hours of GA_3 treatment. Radioactive amino acids were also present during the incubation in order to monitor the average density of newly synthesized proteins by measuring radioactivity profiles in CsCl density gradients. The density shift of newly synthesized proteins represents 100% *de novo* synthesis of the α-amylase timated the precentage of *de novo* synthesis of the α-amylase molecule by comparing the density shift of newly synthesized

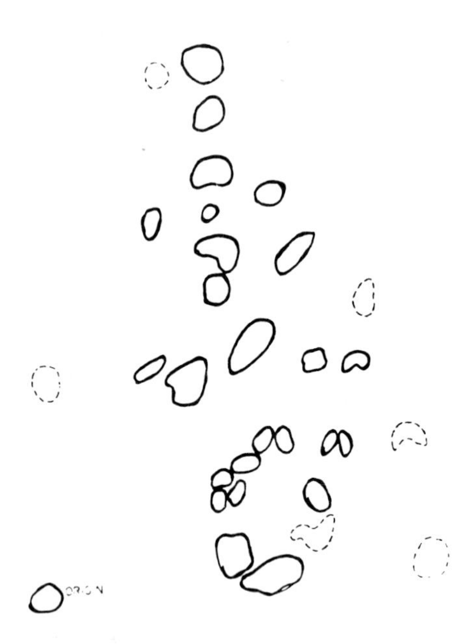

FIGURE 3. Autoradiograph of a fingerprint of α-amylase obtained by trytic digestion of ^{14}C-threonine labeled α-amylase which was isolated from the incubation medium of GA_3 treated embryoless barley half-seeds. The solid lines show the position of ninhydrin-positive spots which coincide with the exposed spots on the radioautogram. The dotted lines are ninhydrin-positive spots which do not contain radioactivity (Varner and Chandra, 1964).

proteins and that of α-amylase. Our estimation was that essentially all the α-amylase molecule was synthesized *de novo* at the time when increased enzyme activity was observed (Ho, 1975).

Following the conclusion that α-amylase was synthesized *de novo*, the next questions were: (1) Since α-amylase is a secretory protein and there is ample evidence in other systems that secretory proteins are synthesized by membrane bound polyribosomes (Siekevitz and Palade, 1960 and 1966), does GA_3 control the synthesis of α-amylase by regulating the availability of membrane and membrane-bound polyribosomes? (2) Does GA_3 cause gene derepression and enhance the synthesis of α-amylase specific mRNA?

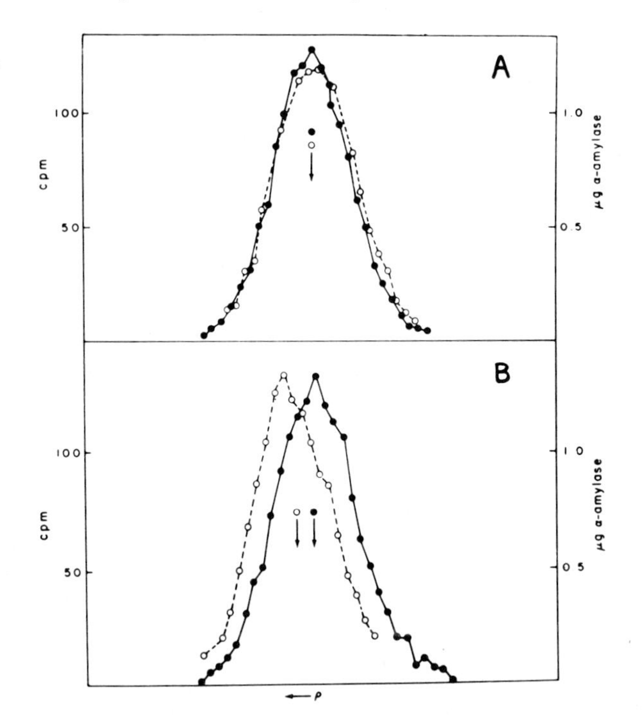

FIGURE 4. Equilibrium distributions of α-amylase in a CsCl density gradient. o-----o α-Amylase activity profile, 25 ug of crude α-amylase prepared from the incubation medium of GA_3 treated aleurone layers. Aleurone layers were incubated in the presence of $H_2{}^{16}O$ (regular water) in (A) and in the presence of $H_2{}^{18}O$ (heavy water) in (B). The buoyant density of α-amylase prepared form $H_2{}^{16}O$ layers is 1.300 g/ml, and that of α-amylase from $H_2{}^{18}O$ layers is 1.314 g/ml. ●——● Radioactivity profile of 2 ug of α-amylase isolated from the incubation medium of GA_3 treated aleurone layers incubated with $H_2{}^{16}O$ (regular) and ^3H-amino acids. Because of the very low quantity of $^3H^{16}O$ labeled α-amylase (2 ug), the α-amylase activity profile in the gradient of (B) represents the distribution of ^{18}O labeled enzyme (Filner and Varner, 1967).

EFFECT OF GA_3 ON THE PROLIFERATION OF MEMBRANE AND MEMBRANE-BOUND POLYSOMES

Direct electron microscopic (EM) observations of GA_3 treated aleurone layers has shown a proliferation of rough endoplasmic reticulum (rough ER) (Jones, 1969a,b; Virgil and Ruddat, 1973). The biochemical determination of the quantity of polysomes

TABLE 1
De Novo Synthesis of α-Amylase As Demonstrated by ^{13}C-Amino Acid Density Labeling

Treatment	Density (g/ml)		Density shift from ^{12}C (0 24 hr)		% of α-Amylase synthesized *de novo*
	α-amylase	radioactivity peak	α-amylase	radioactivity peak	
^{12}C (0 → 24 hr)	1.300 ± 0.001	1.308 ± 0.002			
^{13}C (0 → 24 hr)	1.317 ± 0.002	1.326 ± 0.002	0.017	0.018	>95%
^{12}C (0 → 12 hr) ^{13}C (12 → 24 hr)	1.310 ± 0.001	1.318 ± 0.002	0.010	0.010	100%
^{13}C (0 → 12 hr) ^{12}C (12 → 24 hr)	1.302 ± 0.002	1.311 ± 0.003			

Ten aleurone layers were incubated with either ^{13}C-amino acids or ^{12}C-amino acids (10 mM), radioactive ^{14}C-leucine, buffer, and GA_3 as indicated. After incubation α-amylase was prepared from both layers and incubation medium and the buoyant density of proteins was measured by equilibrium centrifugation in a CsCl density gradient. The radioactivity peak in the gradient represents the distribution of newly synthesized proteins (Ho, 1975).

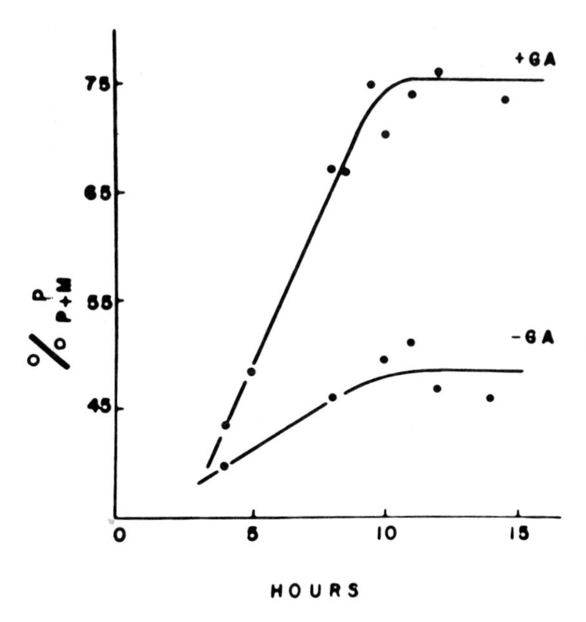

FIGURE 5. The time course of GA_3 enhanced polysome formation expressed by the distribution of ribosomes in polysomes (percent polysomes). The polysomal distribution, % (P/P+M), was calculated by integrating the area of polysome (P) and monosome (M) regions in sucrose density gradients (Evins, 1971).

(Evins, 1971) and of the incorporation of choline and ^{32}P into membrane phospholipid (Evins and Varner, 1971; Koehler and Varner, 1973) has been consistent with EM observations. The GA_3 enhancement of polysome formation starts in about 3 to 4 hours and reaches a maximum at about 12 hours after the addition of hormone (Fig. 5) (Evins, 1971). Aleurone layers treated with GA_3 and incubated in labeled tryptophan and tyrosine possessed polysomes with nascent polypeptides that had a high ratio of tryptophan to tyrosine label. This is characteristic of the amino acid composition of α-amylase and indicates that these polysomes were synthesizing hydrolases (Evins, 1971). However, the GA_3 enhanced α-amylase formation probably does not depend on the synthesis of new ribosomes, because the rRNA synthesis inhibitor, 5-flurouracil, had no effect on α-amylase synthesis (Chrispeels and Varner, 1967).

The GA_3 enhancement of ^{32}P incorporation into phospholipid also started at 3 to 4 hours (Fig. 6) (Koehler and Varner, 1973). Because GA_3 had no effect on glycerol and acetate incorporation (Firn and Kende, 1974; Koehler and Varner, 1973), the enhanced ^{32}P incorporation may represent a turnover of phospholipid, *i.e.* the conversion of storage lipids to membrane lipids. The

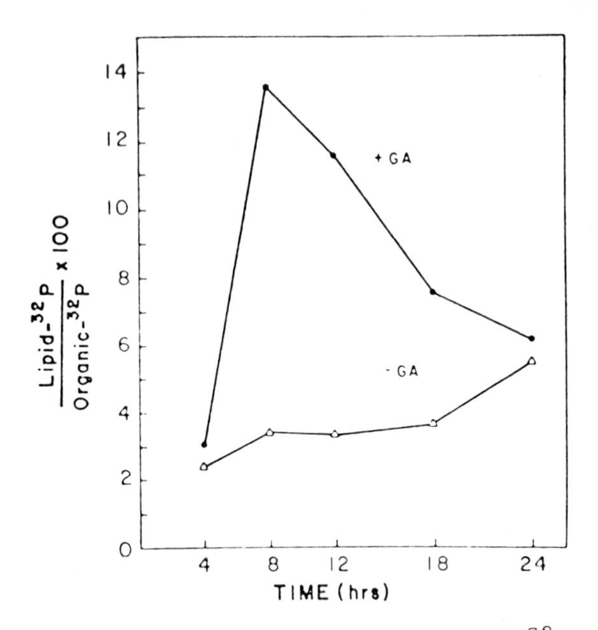

FIGURE 6. The time course of GA_3 enhanced ^{32}P incorporation into phospholipid. Incorporation of ^{32}P is expressed as a percentage of the ^{32}P incorporation into organic phosphates (Koehler and Varner, 1973).

following two observations also support this suggestion. First, the aleurone layer is rich in storage lipids and the major lipid storage particle, the spherosome, disappears after GA_3 treatment (Jones, 1969b). Second, *in vitro* assays indicate a very high phospholipase activity in aleurone cells (Flint and Varner, unpublished observation). Whether this represents *in vivo* activity remains to be tested.

The major component of barley aleurone phospholipid is lecithin (phosphorylcholine) (Koehler and Varner, 1973) and there are three enzymes involved in the formation of lecithin:

$$\text{Choline} + \text{ATP} \xrightarrow{\text{choline kinase}} \text{P-Choline} + \text{ADP}$$

$$\text{P-Choline} + \text{CTP} \xrightarrow[\text{transferase}]{\text{phosphorylcholine-cytidyl}} \text{CDP-Choline} + \text{PPi}$$

$$\text{CDP-Choline} + \text{1,2-diglyceride} \xrightarrow[\text{transferase}]{\text{phosphorylcholine-glyceride}} \text{Lecithin} + \text{CMP}$$

Gibberellic acid enhanced the activity of phosphorylcholine-cytidyl and phosphorylcholine-glyceride transferases (Fig. 7)

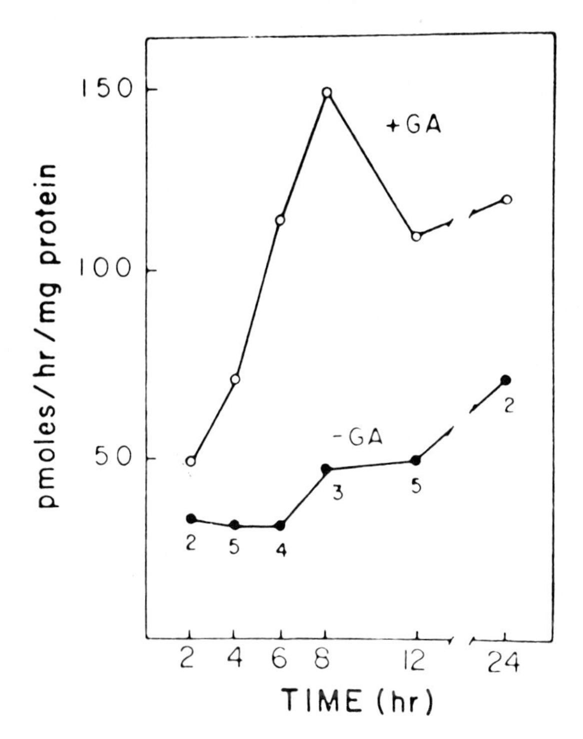

FIGURE 7. The effect of GA_3 on phosphorylcholine glyceride transferase activity. Similar results were obtained with phosphorylcholine cytidyl transferase (Johnson and Kende, 1971).

(Johnson and Kende, 1971). This enhancement has been observed as early as 15 min. after GA treatment (Ben-Tal, 1974). The activity increase of phosphorylcholine-glyceride transferases was apparently due to activation because amino acid analogs and cordycepin did not inhibit this GA_3 effect (Table 2) (Ben-Tal and Varner, 1974). The activity of phosphorylcholine-glyceride transferase can be inhibited by Ca^{++} during the *in vitro* assay. Because aleurone layers contain calcium in the form of phytin, the activation of phosphorylcholine-glyceride transferase could be attributed to redistribution of calcium after GA_3 treatment. This redistribution could be brought about by a changed membrane permeability for Ca^{++} or altered levels of Ca^{++} chelators, such as citrate. Wood and Paleg (1972) demonstrated that GA_3 increased the permeability of artificial model membranes composed of various plant-source lipids.

The activity of membrane bound enzyme, like phosphorylcholine-glyceride transferase, also could be altered by the presence of surfactant. One of the major products of lipid degra-

TABLE 2
Effect of GA_3, Amino Acid Analogs, and Cordycepin On Phosphorylcholine Glyceride Transferase Activity in Barley Aleurone Layers

Treatment during incubation	% of control	Standard deviation
A.		
Control	100	
1 uM GA_3	161	± 38.2
1 uM GA_3 and 7 mM amino acid analogs	167	± 40.4
B.		
Control	100	
1 uM GA_3	164	± 13.4
1 uM GA_3 and 1 mM cordycepin	147	± 18.9

The enzyme was prepared from aleurone layers after 4 hrs. incubation of embryoless half-seeds. The amino acid analogs used were 7-azatryptophan, thioproline, o-fluorophenyalanine, 3,5-diiodotyrosine, ethionine, methyllylglycine, and canavanine, all at 1 mM. Because of the large differences in the absolute pmoles of lecithin formed in different experiments the control in each experiment has been taken arbitrarily as 100 (Ben-Tal and Varner, 1974).

dation, lysolecithin, is a surfactant. Therefore, one could imagine that the breakdown of storage lipid triggers the activation of phospholipid synthesizing machinery.

EFFECT OF GA_3 ON mRNA METABOLISM

Messenger RNA in barley aleurone layers like mRNA in other eucaryotic cells contains a polyadenylic acid (poly A) tail at its 3'-OH end (Ho and Varner, 1974). Therefore, mRNA as well as some of its precursors can be isolated from other major RNA species by oligo-dT cellulose or poly U sepharose column chromatography. Gibberellic acid has enhanced the synthesis of total mRNA starting at 3 to 4 hours after the addition of hormone (Fig. 8) (Ho and Varner, 1974). A maximum was reached which was about 50 to 60% over control at 10 to 12 hours. This increased mRNA was observed when either adenosine or uridine was used as a labeled precursor indicating that the synthesis

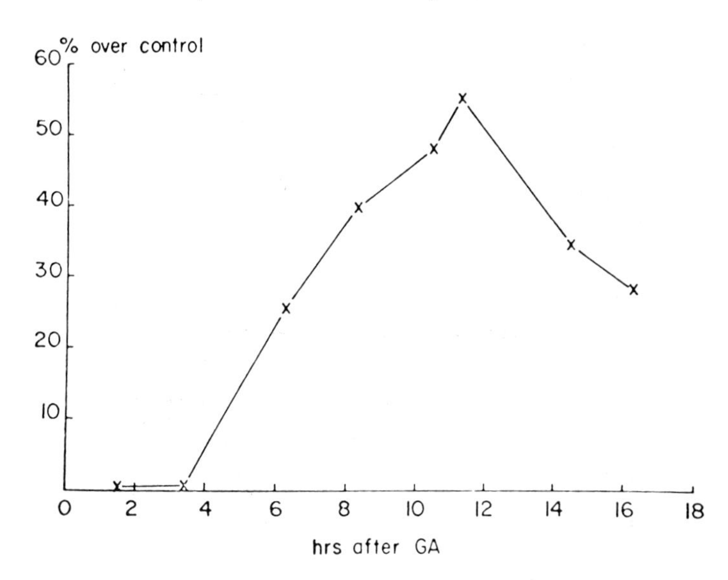

FIGURE 8. The effect of GA_3 on polyA-RNA synthesis. The data are plotted as percentage enhancement over the control without GA_3 (Ho and Varner, 1974).

of both the poly A and the non-poly A regions of mRNA were enhanced by GA_3 (Ho, 1975). In other words the enhancement was not a consequence of adding poly A tails to preexisting inactive forms of mRNA.

The activity of mRNA specific for α-amylase, as measured by its ability to synthesize α-amylase in an *in vitro* protein synthesis system derived from wheat germ, also was enhanced by incubating aleurone layers with GA_3 (Fig. 9) (Higgins *et al.*, 1976). The timing of this enhancement correlated with that for total mRNA synthesis, for polysome formation, and for the rate of α-amylase synthesis suggesting that all these processes were closely related.

Another approach to the study of the transcription dependency of α-amylase was to employ specific RNA synthesis inhibitors. Actinomycin D, the most commonly used RNA synthesis inhibitor, was not effective in barley aleurone layers. However, cordycepin, 3'-deoxyadenosine, which is believed to be a chain terminator of RNA synthesis, effectively inhibited the synthesis of both poly A-RNA and RNA species not containing poly A in barley aleurone layers (Ho, 1975). More importantly, cordycepin inhibited RNA synthesis no matter whether it was added in the absence or presence of GA_3. Cordycepin did not appear to cause any general toxicity to the cells because respiration as measured by oxygen consumption was not affected (Ho, 1975). Cor-

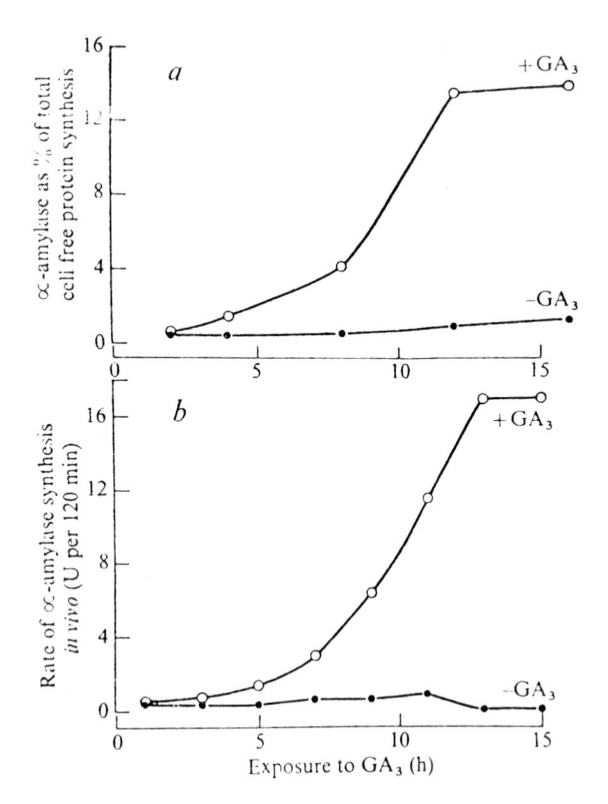

FIGURE 9. a. The time course of GA_3 enhanced level of translatable mRNA for α-amylase. α-Amylase synthesized in a cell-free protein synthesis system derived from wheat germ was estimated from densitometric scan of autoradiographs of SDS gels. α-Amylase migrates as a distinct band on SDS gel. b. The rate of in vivo *α-amylase synthesis of GA_3 treated aleurone layers (Higgins* et al., *1976).*

dycepin inhibited about 90% of α-amylase synthesis if it was added before or at the same time as GA_3 (Ho and Varner, 1974). However, this inhibition decreased progressively when the time of cordycepin addition was delayed (Fig. 10). There was no observable effect of cordycepin on α-amylase production if the inhibitor was added 12 hours or more after the addition of GA_3. Since cordycepin does not influence the degradation of α-amylase (Ho, 1975), the above observations suggest the following three points. (1) The fast α-amylase synthesis after 12 hours of GA_3 treatment was under post-transcriptional control. (2) Messenger RNA for α-amylase must be synthesized be-

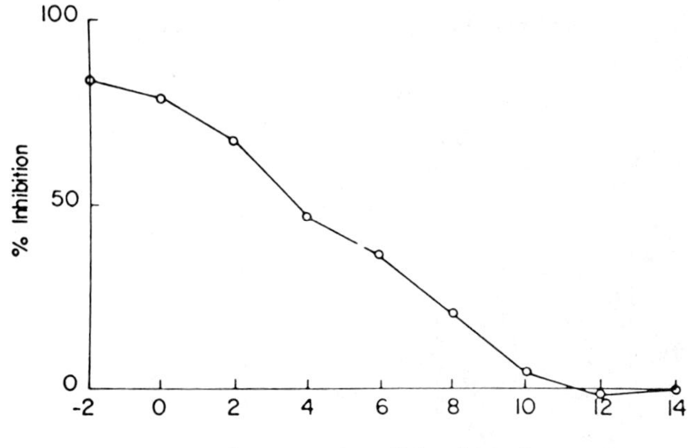

FIGURE 10. Effect of corcydepin on the production of α-amylase. Cordycepin was added at different times after GA_3 as indicated, and the aleurone layers were incubated for a total of 24 hr, when the activity of α-amylase was assayed (Ho and Varner, 1974).

fore or during the first 12 hours of GA_3 treatment. (3) The α-amylase specific mRNA was very stable because although it was formed before or during the first 12 hours of GA_3 treatment it was still functional afterwards.

Although cordycepin did not inhibit α-amylase synthesis if added 12 hours after GA_3, it did inhibit about 45% of total protein synthesis (Table 3) (Ho and Varner, 1974). Since α-

TABLE 3
Effect of Cordycepin on α-Amylase and Total Protein Synthesis

	+ GA_3 only	+ GA_3 with cordycepin
α-amylase (ug/10 layers)	426.7	462.2 (0.1)
^{3}H-leucine incorporation (cpm/10 layers)	247,390	134,170 (46)

Cordycepin (0.1 mM) was added 12 hours after the addition of GA_3. Aleurone layers were incubated with ^{3}H-leucine at 22-24 hours and α-amylase and leucine incorporation were measured at 24 hours after GA_3. Numbers in parenthesis indicate the percent inhibition (Ho and Varner, 1974).

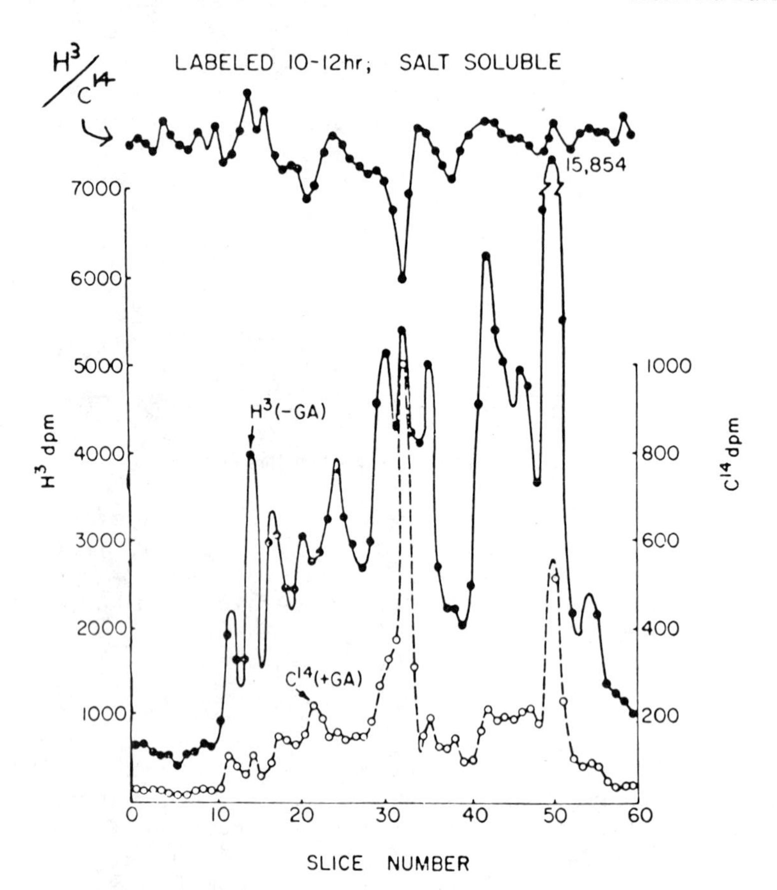

FIGURE 11. Profile of the newly synthesized salt-soluble proteins of control and GA_3 treated aleurone layers. The layers were incubated with labeled amino acids for two hours from 10 to 12 hrs. after the start of the incubation of the layers in buffer, or in buffer with GA_3. The protein profile was analyzed by SDS gel electrophoresis. The major peak in the ^{14}C-profile is α-amylase (Varner, 1975).

amylase is about 40% of the total proteins synthesized at this stage, at least three quarters of all other proteins are apparently translated from relatively short-lived RNAs.

Can the differential stability of mRNAs account for the GA_3 enhanced α-amylase synthesis? Total protein synthesis in barley aleurone layers remains constant after GA_3 treatment suggesting that in order to dramatically increase the synthesis of α-amylase and other hydrolases, the synthesis of other proteins must be decreased. In fact, this was observed by comparing the profile of newly synthesized proteins in the presence or absence of GA_3 with SDS gel electrophoresis (Fig. 11) (Varner, 1975). Ribonuclease activity in barley aleurone layers is high, therefore, the differential stability of α-amylase mRNA versus other mRNAs, as well as the GA_3 enhancement of α-amylase mRNA formation would allow progressive enrichment for α-amylase mRNA. This could then redirect protein synthesis to allow the fast production of α-amylase.

REGULATION OF α-AMYLASE FORMATION BY ABSCISIC ACID

Abscisic acid (ABA) prevents all known GA_3 effects in barley aleurone layers. Although ABA can prevent the response to GA_3, no direct effects of ABA alone in aleurone cells have been observed. The failure of the aleurone cells to respond to GA_3 in the presence of ABA does not result from simple competition between these two hormones because high concentrations of GA_3 do not completely overcome the ABA effect (Chrispeels and Varner, 1966; Jacobsen, 1973). ABA does not influence general

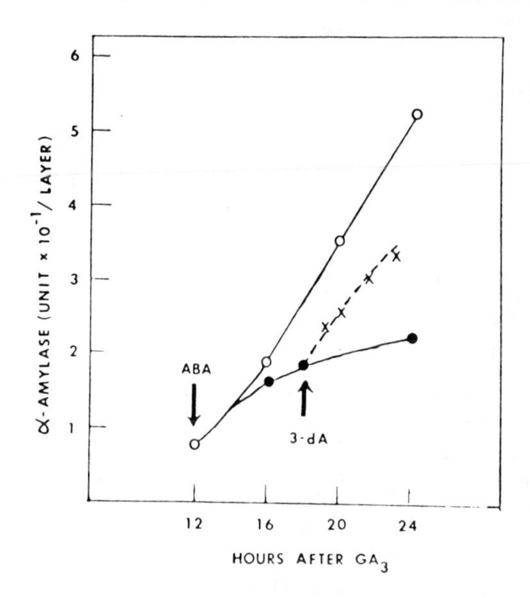

FIGURE 12. Effect of midcourse addition of ABA (25 μM) and cordycepin (0.1 mM) on the synthesis of α-amylase. GA_3 only; GA_3 and ABA; GA_3, ABA, and cordycepin (3'-dA) (Ho and Varner, 1976).

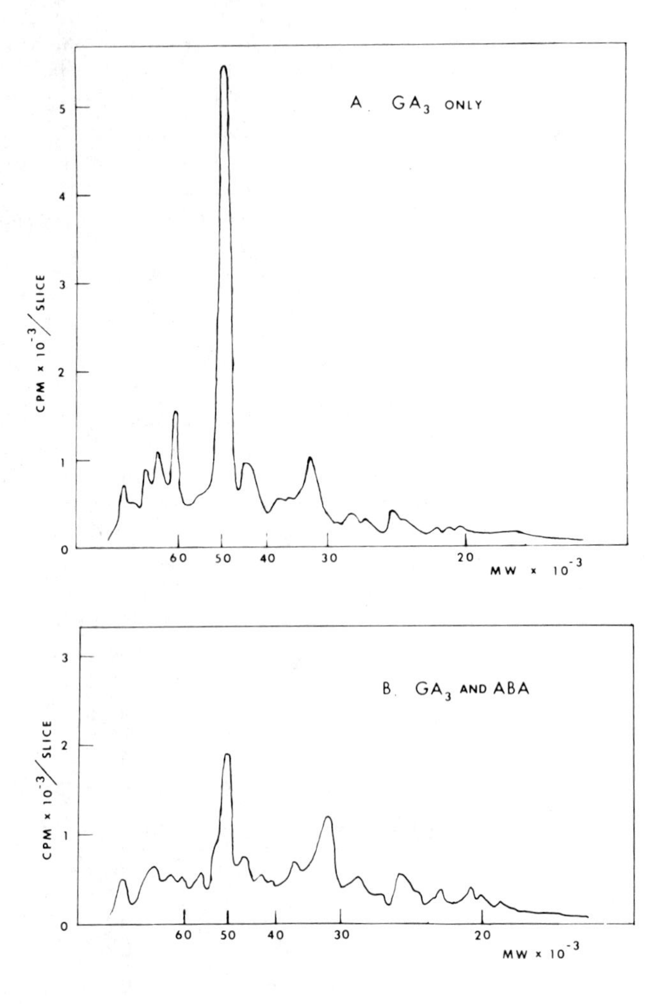

FIGURE 13. Profile of newly synthesized salt-soluble proteins on SDS gel. Abscisic acid was added 12 hours after the addition of GA_3. Aleurone layers were labeled with ^{3}H-leucine for 2 hrs. (18-20 hrs. after GA_3). The major peak in "GA_3 only" profile is α-amylase (Ho and Varner, 1976).

cellular metabolism as measured by O_2 consumption (Chrispeels and Varner, 1966). Since α-amylase synthesis after 12 hours of GA_3 was no longer sensitive to transcription inhibitors such as cordycepin, while ABA at this stage still effectively inhibited α-amylase production, it provided a good opportunity to study the requirement of RNA synthesis for the action of ABA in barley aleurone layers. As shown in Fig. 12 ABA slowed further α-amylase synthesis when it was added 12 hours after GA_3. When ABA was added 12 hours after GA_3 and cordycepin was added at the same time or later, the accumulation of α-amylase either did not stop or was quickly resumed. This indicated that the response of aleurone layers to ABA required the synthesis of a short-lived RNA. Because the rate of α-amylase accumulation of cordycepin addition was close to that of tissue treated with GA_3 alone, it appeared that the amount of α-amylase mRNA was not limiting during the resumption of amylase synthesis, *i.e.* ABA did not reduce the stability of α-amylase mRNA.

Analyzing the profile of the newly synthesized protein by SDS gel electrophoresis, showed that ABA substantially decreased the synthesis of α-amylase while the synthesis of most of the other proteins remained unchanged (Fig. 13). Since α-amylase synthesis after 12 hours of GA_3 was no longer subject to transcriptional control, the inhibition of ABA on α-amylase synthesis must be via a post-transcriptional mechanism, probably the translation of α-amylase mRNA.

EFFECT OF ETHYLENE

Ethylene, also an endogenous hormone in barley seeds, can partially prevent the inhibitory effect of ABA on GA_3-enhanced α-amylase formation in aleurone layers (Jacobsen, 1973). Ethylene together with a high concentration of GA_3 can completely overcome the effect of ABA (Jacobsen, 1973). Ethylene treated aleurone layers always have a higher proportion of α-amylase released into the medium than the layers treated with GA_3 alone (Jacobsen, 1973; Jones, 1968). Barley aleurone cells have very thick walls that bind a significant amount of α-amylase and constitute a major barrier for the release of α-amylase into the medium. Therefore, the fact that ethylene treated aleurone layers have more α-amylase released may be attributed to the following possibilities: (1) Ethylene may further facilitate the partial degradation of cell wall caused by GA_3 treatment. Further breakdown of the wall barrier would enhance enzyme release into the medium. Ethylene has been reported to enhance cell wall degradating enzymes, such as cellulase, polygalacturonase, pectinase and β-1,3-glucanase in other systems (Abeles, 1973). (2) Ethylene may simply enhance the secretion of α-amylase, a process which is defined as transporting material from the interior of the cell across the plasmalemma.

Since secretory processes require membrane recycling, the possible effect of ethylene on secretion could be reflected by the effect of ethylene on ^{32}P incorporation into membrane

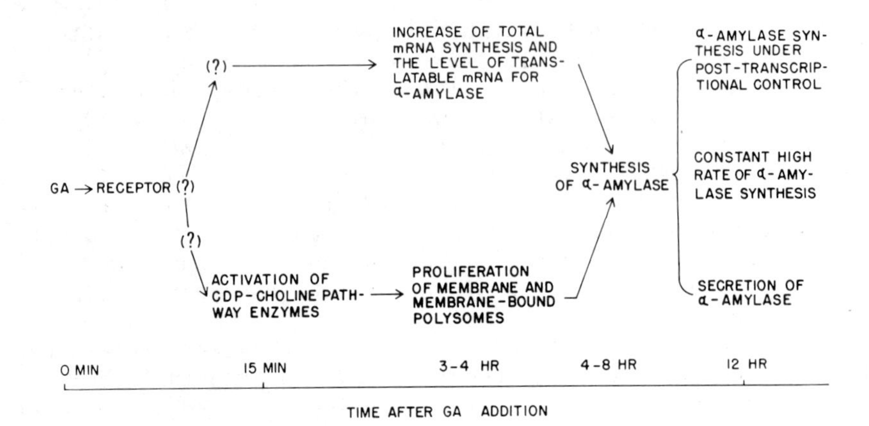

FIGURE 14. Summary of the GA_3 effects in barley aleurone layers.

phospholipids. Recently, however, we have found that ethylene has no significant effect on membrane formation as measured by ^{32}P incorporation into phospholipids (Abroms, Ho and Varner, unpublished observation). This could rule out the possibility that ethylene enhances the secretory process.

SUMMARY AND PROSPECT

The barley aleurone layer is a convenient system to study the mode of action of hormones because this tissue consists of homogeneous, non-dividing cell populations and α-amylase can serve as an excellent biochemical marker for the hormonal action. The synthesis of α-amylase starts after 4 to 8 hours and reaches a maximal steady rate after 12 hours of GA_3 treatment. The increase in α-amylase synthesis is the consequence of two parallel earlier GA_3 effects, *i.e.* an increase in the amount of α-amylase mRNA and the proliferation of membrane, mostly rough ER (Fig. 14). GA_3 enhances total mRNA syntheses and increases the amount of translatable message for α-amylase. The time course of these two events correlate indicating that the increase in α-amylase mRNA is probably due to the *de novo* synthesis of this message. These observations tend to support a previous suggestion that the synthesis of hydrolases may depend on the GA_3 mediated synthesis of their specific mRNAs (Chrispeels and Varner, 1967). Unequivocal proof of this hypothesis would require quantitative measurement of the activity of the α-amylase structural gene at different stages of GA_3 treatment. This could be accomplished by using *in vitro* synthesized, highly labeled, complementary DNA (cDNA) against purified α-amylase mRNA as a hybridization probe to measure the number of α-amylase gene transcripts (mRNA and its precursors). The tDNA would be also useful to measure the number of the α-amylase gene per haploid genome in order to determine whether there is amplification of the α-amylase gene.

Membrane proliferation, as measured by the incorporation of ^{32}P into phospholipid, is proceeded by the activation of CDP-choline glyceride transferase. As mentioned before, Ca^{++} redistribution could be studied either with histochemical techniques or with barley aleurone layers loaded with ^{45}Ca which could be introduced into the seed during seed development.

Membrane proliferation is actually the consequence of phospholipid turnover, *i.e.* from storage lipid to membrane lipid, because total cellular phospholipid does not increase during GA_3 treatment. Therefore, one would imagine that degradation of storage phospholipid, which is controlled by various phospholipases, may precede the proliferation of membranes. A convenient *in vivo* assay of phospholipase would be the incorporation of ^{18}O from $H_2{}^{18}O$ into the carboxylic group of fatty acid. This would detect possible early effects of GA_3 on phospholipase activity. One advantage of this approach, besides convenience and high sensitivity, is that if the fatty acids are reutilized to synthesize new lipid 50% of the ^{18}O atoms will still be retained due to the resonance between the two oxygen atoms on the carboxylic group of free fatty acids.

The rapid synthesis of α-amylase after 12 hours of GA_3 treatment, which is accompanied by the secretion of this enzyme, is apparently under post-transcriptional control because α-amylase synthesis is no longer sensitive to RNA synthesis inhibitors. It appears that the aleurone layers have become fully committed to the synthesis of α-amylase at this stage. Whether GA_3 is still needed at this stage is unclear. However, transferring the aleurone layer to GA_3 free medium at this stage does not decrease α-amylase production (Ho and Varner, unpublished observation).

Abscisic acid is still able to specifically prevent the synthesis of α-amylase if it is added 12 hours after the addition of GA_3. ABA must therefore act on a post-transcriptional site to regulate the synthesis of α-amylase. Since ABA apparently does not effect the stability of α-amylase mRNA, it is possible that ABA specifically prevents the translation of α-amylase mRNA. However, the effect of ABA itself depends on the synthesis of a short-lived RNA (tentatively called "regulator RNA"). Whether this "regulator RNA" or its translational product ("regulator protein) can prevent the synthesis of α-amylase is still unknown. In human reticulocytes (Fuhr and

Overton, 1975) and embryonic chick muscle (Bester *et al.*, 1975) a new species of low molecular weight RNA has been reported to be able to directly regulate the translation of specific mRNA. Since α-amylase mRNA can be successfully translated in a cell free protein synthesis system from wheat germ, this *in vitro* α-amylase synthesis could be employed as a sensitive assay for the "regulator RNA". RNA fractions isolated from ABA treated aleurone layers could be assayed for regulator RNA activity by their ability to prevent *in vitro* α-amylase synthesis. This assay could eventually lead to the isolation of this "regulator RNA" which would be useful to study the action of ABA.

What is the primary effect of hormones in barley aleurone layers? A logical answer would be to bind to specific hormone receptors. It appears that only protein molecules can provide the high degree of specificity that a hormone receptor must possess. Although attempts have been made to look for the GA_3 receptors in barley aleurone layers, no helpful information is yet available. Gibberellins with high specific radioactivity are now available (Nadeav and Rappaport, 1974) and the preparation of gibberllin affinity absorbent has been reported (Knöfel *et al.*, 1975). With the help of these new materials as well as photoaffinity labeled hormones, the search for hormone receptors might be facilitated.

Animal peptide hormones, such as insulin, require cyclic AMP (cAMP) as a second messenger to carry out the various divergent secondary effects. Despite those reports using very high concentration of exogenous cAMP to partially mimic the action of GA_3, there is no evidence for the presence of cAMP in barley aleurone layers (Keates, 1973). The technique used allowed detection of cAMP levels of 2 pmoles/g fresh weight which is less than one hundredth of the cAMP concentration of animal tissues without hormone treatment. Therefore, it is unlikely that cAMP mediates the action of GA_3 in barley aleurone layers.

Finally, can the information from barley aleurone layers be used to extrapolate the mode of action of hormones in other systems? For the time being, this question only can be answered indirectly by comparing the information from barley aleurone layers with those from other systems.

Similar effects of GA on hydrolases production and other processes have been reported in the aleurone layers of various cereal grains, including wheat, oat, maize, and rice (Jones, 1975). Another well known GA effect is to cause cell elongation, sometimes accompanied by cell duplication, in various plant tissue, such as lettuce and cucumber hypocotyls, pea epicotyl, and Avena internode (Jones, 1975). The GA_3 enhanced elongation of lettuce hypocotyls has been studied recently by

Srivastava *et al.* (1975). They found that GA_3 enhanced elongation was accompanied by increased cell wall dry weight which apparently resulted from increased synthesis and secretion of wall materials. Their electron microscopic observations revealed that cell elongation was preceded by a proliferation of membrane and membrane bound polysomes. Therefore, although the most noticable GA_3 effect in lettuce hypocotyls (cell elongation) is very different from that in barley aleurone layers (synthesis and secretion of hydrolases), the mode of action of GA_3 in these two systems appears to be similar.

Ihle and Dure (1970) working with precociously germinating cotton embryos, found that ABA is able to inhibit the translation of carboxypeptidase mRNA. Because actinomycin D prevented the ABA inhibition, they suggested that the effect of ABA may depend on the synthesis of a suppressor molecule. Again, this is very similar to what we observed in barley aleurone layers. Translational control by ABA has also been observed in wheat embryos (Chen and Osborne, 1970) and barley leaves (Bonnafores *et al.*, 1973).

ACKNOWLEDGMENT

I wish to express my thanks to Professor J. E. Varner for reading this manuscript critically. This work was supported in part by grants from the National Science Foundation (GB-39944 and PCM 78-16143).

REFERENCES

Abeles, F. B. (1973). *Ethylene in plant biology*. Academic Press, New York.

Ben-Tal, Y. (1974). Activation of phosphorylcholine glyceride transferase by gibberellic acid in barley aleurone cells. Thesis, Michigan State University, East Lansing, Michigan.

Ben-Tal, Y., and Varner, J. E. (1974). An early response to gibberellic acid not requiring protein synthesis. *Plant Physiol. 54*, 813-816.

Bester, A. J., Kennedy, D. S., and Heywood, S. M. (1975). Two classes of translational control RNA: their role in the regulation of protein synthesis. *Proc. Nat. Acad. Sci. U.S. 72*, 1523-1527.

Bonnafous, J. C., Mousseron-Canet, M., and Olivé, J. L. (1973). Translational control in barley coleoptiles by abscisic acid. *Biochim. Biophys. Acta 312*, 165-171.

Chen, D., and Osborne, D. J. (1970). Hormones in the translational control of early germination in wheat embryos. *Nature 226*, 1157-1160.

Chrispeels, M. J., and Varner, J. E. (1966). Inhibition of gibberellic acid induced formation of α-amylase by abscisin II. *Nature 212*, 1066-1067.

Chrispeels, M. J., and Varner, J. E. (1967). Hormonal control of enzyme synthesis: on the mode of action of gibberellic acid and abscisin in aleurone layers of barley. *Plant Physiol. 42*, 1008-1016.

Evins, W. H. (1971). Enhancement of polyribosome formation and induction of tryptophan-rich proteins by gibberellic acid. *Biochem. 10*, 4295-4303.

Evins, W. H., and Varner, J. E. (1971). Hormone-controlled synthesis of endoplasmic reticulum in barley aleurone cells. *Proc. Nat. Acad. Sci. U.S. 68*, 1631-1633.

Filner, P., and Varner, J. E. (1967). A test for *de novo* synthesis of enzymes: density labeling with H_2O^{18} of barley α-amylase induced by gibberellic acid. *Proc. Nat. Acad. Sci. U.S. 58*, 1520-1526.

Firn, R. D., and Kende, H. (1974). Some effects of applied gibberellic acid on the synthesis and degradation of lipids in isolated barley aleurone layers. *Plant Physiol. 54*, 911-915.

Fuhr, J. E., and Overton, M. (1975). Translational control of globin synthesis by low molecular weight RNA. *Biochem. Biophys. Res. Commun. 63*, 742-747.

Higgins, T. J. V., Zwar, J. A., and Jacobsen, J. V. (1976). Gibberellic acid enhances the level of translatable mRNA for α-amylase in barley aleurone layers. *Nature 260*, 166-169.

Ho, D. T.-H. (1976). On the mechanism of hormone controlled enzyme formation in barley aleurone layers. Thesis, Michigan State University, East Lansing, Michigan.

Ho, D., T.-H., and Varner, J. E. (1974). Hormonal control of messenger ribonucleic acid metabolism in barley aleurone layers. *Proc. Nat. Acad. Sci. U.S. 71*, 4783-4786.

Ho, D. T.-H., and Varner, J. E. (1976). Response of barley aleurone layers to abscisic acid. *Plant Physiol. 57*, 175-178.

Ihle, J. N., and Dure, L. III. (1970). Hormonal regulation of translation inhibition requiring RNA synthesis. *Biochem. Biophys. Res. Commun. 38*, 995-1001.

Jacobsen, J. V. (1973). Interactions between gibberellic acid, ethylene, and abscisic acid in control of amylase synthesis in barley aleurone layers. *Plant Physiol. 51*, 198-202.

Jacobsen, J. V., and Zwar, J. A. (1974). Gibberellic acid causes increased synthesis of RNA which contains poly(A) in barley aleurone tissue. *Proc. Nat. Acad. Sci. U.S. 71*, 3290-3293.

Johnson, K. D., and Kende, H. (1971). Hormonal control of lecithin synthesis in barley aleurone cells: regulation of the CDP-choline pathway by gibberellin. *Proc. Nat. Acad. Sci. U.S. 68*, 2674-2677.

Jones, R. L. (1968). Ethylene enhanced release of α-amylase from barley aleurone cells. *Plant Physiol. 43*, 442-444.

Jones, R. L. (1969a). Gibberellic and the structure of barley aleurone cells I, changes during lag phase of α-amylase synthesis. *Planta 85*, 359-375.

Jones, R. L. (1969b). The fine structure of barley aleurone cells. *Planta 88*, 73-86.

Jones, R. L. (1973). Gibberellins: their physiological role. *Ann. Rev. Plant Physiol. 24*, 571-598.

Keates, R. A. B. (1973). Evidence that cyclic AMP does not mediate the action of gibberellic acid. *Nature 244*, 355-357.

Knöfel, H. D., Müller, P., Kramell, R., and Sembdner, G. (1975). Preparation of gibberellin affinity adsorbents. *FEBS Letters 60*, 39-41.

Koehler, D. E., and Varner, J. E. (1973). Hormonal control of orthophosphate incorporation into phospholipids of barley aleurone layers. *Plant Physiol. 52*, 208-214.

Nadeau, R., and Rappaport, L. (1974). The synthesis of ^3H-gibberellin A_3 and palladium-catalyzed actions of carrier free tritium on gibberellin A_3. *Phytochem. 13*. 1537-1545.

Siekevitz, P., and Palade, G. E. (1960). A cytochemical study on the pancreas of the guinea pig. V. *in vivo* in incorporation of leucine-1-^{14}C into the chymotrypsinogen of various cell fractions. *J. Biophys. Biochem. Cytol (J. Cell Biol.) 7*, 619-630.

Siekevitz, P., and Palade, G. E. (1966). Distribution of newly synthesized amylase in microsomal subfractions of guinea pig pancreas. *J. Cell Biol. 30*, 519-530.

Srivastava, L. M., Sawhney, V. K., and Taylor, I. E. P. (1975). Gibberellic-acid-induced cell elongation in lettuce hypocotyls. *Proc. Nat. Acad. Sci. U.S. 72*, 1107-1111.

Varner, J. E. (1975). Hormone mediated integration of seedling physiology. *Advan. Exp. Med. Biol. 62*, 65-78.

Varner, J. E., and Chandra, G. R. (1964). Hormonal control of enzyme synthesis in barley endosperm. *Proc. Nat. Acad. Sci. U.S. 52*, 100-106.

Varner, J. E., and Ho, D. T.-H. (1976). The role of hormones in the integration of seedling growth. In *Molecular Biology of Hormone Action* (Papaconstantinou, J., ed.), Academic Press, N.Y.

Vigil, E. L., and Ruddat, M. (1973). Effect of gibberellic acid and actinomycin D on the formation and distribution of rough endoplasmic reticulum in barley aleurone cells. *Plant Physiol. 51*, 549-558.

Wood, A., and Paleg, L. G. (1972). The influence of gibberellic acid on the permeability of model membrane systems. *Plant Physiol. 50*, 103-108.

Yomo, H., and Varner, J. E. (1971). Hormonal control of a secretory tissue. In *Current Topics in Developmental Biology, Vol. 6* (Moscona, A. A., and Monroy, A., eds.), pp. 111-144, Academic Press, N.Y.

SECTION III
Plant Viruses and Bacterial Agents

REPLICATION OF RNA PLANT VIRUSES

George Bruening
Roger Beachy
Milton Zaitlin

Department of Biochemistry and Biophysics
University of California
Davis, California

Among virologists it is the plant virologists who are most often privileged to study organisms with catchy names evoking firm visual images, such as:

- artichoke mottle crinkle virus
- barley stripe mosaic virus
- chicory blotch virus
- cowpea mosaic virus
- cucumber green mottle mosaic virus
- grapevine fanleaf virus
- maize rough dwarf virus
- odontoglossum ringspot virus
- pea ēnation mosaic virus
- pelargonium leaf curl virus
- plum pox virus
- potato yellow dwarf virus
- sowthistle yellow vein virus
- tobacco rattle virus
- tomato bushy stunt virus
- tomato spotted wilt virus

Currently, twenty groups of plant viruses are recognized (Harrison *et al.*, 1971; Fenner, 1976), though four of these are considered to have only one member each. Other viruses, well characterized enough to be excluded from the twenty recognized groups, have not yet been assigned to groups. Plant viruses with single-stranded RNA, with double-stranded RNA and with double stranded DNA in the virus particles (virions) are known; a few have a membrane. However, for most (e.g., members of eighteen out of the twenty recognized groups) plant viruses the macromolecular constituents of the virion are single stranded RNA and protein. Usually only one kind of polypeptide is present and this is designated the "coat protein" or the "capsid protein".

ISBN 012-601950-9

PLUS-STRAND RNA VIRUSES

It is the "simple RNA viruses" with a genome of single-stranded RNA which have been studied most thoroughly. Some plant viruses achieve high concentrations of virions in the infected tissue (up to 9 g/l of expressed sap for tobacco mosaic virus (TMV) in tobacco), which favors study of the chemical structure of the virions; the lack, until recently, of suitable host systems for synchronous infection has retarded study of plant virus replication. Infectious RNA has been isolated from many of these viruses, following the initial successes of Gierer and Schramm (1956) and Fraenkel-Conrat *et al.* (1957) with tobacco mosaic virus; this implies, but does not prove, that the virion RNA has nucleotide sequences which, with proper initiation and termination, could be directly translated into virus-specified proteins. That is, these viruses are probably "plus-strand viruses" with particles containing the plus (+) strand or "sense" strand of RNA rather than a complementary negative (-) strand RNA which presumably is not translated. That TMV is a plus strand virus, at least for the capsid protein gene, was implied by comparisons of mutational events (amino acid substitutions) and the genetic code for "spontaneous" and induced TMV capsid protein mutants (Funatsu and Fraenkel-Conrat, 1964; Wittman and Wittman-Liebold, 1966). That the capsid protein gene is a (+) strand has been demonstrated unequivocally by Guilley *et al.* (1975a). They found among the large oligonucleotides formed by a partial digestion of TMV RNA with T1 ribonuclease one which bound TMV capsid protein (and could therefore be separated from other large oligonucleotides) and which had the nucleotide sequence specifying residues 95 to 129 of the capsid protein according to a plus strand code. Garfin and Mandeles (1975) isolated an oligonucleotide of 16 residues from TMV RNA which corresponds in sequence to residues 53 to 57 of the capsid protein. Other evidence for (+) strand RNA in TMV is given in the discussion of the translation products, below.

Some eucaryotic messenger RNAs (mRNAs) have a single nucleoside residue at the 5'-terminus which is inverted with respect to the 5'→3' orientation of the other residues in the polynucleotide chain and is connected to it by a triphosphate bridge (Shatkin and Both, 1976). The inverted terminal nucleoside residue is 7-methylguanosine and the terminal ("capped") sequence may be abbreviated $m^7G(5')ppp(5')N$...where N stands for a purine nucleoside residue in most cases and occasionally for a pyrimidine. Other eucaryotic mRNAs have polyadenylate sequences at the 3'-terminus; some mRNAs have both a cap and polyadenylate. The sequence $m^7G(5')ppp(5')Gp$ has been found in ^{32}P-labeled TMV RNA and 50 to 80% of the 5'-terminals appear to be capped with this sequence (Zimmern, 1975; Keith and Fraenkel-Conrat, 1975). Brome mosaic virus (BMV) (Dasgupta *et al.*, 1976), alfalfa mosaic virus (AMV) (Pinck, 1976) and possibly cucumber mosaic virus (Symons, 1976) RNAs bear the same 5'-terminal sequence. Cowpea mosaic virus RNA has 3'-terminal polyadenylate sequences of about 200 residues (El Manna and Bruening, 1973) and the RNAs of two viruses serologically related to cowpea mosaic virus also contain polyadenylate, presumably 3'-terminal (Semancik, 1974; Oxefelt, 1976). These facts are indirect evidence for the RNAs of these viruses being (+) strands.

The single-stranded RNA of lettuce necrotic yellow virus has not been demonstrated to be infectious; when virions were incubated with a non-ionic detergent and the four ribonucleoside triphosphates, RNA was synthesized which is complementary to (i.e., can form molecular hybrids with) the virion RNA (Francki and Randles, 1973). Presumably the RNA synthesized by the virion-bound polymerase ("transcriptase") has messenger sequences and lettuce necrotic yellows virus is a negative strand virus. A virion-bound polymerase has also been found (Ikegami and Francki, 1976) in sub-viral particles of Fiji disease virus, which has a genome of double-stranded RNA.

Modes of Translation

Even for the plus strand RNA viruses, the plant viruses about which we know most, it is not yet possible to write down a step-by-step replication mechanism based entirely on experimental observations. Fig. 1 presents an outline of the events (not all necessarily sequential) which can reasonably be expected to take place. What evidence is there for the entities and events illustrated in Fig. 1? It is assumed that a crucial early event after exposure of the RNA to the cytoplasm must be the translation of some portion of the virion RNA to form one or more proteins which catalyze the synthesis of more copies of virus genome RNA (step 3 of Fig. 1). Presumably the proper polymerase (replicase) activity would not be present before inoculation. However, even this assumption is open to question. RNA polymerase activities which are resistant to actinomycin D (which binds to DNA and thereby inhibits DNA-dependent RNA polymerases), which are stimulated by added viral RNA and which require all four ribonucleoside triphosphates (but no ribonucleoside diphosphates), have been observed in several uninoculated plant tissues (Astier-Manifacier and Cornuet, 1971; Duda *et al.*, 1973; Geelen *et al.*, 1976; Fraenkel-Conrat, 1976). Double-stranded RNA most probably is the pro-

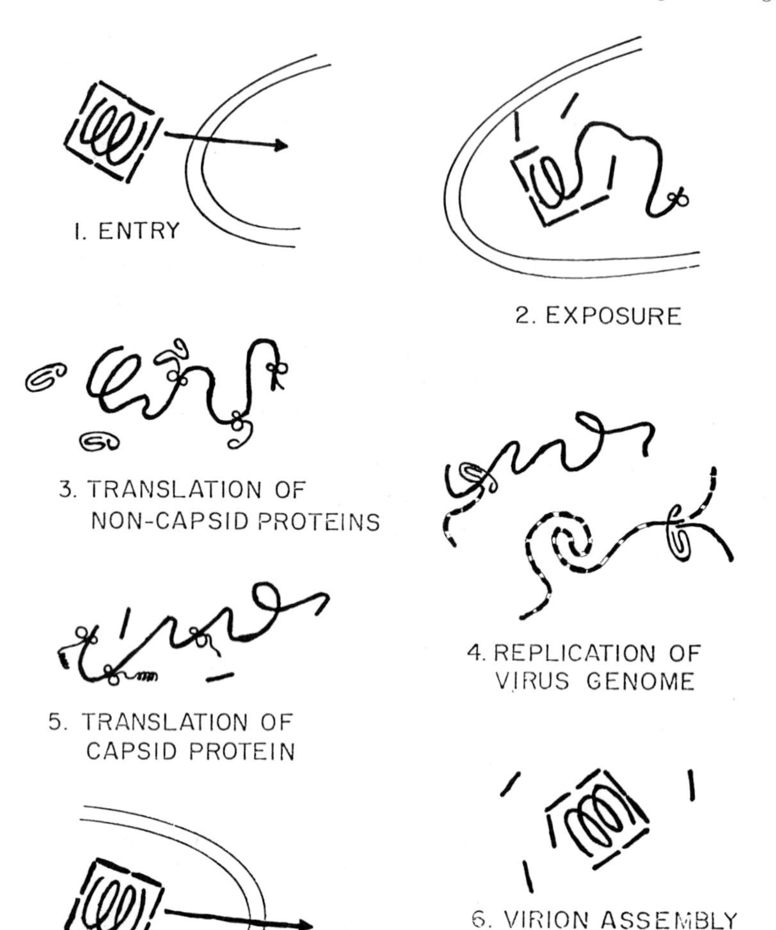

FIGURE 1. Expected events in the life cycle of a (+) strand RNA virus. The infecting virion is represented by a coil of (+) strand RNA in a capsid composed of only eight molecules of capsid protein, a particular capsid structure corresponding to no known virus! After entry of virion RNA, with or without the capsid protein or other constituents, the RNA must be exposed to the cytoplasm. A ribosome (∞) is shown attached to the virion RNA in step 2 because translation of at least part of the virion RNA is presumed to be an essential early event. Step 3 illustrates the synthesis of a specific non-capsid protein, an RNA-dependent RNA polymerase. In step 4 the polymerase ("replicase") is shown replicating the virion RNA through an intermediate, the (-) strand RNA. The (-)

duct; that the added RNA can act as a template is supported by some data of Astier-Manifacier and Cornuet (1971) for Chinese cabbage but contraindicated for tobacco by data of Duda *et al.* (1973). However, it is likely that (+) strand plant viruses specify a polypeptide(s) which participates in the RNA replication reactions. Zaitlin *et al.* (1976) and Brishammer (1976) have discussed the relationship of virus specified polypeptides of TMV to the RNA polymerase activities of infected tissue.

Double-stranded RNA has been isolated from TMV-infected tobacco (Shipp and Hazelkorn, 1964; Burdon *et al.*, 1964) and later was found to be associated with infections by other (+) strand plant viruses. The double stranded RNAs are routinely demonstrated to be virus related by melting them and incubating with radioactive virion RNA to form molecular hybrids. TMV double-stranded RNA yields infectious TMV RNA upon melting (Jackson *et al.*, 1971) and is referred to as RF, for "replicative form". RNA from infected cells which has significant amounts of single-stranded as well as double-stranded RNA is referred to as RI, for "replicative intermediate". Both RI and RF lose radioactivity, apparently to virion RNA, in pulse-and-chase experiments (Jackson *et al.*, 1972). All these results strongly implicate (-) strand RNA (step 4 of Fig. 1) as an intermediate in the synthesis of virion RNA. However, the (-) strands may not be present in double-stranded complexes *in vivo*. Nilsson-Tillgren and Kielland-Brandt (1976) mixed TMV containing radioactively labeled RNA with unlabeled, infected tissue and isolated RF and RI from the mixture. Both the RF and the RI were found to be radioactive, implying that free (+) strands and free (-) strands can form molecular hybrids during the isolation procedure. Because the replication of several (+) strand RNA plant viruses is not inhibited or is inhibited very little by actinomycin D, DNA of the host is assumed to have no direct role in the replication process.

---

strand is represented by (▬▭▬) and polypeptide strands by (▬▬). Formation of the capsid protein molecules, virion assembly and release complete this abbreviated version of the virus life cycle. Steps 3, 4 and 5 are not necessarily sequential. For plant viruses steps 1 and 7 are vector mediated, though there is some evidence that the capsid may influence the host range of a virus and therefore could be involved in a recognition phenomenon at entry (Atabekov, 1975).

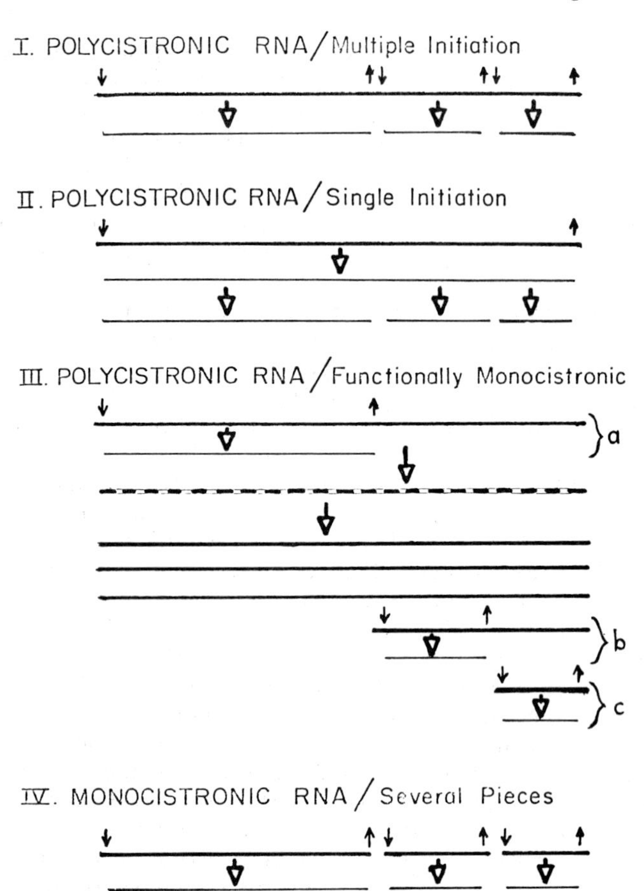

FIGURE 2. Modes of translation for (+) strand viruses. In parts I, II and III the upper heavy line represents the virion RNA. Polypeptide chains are represented by narrow lines and (-) strand RNA by a dashed line. Downward and upward pointing arrows represent initiation sites and termination sites, respectively, for the synthesis of polypeptide chains. In part II the small, presumably functional polypeptides are formed by the action ("protein processing") of a specific protease(s) on the single large polypeptide which was initiated at the single site on the virion RNA. In part III the virion RNA is functionally monocistronic (a) and the polypeptides corresponding to all but the first gene cannot be translated from the virion RNA (because of specific folding of the RNA molecule or the requirement for an adjacent 5'-terminus, for example). The other genes are translated (b, c) from specific fragments of the virion RNA which may behave as functionally or actually monocistronic mRNAs. Portions of (+) strand

That the capsid protein is specified by the virus (step 5 of Fig. 1) has been proved by the *in vitro* translation of the protein from virion RNA or RNA related to it, as will be discussed in greater detail below. Assembly (step 6) of several (+) strand viruses has been demonstrated *in vitro* (Casjens and King, 1975).

In this review we focus on some alternative modes for synthesizing the proteins coded for by the virus genome because substantial progress in this area has been made recently, and it is now possible to compare the replication "strategies" of several plant viruses. In Fig. 2 some likely alternatives are diagrammed. In each of the four modes shown the expression of the virus genome, consisting of several genes, by means of several polypeptide chains (three, for the cases illustrated), is accomplished in a different way. For modes I, II and III the virus is monochromosomal. Mode I has been confirmed (though with elaborate regulatory features not shown in the simplified diagram of Fig. 2) for the RNA bacteriophages (Weissman, 1974); each gene has a translation initiation site. Mode II is well represented by the replication of certain small (picornaviruses, e.g., poliovirus) and also some more complex RNA animal viruses, though the specific cleavage of polypeptide chains (protein processing) is also well established for certain DNA bacteriophages (Hershko and Fry, 1975).

Mode III resembles mode II in that only a single initiation site on the virion RNA is functional. However, the intact genome RNA is not entirely translated in mode III. RNAs with nucleotide sequences corresponding to fragments of the virion RNA serve as the actual mRNAs for some of the virus-specified polypeptides. The (-) strand RNA is shown as an intermediate because cleavage of the infecting virion RNA molecule probably must be avoided since enzymatic mechanisms for specifically joining large RNA fragments are not known. The small viral mRNAs could be derived from full length (-) strand RNA by cleavage of full length (+) strand copies (RNA processing) or by partial transcription or they could be transcribed from processed fragments of (-) strand RNA. Mode III obviously is complex but does provide potential mechanisms for close control over protein synthesis. It is made tenable

---

drawn directly above or below one another in this diagram represent identical nucleotide sequences. In part IV the virion RNA consists of several monocistronic pieces each independently translated. The virus is considered to be "multichromosomal" or to have a "divided genome".

a priori by the discovery of numerous RNA processing mechanisms in both procaryotic and eucaryotic systems (Perry, 1976; note that RNA processing is a general term used to describe post-transcription covalent modifications, both nucleolytic and others, such as "capping"). The animal viruses of the togavirus group have a high molecular weight virion RNA which apparently is only partially translated. The polypeptide, which is a precursor to the virion proteins, is translated from a smaller, "subgenomic" RNA which has the same polarity as the virion RNA and has the 3'-one-third of the nucleotide sequences of the virion RNA. It can be recovered from the polysomes of infected cells (Simmons and Strauss, 1974; Clegg and Kennedy, 1975; Wengler and Wengler, 1976). Since the precursor polypeptide must be cleaved to form the virion porteins, the replication of togaviruses seems to involve both mode III and mode II.

It is common among plant viruses (and has been demonstrated or seems very likely to hold for nine out of the twenty recognized groups of plant viruses) to have genes distributed among several pieces of RNA and to have the different RNAs housed in different capsids. More than one particle or RNA is required to initiate an infection. The existence of these multicomponent, multichromosomal (or "divided genome" or "covirus") systems favors repeated discovery of replication mechanisms corresponding to mode IV among the (+) strand RNA plant viruses. The known (+) strand multichromosome viruses have either two (Bruening, 1977) or three (van Vloten-Doting and Jaspars, 1977) "genomic" RNAs in the virions.

In the following sections the idealized modes of translation which are summarized in Fig. 2 will be related to the replication mechanisms of three viruses. The idealized modes represent extremes; some viruses can be expected to utilize an amalgam of the modes. Evidence for two (and possibly three) of the four idealized modes has been obtained for the three viruses: tobacco mosaic virus (TMV), brome mosaic virus (BMV) and alfalfa mosaic virus (AMV).

Translation of TMV RNAs

Circumstantial evidence against the first mode of translation (Fig. 2) and in favor of the third mode in the TMV replication cycle has been available for a few years. The capsid protein of TMV can be detected easily in TMV-infected tissue and probably is the most abundant virus-specified macromolecule. If TMV-RNA acts as a polycictronic messenger RNA with multiple initiation and termination sites then, by analogy with the RNA bacteriophage systems (Weissman, 1974; Davies and Kaesberg, 1973), the capsid protein ought to be among the products synthesized *in vitro* in the efficient cell-free protein synthesizing systems derived from wheat embryo or germ if TMV-RNA is added to them. Wheat germ systems apparently are capable of translating di- and tri-cistronic messenger RNAs (Davies and Kaesberg, 1973; Davies and Kaesberg, 1974; Schwinghamer and Symons, 1975), though generally at a lower efficiency or with a lesser preference than for monocistronic messengers. Efron and Marcus (1973) and Roberts *et al.* (1974) have demonstrated convincingly, by analysis of tryptic peptides, that the capsid protein (or at least large polypeptide fragments corresponding to it) can be synthesized *in vitro* under the direction of TMV-RNA. However, the proportion of radioactivity associated with the capsid protein-related products amounted to only a small proportion (probably less than 1%) of the total of the polypeptides synthesized. In these systems the radioactivity was incorporated into polypeptides of apparent molecular weights ranging from less than 10,000 to more than 100,000. A spectrum of polypeptides also was formed when TMV-RNA was injected into frog oocytes which were subsequently incubated with radioactive amino acids (Knowland, 1974). However, the principal product had an apparent molecular weight of about 140,000. Capsid protein polypeptides were not detected in trypsin digests of the incorporation products. (Such peptides would of course be expected if, as under translation mode II, Fig. 2, the large polypeptide were a precursor of the capsid protein.) The approximately 140,000 dalton polypeptide also was formed in a cell-free protein synthesizing system derived from reticulocytes (Knowland, 1974). The significance of the high molecular weight polypeptide(s) is discussed later in this section. The lack of efficient synthesis of capsid protein in the cell-free and oocyte systems implies that TMV-RNA does not behave as a polycistronic messenger RNA *in vivo*.

Babos (1969, 1971) found RNA in TMV infected, but not in uninoculated, tobacco leaves which sedimented more slowly than TMV-RNA and which continued to be synthesized in the presence of actinomycin D. The RNA was associated with ribosomes and could be hybridized to double-stranded RNA from infected tissue but not to TMV-RNA itself; these are all characteristics of RNA expected to be found if mode III of translation is operative. Jackson *et al.* (1972) and Siegel *et al.* (1973) studied virus-related (as judged by RNA-RNA molecular hybridization tests) RNA of infected cells and of infected leaf slices using polyacrylamide gel electrophoresis to obtain higher resolution separations. The predominant and most easily observed virus-related, single-stranded RNA was named LMC (for "low molecular weight component") and is considered to have a molecular weight in the range 250,000 to 300,000. It was found to be metabolically more stable than the virus-related double-stranded RNAs of the same cells. Evidence of lesser amounts of other virus-related RNAs, larger in molecular weight than LMC but smaller than the virion RNA, also was obtained. Bourque *et al.* (1975) and Siegel *et al.* (1976) demonstrated that LMC is formed before or in the phase of rapid accumulation of virions and can be detected after only two to four hours of labeling with isotopic precursor. It is recovered under conditions which apparently preserve 23S RNA of organelle ribosomes, the major cellular RNA which seems most subject to loss through breakdown before or during isolation. These observations support the idea that LMC is not a non-specific breakdown product of TMV-RNA and that it has a role in the virus replication cycle. It can be recovered from polysomes (Babos, 1971; Beachy and Zaitlin, 1975) and therefore presumably serves as a mRNA.

Both the LMC RNA of the common strain of TMV and a similar RNA from a TMV strain which infects legumes behave as mRNAs for their respective capsid proteins. We will discuss first the results for common TMV. Hunter *et al.* (1976) recovered RNA from infected tobacco tissue by phenol extraction. The RNA was roughly fractionated by centrifugation on sucrose density gradients into 6-18S, 18S, 25S and >25S fractions (TMV virion RNA is approximately 28S). RNA from each fraction was added to a cell-free polypeptide-synthesizing system derived from wheat germ. The 25S and >25S RNA fractions stimulated the synthesis of a spectrum of polypeptides which is very similar to that stimulated by TMV RNA whereas the 6-18S fraction engendered strong incorporation into a polypeptide which migrated with capsid protein during electrophoresis in gels. Radioactive histidine and methionine, two amino acids which are absent from the capsid protein of common TMV, were not incorporated into this polypeptide made *in vitro*. The same *in vitro* product (labeled with leucine) participated in a reconstitution reaction with TMV RNA forming a labeled product with the buoyant density of TMV (in a centrifuged CsCl gradient). When the 6-18S RNA fraction was injected into frog oocytes a polypeptide of the same apparent size was synthesized. Digestion of this polypeptide with trypsin gave peptides which had the electrophoretic and chromatographic properties of authentic capsid protein peptides. As expected, the 6-18S fraction consisted of a collection of RNAs. Only RNA from a zone with an apparent molecular weight of about 250,000 (as expected for LMC) caused synthesis of the capsid protein-sized polypeptide. Thus, the LMC RNA appeared to direct the synthesis of a polypeptide which either is, or is within a few amino acids of being, the capsid protein. Siegel *et al.* (1976) obtained similar results. They labeled LMC with

^{3}H-uridine in the presence of actinomycin D. The drug not only suppressed the synthesis of host RNA but seemed to stimulate LMC synthesis. The LMC RNA was purified by gradient centrifugation and gel electrophoresis. It engendered in a wheat cell-free system a polypeptide with electrophoretic and serological properties expected for capsid protein.

The studies of both TMV particles and TMV replication have been aided by the availability of a collection of virus strains. The common strain of TMV does not systemically infect legumes (though certain legumes serve as useful local lesion hosts for many TMV strains). Lister and Thresh (1955) discovered a virus which systemically infects cowpeas and various beans and which is morphologically similar and serologically related to common TMV. Since that time other legume TMVs have been isolated, e.g., dolichos enation mosaic virus and sunn-hemp mosaic virus. These seem to be serologically similar to the virus isolated by Lister and Thresh and to share the property of forming, in addition to the approximately 300 nm rods which are characteristic of TMV, discrete classes of shorter rods (Kassanis and Varma, 1975). Dunn and Hitchborn (1965), Kassanis and McCarthy (1967) and Morris (1974) independently observed in preparations of these viruses a prominent class of rods which were about 40 nm in length and sediment at approximately 30S (compared to approximately 190S for the rull length rods). Morris postulated that the RNA of these short rods was analogous to LMC and mRNA for the capsid protein.

The virus isolated by Lister and Thresh (1955), and referred to here as C_{c}TMV, has been most intensely studied among the legume TMVs. The amino acid sequence of the capsid protein has been determined (Rees and Short, 1975). It has 161 amino acid residues, versus 158 in common TMV. Maximum alignment of the C_{c}TMV sequences to the common TMV sequences can be achieved by numbering both sequences directly from the (acetylated) amino terminal residue, except that one residue of C_{c}TMV protein which is between residues 64 and 69 of TMV and the two residues at and next to the carboxyl terminus of C_{c}TMV remain without numbers. Sixty-six of the 158 residues are then similarly located in the two structures. One may ask whether C_{c}TMV is really not a TMV strain. Among the 66 common residues are 18 which apparently are invariant among all other TMV strains and mutants for which the amino acid sequences are known. These 18 include the sequence Asp-Thr-Arg-Asn-Arg and the four dicarboxylic (Asp + Glu) residues which are considered to be those forming the two carboxyl pairs which have a critical role in virion assembly (Kaper, 1975; Rees and Short, 1975). Evidence of similar viral RNA translation products, discussed below, further validates direct comparison of C_{c}TMV and TMV.

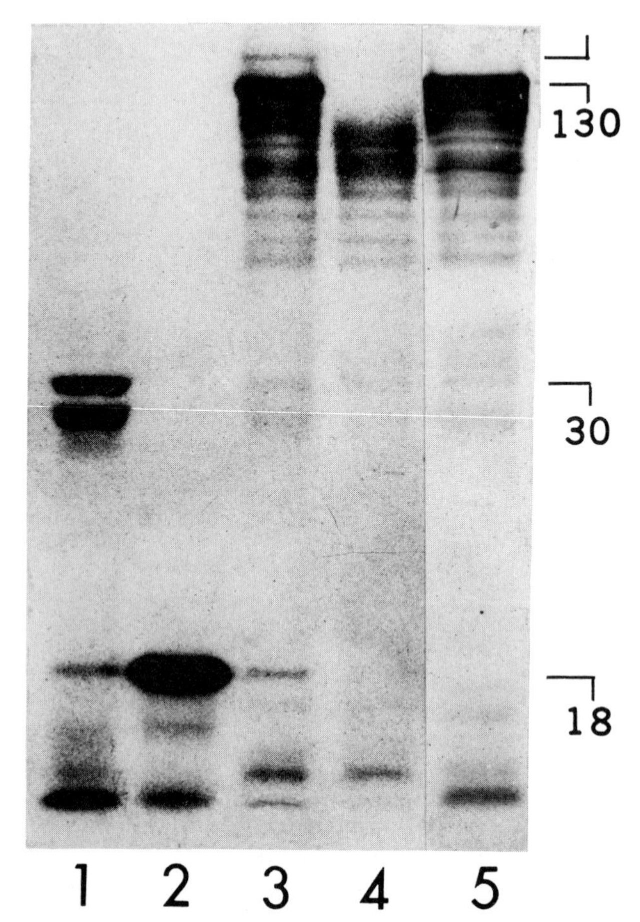

FIGURE 3. Translation of C_cTMV RNAs in a wheat germ cell-free system. The conditions for isolating the RNAs, preparing the cell-free polypeptide synthesizing system and analyzing the products by electrophoresis on 12% polyacrylamide gels are all described by Bruening et al. (1976). The migration of the ^{3}H-leucine labeled, sodium dodecyl sulfate-denatured polypeptides is from top to bottom and the zones were detected by fluorography (Laskey and Mills, 1975). The RNAs added to the system are: track 1, I_2-RNA; track 2, S-RNA; track 3, L-RNA; track 4, PROT-RNA of molecular weight 1.1 × 10^6; and track 5, PROT-RNA of molecular weight 1.4 × 10^6. PROT-RNA is derived from L-RNA by nuclease digestion and removal of 3'-terminal portions of the molecules as described in the text and in the legend to Fig. 4. If no RNA is added to the cell-free system, only a very few, very faint zones are ob-

Whitfeld and Higgins (1976) and Bruening *et al.* (1976) found in C_cTMV not only the short rods (S rods) of about 40 nm length but also rods (I_2 rods) of an intermediate length of about 100 nm. The estimated molecular weights (from polyacrylamide gel electrophoresis) were approximately 280,000 and 650,000 for S-RNA and I_2-RNA, respectively. (The long rods were about 310 nm in length; the L-RNA derived from them could not be separated from common TMV RNA by polyacrylamide gel electrophoresis and so is considered to have the molecular weight of common TMV RNA, 2.1×10^6,) The L rods and S rods and their RNAs were purified by multiple step procedures. S rods at 150 μg/ml produced no symptoms on beans or tobacco and no recoverable virus whereas L rods were infectious at 1 μg/ml (Bruening *et al.*, 1976). S rods neither stimulated nor inhibited the infectivity of L rods. L rods, I_2 rods and S rods were all recovered from tissue infected with very highly purified L-RNA. These observations discount the possibility that C_cTMV is a multichromosomal virus (Fig. 2, mole IV). It is a multicomponent system but not a multichromosomal or divided genome system because L-RNA has all the genetic information to induce an infection which cannot be distinguished from that induced by inoculating L rods plus I_2 rods plus S rods. S rods could not be detected in common TMV preparations (Bruening *et al.*, 1976; Whitfeld and Higgins, 1976). However, I_2 rods have been (Beachy and Zaitlin, unpublished).

L-RNA was effective in stimulating the incorporation of amino acids into polypeptides by wheat germ systems and, as was the case with common TMV, high molecular weight polypeptides were formed (Higgins *et al.*, 1976; Bruening *et al.*, 1976). In many experiments no material corresponding to capsid protein was detected. In some experiments, however, traces of what probably was capsid protein were seen (Fig. 3, track 3), even with highly purified L-RNA was the mRNA. Bruening *et al.* (1976) found that the two most slowly migrating zones among the L-RNA products (apparent molecular weights 165,000 and 130,000) co-migrated with zones found in the analysis of proteins formed *in vivo* in C_cTMV-infected tissue. The *in vivo* polypeptides were synthesized in the presence of high concentrations of actinomycin D (60 μg/ml) and chloramphenicol (1 mg/ml), which had been vacuum-infiltrated into leaf slices, and the polypeptides were not observed in material from uninoculated tissue. Pairs of high molecular weight, apparently

---

served. The principal zone in track 2 migrated at the rate of C_cTMV capsid protein, $M \sim 18{,}000$. The "165,000" and "130,000" molecular weight polypeptides are the top two zones (lightly and heavily exposed, respectively) of track 3.

infection-related polypeptides have been observed for common TMV in several systems [leaves and separated leaf cells: Zaitlin and Hariharasubramanian (1972), Scalla *et al.* (1976); protoplasts: Sakai and Takebe (1974), Paterson and Knight (1975); frog oocytes injected with TMV RNA: Knowland (1974); rabbit reticulocyte and wheat germ cell-free polypeptide synthesizing systems: Knowland *et al.* (1975)]. The molecular weights assigned to the two zones have varied (130,000 to 155,000 for one and 150,000 to 195,000 for the other). It is likely, however, that the same or closely related polypeptides have been observed by the various workers. The lower molecular weight polypeptides (Fig. 3, track 3) may be due to premature termination or ribosomes packed together near the termination signal because of faulty termination (Davies and Samuel, 1975). Replicase activity may be associated with a protein of molecular weight approximately 130,000, though this conclusion is not certain (Brishammer and Junti, 1974; Zaitlin *et al.*, 1976).

As is indicated by Fig. 3, track 2, the principal product from wheat germ cell-free systems directed by S-RNA is a polypeptide which migrates with C_cTMV capsid protein during electrophoresis (Bruening *et al.*, 1976; Higgins *et al.*, 1976). The close relationship of the *in vitro* product to capsid protein was confirmed by reconstitution experiments, serological reaction and peptide mapping. Thus, LMC and S-RNA seems functionally indistinguishable except that S-RNA becomes encapsidated by C_cTMV protein in the course of the infection whereas LMC remains free and does not reconstitute with common TMV protein *in vitro* (Siegel *et al.*, 1973).

The principal products from incorporations directed by I_2-RNA (Fig. 3, track 1) were a pair of polypeptides with molecular weights of approximately 30,000 (Bruening *et al.*, 1976; Higgins *et al.*, 1976). Because RNA-RNA molecular hybridization experiments (Bruening *et al.*, 1976) had demonstrated that S-RNA nucleotide sequences were present in I_2-RNA and because the difference in the molecular weights of the two RNAs (650,000 - 280,000 = 370,000) would not allow enough genetic information to specify two polypeptides of approximately 30,000 daltons the two polypeptides probably share amino acid sequences (assuming that I_2-RNA represents a single species). There is only one report (Zaitlin and Hariharasubramanian, 1972) of an infection-related (common TMV) protein of similar size. The function of the I_2-RNA translation product is unknown.

If the I_2-RNA product of C_cTMV is counted as a single polypeptide [the I_2-RNA of common TMV also directs the synthesis of a polypeptide of approximately 30,000 daltons (Beachy and Zaitlin, unpublished)], then four TMV and C_cTMV gene products have been identified in the above discussion: the 165,000 and 130,000 molecular weight polypeptides, the I_2 product and the capsid protein. Considerations of coding capacity (RNA is able to code for about one-ninth its weight in protein) make it likely that the two large polypeptides cannot be separately coded for by L-RNA and that they must be translated at least partly from a common nucleotide sequence. RNA-RNA molecular hybridization tests have shown that the nucleotide sequences of both S-RNA and I_2-RNA were contained in L-RNA of C_cTMV (Bruening *et al.*, 1976); Hunter *et al.* (1976) have demonstrated that specific large oligonucleotides from ribonuclease digestions of LMC were present in TMV virion RNA. The available data on TMV and C_cTMV favor mode III (Fig. 2) as a close approximation of the principal translation "strategy" of these viruses.

Genetic Map of TMV

Work with both C_cTMV and common TMV allows a preliminary genetic map to be constructed. An experimentally valuable property of many rod-shaped plant viruses is their ability to form "partially stripped virus" (PSV) when treated with protein denaturing agents under mild conditions (Fig. 4). Agents which have been used are the detergent sodium dodecyl sulfate (SDS), dilute alkali (pH 10.5) and dimethyl sulfoxide (DMSO). All of these agents cause TMV to be converted into "firecrackers", virus-like particles shorter than virions and with an RNA "fuse" at one end. If capsid protein is stripped from only one end of any particle to form the PSV, then it is reasonable to assume that the same end of the virion RNA is exposed on every PSV particle. There has been some confusion as to which end, the 3' or the 5', is exposed in PSV of common TMV. We will now discuss some experiments with the PSV of C_cTMV and TMV.

The virion RNAs of several plant viruses have been shown to participate in a reaction which is parallel to the reaction of tRNA with ATP and amino acids. That is, the virion RNA can be aminoacylated with a specific amino acid in the presence of crude enzyme preparations from a variety of sources (Kohl and Hall, 1974). Common TMV accepts histidine (Öberg and Philipson, 1972) and the site of acceptance has been shown to be the 3'-terminus of the RNA (Salomon *et al.*, 1976). Beachy *et al.* (1976) found that C_cTMV L-RNA and PSV accept valine. The RNA fragment which remains protected by the capsid protein after digestion with ribonuclease (PROT RNA, Fig. 4) did not accept valine. The results imply that it was the 3'-terminal sequences of L-RNA which were exposed in PSV of C_cTMV. Since

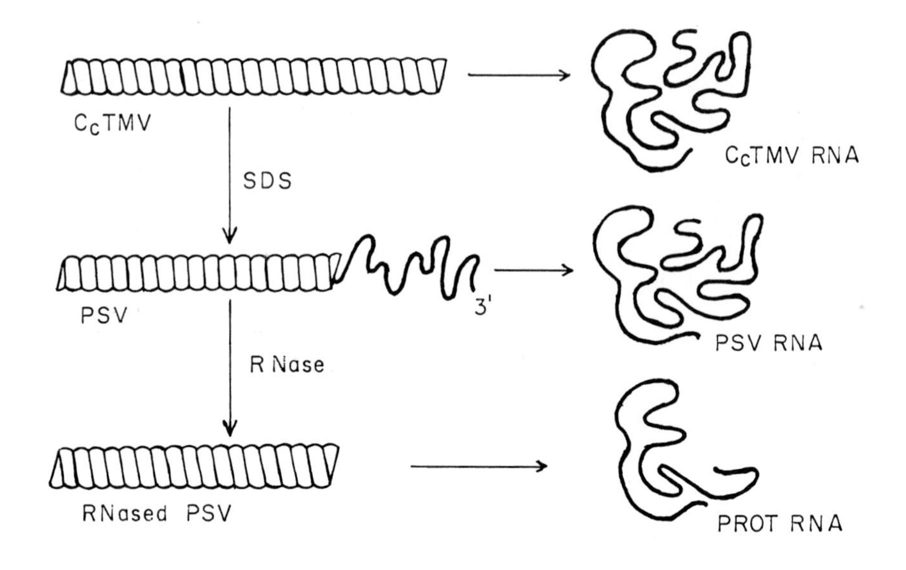

FIGURE 4. Recovery of specific large fragments from the 5'-terminal portion of C_cTMV L-RNA. When purified L rods were incubated for 2 to 5 min at 60°C with a 3 to 10-fold excess (by weight) of sodium dodecyl sulfate (SDS), partially stripped virus (PSV) was formed by the polar removal of some coat protein molecules from the RNA, beginning at the 3'-terminus of the RNA. The PSV was separated from residual intact rods by sucrose density gradient centrifugation and was treated with ribonuclease to remove the 3'-terminal sequences. The RNA protected from ribonuclease digestion (PROT-RNA) by the unstripped capsid protein was recovered by extraction of the protein into a phenol phase. That only one end of any rod was stripped by treatment with SDS was established by examination of PSV in the electron microscope; it is likely that the 3'-end is stripped because L-RNA and PSV but not PROT RNA can be enzymatically aminoacylated (with valine), a property expected to be specific for the 3'-terminus (Beachy et al., *1976).*

S-RNA and I_2-RNA also accept valine, those RNAs probably have nucleotide sequences which correspond to the 3'-terminal sequences of L-RNA whereas PROT RNA might be expected to lack the S-RNA and I_2-RNA sequences.

The postulated sequence relationships between S-RNA, I_2-RNA, L-RNA and PROT RNA were confirmed by Beachy *et al.* (1976) by means of RNA-RNA molecular hybridization and *in vitro* protein synthesis experiments. Tritium labeled S-RNA was hybridized to melted, double-stranded RNA from C_cTMV-infected tissue and in the presence of increasing concentrations of L-RNA,

S-RNA, PSV or PROT RNA. The amount of tritium labeled S-RNA incorporated into molecular hybrids was detected as acid-insoluble radioactivity remaining after incubation with ribonucleases in 0.1 M sodium phosphate, 0.1 M NaCl, pH 7.1 (a solution in which double-stranded RNA is resistant to attack by low concentrations of ribonuclease). As expected, the ribonuclease-resistant radioactivity was greatest in the absence of added unlabeled RNA or PSV. The recovered radioactivity declined with increasing amounts of unlabeled L-RNA, S-RNA or PSV in the reaction mixtures, and the data were consistent with *all* of the tritium labeled S-RNA sequences being represented in the unlabeled L-RNA, S-RNA and PSV. In contrast, the PROT RNA failed to compete with tritium labeled S-RNA in the hybridization reaction. Therefore, the S-RNA nucleotide sequences (and, obviously, the gene for the capsid protein) must be located near the 3'-end of L-RNA. Previously, I_2-RNA was found to compete with S-RNA (Bruening *et al.*, 1976). Therefore, the gene for the I_2-RNA product must be close to the capsid protein gene. Since both I_2-RNA and S-RNA accept valine (and presumably have the same 3'-terminal sequences) and S-RNA is too low in molecular weight to specify both the I_2 product and capsid protein, the gene for the I_2-RNA product seen *in vitro* is almost certainly on the 5'-side of the capsid protein gene in I_2-RNA and L-RNA. The gene for the 165,000 and 130,000 dalton polypeptides must be on the 5'-side of the gene for the I_2-RNA product on L-RNA. This set of genes probably exhausts the coding capacity of L-RNA. The sum of the molecular weights of the three gene products (neglecting the 130,000 dalton polypeptide on the assumption that information specifying it resides completely within the gene for the 165,000 dalton polypeptide) is 165,000 + 30,000 + 18,000 = 213,000. An RNA of molecular weight of at least $\sim 1.92 \times 10^6$ would be required to specify the three products independently, compared to a molecular weight of approximately 2.1×10^6 for L-RNA. Thus, in the 5'-to-3' direction, the genetic map of C_cTMV appears to be:

165,000/130,000 dalton polypeptide;
I_2-RNA product; capsid protein.

The entire L-RNA seems to be a (+) strand since all of the gene products listed above have been formed *in vitro* from L-RNA and RNAs having the same polarity as L-RNA.

Beachy *et al.* (1976) obtained more direct evidence that nucleotide sequences in the 5'-portion of L-RNA specify the 130,000 dalton product. PROT RNA failed to accept valine in the aminoacylation reaction and failed to compete with S-RNA sequences in molecular hybridization reactions, as discussed above. However, PROT RNA did direct the synthesis of polypeptides in a wheat germ system. From Fig. 3, tracks 3 and 5, it is apparent that PROT RNA of molecular weight approximately 1.4×10^6 and L-RNA specify a similar pattern of polypeptides (including a prominent zone at 130,000), except that the weak zones for capsid protein and 165,000 molecular weight polypeptide apparently are not present among the PROT RNA products. An RNA of molecular weight 1.4×10^6 would be expected to be unable to specify a polypeptide of molecular weight 165,000, just as PROT RNA of molecular weight approximately 1.1×10^6 failed to direct the synthesis of the 130,000 molecular weight polypeptide (track 4 of Fig. 3). If PROT RNA of weight 1.4×10^6 does specify the same 130,000 molecular weight polypeptide as does L-RNA, then (from considerations of coding capacity) the 130,000 and 165,000 molecular weight polypeptides probably share an initiation site. As has been discussed above, the high molecular weight polypeptides probably are involved in replicating the virion RNA.

Diagram III in Fig. 2 represents approximately our current concept of the principal modes of expression of C_cTMV (and presumably common TMV) genes, though without indicating the locations of small untranslated regions which must exist or explicitly showing the synthesis of *two* high molecular weight polypeptides. A consistent pattern of translation is postulated with L-RNA, I_2-RNA and S-RNA having one active initiation site each (corresponding to a, b and c of the diagram, respectively). However, some of the *in vitro* products indicated partial deviation from diagram III. Small amounts of capsid protein were generated in some *in vitro* polypeptide synthesis assays directed by highly purified preparations of L-RNA and I_2-RNA (tracks 3 and 1, Fig. 3). Whether this capsid protein was translated directly from L-RNA and I_2-RNA at low efficiency or whether some of the RNA molecules were cleaved to form small amounts of S-RNA or an S-RNA-like molecule is unknown. Similarly, a trace of I_2-RNA products appeared to be present among the L-RNA products (track 3, Fig. 3) and could be similarly explained. What was more difficult to explain was what appeared to be a trace amount of I_2-RNA product in the reaction mixtures directed by PROT RNA (tracks 4 and 5).

The preponderance of evidence concerning the location of the capsid gene of common TMV also favors a position close to the 3'-terminus of the virion RNA. Richards *et al.* (1975) prepared PSV with a distribution of lengths of the remaining encapsidated portion ranging around 190 nm (corresponding to a PROT RNA molecular weight of approximately 1.3×10^6) by incubating virions in 75% DMSO at 24°C for 30 minutes. The evidence favoring the 3'-end as the end stripped by DMSO is as follows: The reconstitution reaction, in which capsid protein and RNA react to form virion-like particles, seems to protect first the 5'-terminal region. The 3'-terminus, which had been chemically labeled with a reagent considered to be specific for the 3'-terminus, remained unprotected (Thouvenel *et al.*, 1971). A similar conclusion has been drawn using other means to detect specific termini (Butler and Klug, 1971; Ohno *et al.*, 1971). [It must be noted here that since a "capped" TMV RNA has nucleoside residues at both ends which are oriented as expected for a 3'-terminus, it is difficult to understand how the level of labeling of the two RNA termini could be so clearly distinguished in the experiments of Thouvenel *et al.* (1971). Most other approaches to distinguishing the termini also suffer from some ambiguity.] Nicolaieff *et al.* (1975) showed that PROT RNA from DMSO-produced PSV would reconstitute and that partially reconstituted virus and PSV therefore presumably had the same end of the virion RNA protected. Richards *et al.* (1975) separately digested TMV RNA, PROT RNA from the 190 nm particles and PSV with T1 ribonuclease and recovered those oligonucleotides which formed a complex with capsid protein, according to the procedure of Guilley *et al.* (1975a). These oligonucleotides (from nucleotide sequences which code for part of the capsid protein) were present in PSV and TMV RNA but absent from the PROT RNA. Therefore, if the polarity of stripping by DMSO is 3' to 5', the capsid protein gene must be located in the 3'-most third of the virion RNA.

PROT RNA from PSV prepared by incubation at high pH can be aminoacylated with histidine and has been shown chemically to contain the 3'-terminus of the virion RNA (Perham and Wilson, 1976; Ohno and Okada, 1976). Therefore, alkali-stripping of TMV must be 5' to 3'. Hunter *et al.* (1976) showed that a PROT RNA of molecular weight approximately 350,000 gave a pattern of oligonucleotides (after digestion with T1 ribonuclease) very similar to that of LMC RNA; both patterns had oligonucleotides previously recognized to be derived from the capsid protein gene. Therefore, the capsid protein gene must be located in the 3'-most sixth of the virion RNA. [The size of LMC RNA (and S-RNA) limits the gene to the 3'-most seventh if LMC RNA and virion RNA have the same 3'-terminus.] However, the 480 or so nucleotides (approximately 7.5% of the virion RNA) which must code for the capsid protein do not coincide with the 3'-terminal 480 residues of the virion RNA because the 71 3'-most nucleotides cannot code for the carboxyl terminus of the capsid protein (Guilley *et al.*, 1975a).

Kado and Knight (1968) were the first to try to localize the capsid protein gene of TMV. They inoculated ribonuclease-treated PSV (obtained by SDS treatment) of common TMV together with low concentrations of a serologically distinct TMV strain.

In 5 out of 21 trials with treated common TMV rods about half the full length, common TMV capsid protein was detected serologically. Since it was believed at that time that SDS stripping of common TMV proceeded from the 3'-end of the RNA, they concluded that the capsid protein gene lay in the 5'-half of the virion RNA. Rods which were 30% of the full length failed to give the effect. Recent work of Ohno and Okada (1976) strongly supports the 5'-terminus as the initial site for SDS-stripping of common TMV [in contrast to the results of Beachy *et al.* (1976) with C_cTMV], so that the conclusion of Kado and Knight (1968) can be modified to locate the gene in the 3'-half. However, the failures of the 30% rods to give the effect and the technical difficulty of producing 50% rods absolutely free of full length rods make it difficult to draw conclusions from this pioneering work.

Translation of BMV RNAs

The properties of brome mosaic virus (BMV) and similar viruses have been reviewed by Lane (1974), Kaesberg (1975) and Davies (1976). Three distinct particles, which nevertheless are very similar in chemical and physical properties, are formed during BMV infection (Fig. 5). The particles are isometric, in fact nearly spherical, rather than rod-shaped. The capsid is composed of 180 copies of a single kind of polypeptide of molecular weight 20,000. An infection can be initiated by particles H, M and L or RNAs 1, 2 and 3. No pair from the set of RNAs 1, 2 and 3 is sufficient for infectivity, and RNA 4 neither stimulates nor inhibits significantly the infectivity of 1, 2 and 3 (Lane and Kaesberg, 1971). "Pseudo-recombinant" brome mosaic viruses have been constructed by inoculating a mixture of two RNAs (from the set 1, 2 and 3) from one virus strain or variant and one RNA from another. By relating the kind of capsid protein formed by the pseudohybrid with the sources of the viral RNAs which were inoculated, it was possible to show that RNA 3 specified the capsid protein. Infections initiated with RNAs 1, 2 and 3 yielded all four RNAs. Therefore, the nucleotide sequences of RNA 4 ought to be present in one of the three "genomic" RNAs (1, 2 or 3). Since the oligonucleotides derived by digestion of RNA 4 with specific nucleases appear to be a subset of the oligonucleotides from RNA 3, but not of those from RNAs 1 or 2, RNA 3 must have the entire nucleotide sequence of RNA 4 (Shih *et al.*, 1972). When RNA 4 of one strain was added to RNAs 1, 2 and 3 of another, the capsid protein of the progeny was controlled by the source of RNA 3 (Lane, 1974). Therefore, it seems likely that RNA 4 is not replicated directly and is not part of the

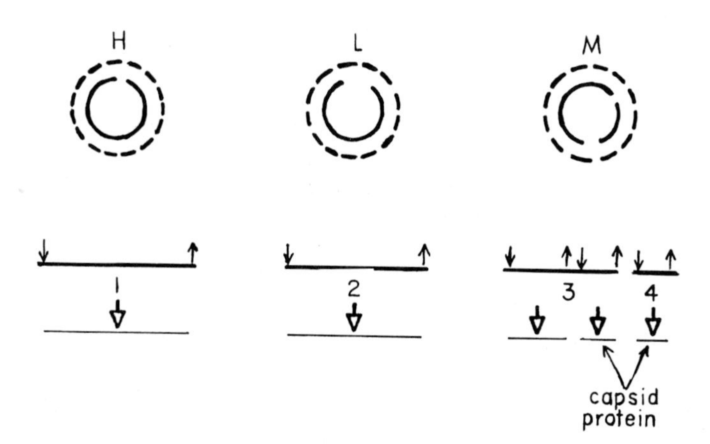

FIGURE 5. Translation pattern for BMV RNAs and their distribution in virions. Three kinds of particles are required to induce a brome mosaic virus infection, designated H, M and L in order of decreasing RNA content. The capside are represented by circles of capsid protein molecules (▬ ▬). The virion RNAs 1, 2, 3 and 4 have molecular weights of about 1.05 to 1.1, 1.0, 0.7 to 0.75 and 0.28 to 0.3 × 10^6, respectively. RNAs 1 and 2 (from virions H and L, respectively) direct in vitro *the synthesis of a spectrum of polypeptides, the largest and most abundant of which is large enough in each case to account for the coding capacity of the RNA. RNA 4 is monocistronic messenger RNA for capsid protein. RNA 3 may be dicistronic because it directs the synthesis of "protein 3a" and small amounts of capsid protein in a wheat germ cell-free system. However, it is not clear whether the capsid protein is directly translated from RNA 3 or from a fragment of RNA 3 which may be formed during the incubation in the cell-free system and acts as does RNA 4.*

virus genome. The search for a double-stranded RNA [and therefore a (-) strand RNA] corresponding in size to RNA 4 has been unsuccessful in one case (Phillips *et al.*, 1974) and successful in another (Bastin and Kaesberg, 1976). Since the (+) strand of TMV double-stranded RNA can be derived from virion RNA in the process of isolation (Nilsson-Tillgren and Kielland-Brandt, 1976), the demonstration of a double-stranded form of BMV RNA 4 does not necessarily indicate that RNA 4 is replicated independently rather than derived from RNA 3. That is, the RNA 4 RF (replication form) could be generated from a (-) strand of RNA 3 and a (+) strand of RNA 4 with subsequent nuclease removal of the (-) strand tail, all in the course of isolating the double-stranded RNA (Bastin and Kaesberg, 1976).

RNA 4 clearly is a monocistronic messenger RNA for the capsid protein as judged by analyses of the product from a wheat embryo protein synthesizing system (Shih and Kaesberg, 1973) and the sequence of a "capped" 52 nucleotide fragment from the 5'-end of RNA 4 which has an initiator codon (AUG) followed by a sequence specifying the first 14 amino acids of the capsid protein (Kaesberg, 1975). As is indicated in Fig. 5, RNAs 1 and 2 appear to be monocistronic since they can direct the synthesis of polypeptides of apparent molecular weights approximately 120,000 and 110,000 (Shih and Kaesberg, 1976), accounting for the respective coding capacities. RNA 3 serves efficiently as a messenger for an approximately 34,000 molecular weight polypeptide of unknown function designated "protein 3a." This represents less than half the coding capacity of RNA 3. Protein 3a and capsid protein have no common tryptic peptides, so the 3a and capsid protein genes probably are encoded in different parts of RNA 3. In some experiments small amounts of capsid protein were synthesized when RNA 3 is added to a wheat embryo protein synthesis system. This could represent direct translation. Alternatively, RNA 3 may be cleaved (to a limited extent) to form RNA 4 or an RNA 4-like molecule after it has been added to the cell-free protein synthesizing system, or traces of RNA 4 may be present as a contaminant in an RNA 3 preparation. [RNA 4 is known to compete effectively with RNA 3 (Shih and Kaesberg, 1976)]. None of these alternatives have been experimentally eliminated (Davies, 1976). The "genomic" RNAs 1, 2 and 3 seem to be expressed by a combination of translation mode IV (for RNAs 1 and 2) and mode III (for RNA 3) with some possibility of some mode I translation for RNA 3.

Limited digestion of each of the four BMV RNAs with T1 ribonuclease releases an approximately 150 nucleotide fragment which is very similar in sequence for each of the four RNAs and is the 3'-terminal fragment of each. The fragments and the intact virion RNAs can be aminoacylated with tyrosine. The nucleotide sequences of the 3'-terminal fragments from RNAs 3 and 4 apparently are identical and those sequences occur only once in each molecule. Since RNA 4 is derived from RNA 3 *in vivo* and a common, but unique, sequence is the 3'-terminal fragment of both, the obvious conclusion is that RNA 4 nucleotide sequences are the same as the 3'-most sequences of RNA 3 and that the 5'-to-3' gene order on RNA 3 (P. Kaesberg, personal communication) is:

3a protein; capsid protein

Though BMV is principally a "mode III and IV virus" and TMV a "mode III virus" with regard to translation of viral RNAs, there are striking similarities between the two systems. RNA 4 and LMC RNA/S-RNA both are monocistronic capsid protein messenger RNAs. RNA 3 and I_2-RNA both seem to be dicistronic messenger RNAs which efficiently direct the synthesis of a polypeptide of molecular weight 30,000 to 35,000 and the synthesis of capsid protein at a low efficiency or perhaps not at all, if RNA cleavage is the explanation for the capsid protein which *is* synthesized. For both I_2-RNA and RNA 3 the capsid protein gene is 3'-most. Both viruses also seem to have associated with them two polypeptides with molecular weights exceeding 100,000. However, the two BMV large polypeptides could have entirely distinct amino acid sequences (since they are translated from different RNAs) whereas the TMV polypeptides probably share most of their amino acid sequences.

Translation of AMV RNAs

Alfalfa mosaic virus (AMV), like BMV, has four major size classes of RNAs in the virions. In order of decreasing size these are generally referred to as B-RNA, M-RNA, T_b-RNA and T_a-RNA. For purposes of comparison with the BMV system the four AMV RNAs will be designated 1, 2, 3 and 4 because they have molecular weights close to, though not equal to, those of the BMV RNAs and, as discussed below, some functional similarity. There are important distinctions between the AMV and BMV systems (reviewed by van Vloten-Doting, 1976; van Vloten-Dotin and Jaspars, 1977). The AMV virions are bacilliform in shape. Particles B, M and T_b have one molecule of RNAs 1, 2 and 3, respectively. Particle T_a has two molecules of RNA 4. All of the particles have similar RNA/protein ratios and vary in the length of the bacilliform rodlets according to the mass of RNA present. Thus in order of decreasing length they are B, M, T_b and T_a. AMV infections can be initiated by particles B, M and T_b or by RNAs 1, 2 and 3 plus a few molecules of capsid protein or RNAs 1, 2, 3 and 4. RNAs 1, 2 and 3 alone are not sufficient. When RNAs 1 through 4 are the inoculum, RNA 4 apparently serves principally as a generator of capsid protein since it can efficiently direct capsid protein synthesis *in vitro* but does not control the kind of capsid protein found in the progeny from mixed inoculation experiments (Bol and van Vloten-Doting, 1973). That is, AMV RNA 4, like BMV RNA 4, is not genomic and the genome consists of three pieces of RNA. The role of the capsid protein in initiating an AMV infection is not clear. It is known that

AMV RNAs have high-affinity binding sites for capsid protein and that capsid protein from certain other "3-component" viruses, which also require capsid protein for infectivity, will substitute for AMV protein but BMV capsid protein will not (van Vloten-Doting, 1976).

AMV RNA 3 directs the *in vitro* synthesis of a "3a-like" protein of molecular weight 35,000 and AMV capsid protein, as expected. However, a polypeptide of molecular weight 65,000 is also formed in a wheat germ system to which RNA 3 was added (van Vloten-Doting, 1976). This polypeptide is recognized by antiserum to AMV and therefore probably has capsid protein amino acid sequences. It almost certainly is a "read through" product which has both "3a-like" and capsid protein amino acid sequences since it is large enough to be a complete translation of RNA 3. Whether the 65,000 molecular weight polypeptide, or even 3a-like protein, are formed *in vivo* is not known. One might expect that AMV RNAs 1 and 2, by analogy with BMV, would be monocistronic. However, RNAs which seem to be specific fragments of RNAs 1 and 2 have been observed in polysomes from infected tissue and as minor components of virion RNA preparations. Bol *et al.* (1976) have suggested that each of the three genomic RNAs of AMV *may* be dicistronic with the 3'-most gene being expressed principally by means of a subgenomic RNA. That is, AMV may utilize translation mode III in triplicate.

SUMMARY

We have discussed the replication of (+) strand RNA plant viruses, viruses which have in the virion(s) RNA(s) of the polarity capable of being translated into protein. The emphasis of this article is on modes of translation. Four possible modes (Fig. 2) are identified from evidence on plant and other viruses and on protein synthesis is uninfected cells. The modes are:

I. The virion RNA is polycistronic and several separate initiation and termination sites allow the direct translation of the virus-specified proteins from the virion RNA.

II. The virion RNA is polycistronic but only a single initiation and termination site are present. The "polyprotein" translated from the virion RNA is cleaved by specific proteases to form the virus-specified proteins.

III. The virion RNA is polycistronic but functionally monocistronic. That is, only the 5'-most gene is translated directly from the virion RNA. The other virus-specified proteins are translated from "sub-genomic" RNAs formed during the infection and representing the nucleotide sequences of the virion RNA which are not close to the 5'-terminus.

IV. The virus has several virion RNAs composing the genome. These are monocistronic messenger RNAs which can be directly translated.

Available data indicate that these idealized modes only approximate the replication mechanism of a given virus. However, mode III is consistent with the experimental evidence for tobacco mosaic virus replication. Brome mosaic virus has a genome composed of three RNAs. Two of these seem to be monocistronic (mode IV), whereas the third RNA seems to be dicistronic and to be expressed by mode III. Alfalfa mosaic virus also has genes distributed among three virion RNAs. Each of these may be expressed by mode III, though the conclusion is not firm. The capsid protein and a polypeptide of molecular weight 30,000 to 35,000 apparently are synthesized in a strikingly similar manner during the replication of all three of these viruses. In each case an RNA of molecular weight approximately 0.3×10^6 is a monocistronic messenger RNA for the capsid protein and an RNA of molecular weight approximately 0.7×10^6 has nucleotide sequences which encode the capsid protein gene and others which encode the gene for the larger polypeptide. The principal product from the *in vitro* translation of the approximately 0.7×10^6 molecular weight RNA is in each case the larger polypeptide. For tobacco mosaic virus and brome mosaic virus the capsid protein gene is closest to the 3'-terminus of this RNA.

REFERENCES

Astier-Manifacier, S., and Cornuet, P. (1971). RNA-dependent RNA polymerase in Chinese cabbage. *Biochem. Biophys. Acta 232*, 484-493.

Atabekov, J. B. (1975). Host specificity of plant viruses. *Ann. Rev. Phytopathol. 13*, 127-145.

Babos, P. (1969). Rapidly labeled RNA associated with ribosomes of tobacco leaves infected with tobacco mosaic virus. *Virology 39*, 893-900.

Babos, P. (1971). TMV-RNA associated with ribosomes of tobacco leaves infected with TMV. *Virology 43*, 597-606.

Bastin, M., and Kaesberg, P. (1976). A possible replicative form of brome mosaic virus RNA 4. *Virology 72*, 536-539.

Beachy, R. N., and Zaitlin, M. (1975). Replication of tobacco mosaic virus. VI. Replicative intermediate and TMV-RNA related RNAs associated with polyribosomes. *Virology 63*, 84-97.

Beachy, R. N., Zaitlin, M., Bruening, G., and Israel, H. W. (1976). A genetic map for the cowpea strain of TMV. *Virology 73*, 498-507.

Bol, J. F., and van Vloten-Doting, L. (1973). Function of top component a RNA in the initiation of infection by alfalfa mosaic virus. *Virology 51*, 102-108.

Bol, J. F., Bakhuizen, C. E. G. C., and Rutgers, T. (1976). Alfalfa mosaic virus polyribosomes: composition and biosynthetic activity. *Virology 75*, 1-17.

Bourque, D. P., Hagiladi, A., and Wildman, S. G. (1975). Experimental test of a TMV replication model. Attempts to identify molecular precursors of TMV-RNA and determine the time required to synthesize a TMV-RNA molecule. *Virology 63*, 135-146.

Brishammer, S. (1976). Separation studies of TMV replicase. *Ann. Microbiol. (Inst. Pasteur) 127A*, 25-31.

Brishammer, S., and Juntti, N. (1974). Partial purification and characterization of soluble TMV replicase. *Virology 59*, 245-253.

Bruening, G. (1977). Plant covirus systems. Two component systems. In *Comprehensive Virology* (Fraenkel-Conrat, H., and Wagner, R. R., eds.), Plenum Press, New York. *Vol. II*, pp. 55-141.

Bruening, G., Beachy, R. N., Scalla, R., and Zaitlin, M. (1976). *In vitro* and *in vivo* translation of the ribonucleic acids of a cowpea strain of tobacco mosaic virus. *Virology 71*, 498-517.

Burdon, R. H., Billeter, M. A., Weissman, C., Warner, R. C., Ochoa, S., and Knight, C. A. (1964). Replication of viral RNA. V. Presence of a virus-specific double-stranded RNA in leaves infected with tobacco mosaic virus. *Proc. Nat. Acad. Sci. U.S. 52*, 768-775.

Butler, P. J. G., and Klug, A. (1971). Assembly of the particle of tobacco mosaic virus from RNA and disks of protein. *Nature, New Biology 229*, 47-50.

Casjens, S., and King, J. (1975). Virus assembly. *Ann. Rev. Biochem. 44*, 555-611.

Clegg, C., and Kennedy, I. (1975). Translation of Semliki-Forest-virus intracellular 27-S RNA. Characterization of the products synthesized *in vitro*. *Eur. J. Biochem. 53*, 175-183.

Dasgupta, R., Harada, F., and Kaesberg, P. (1976). Blocked 5'-termini in brome mosaic virus RNA. *J. Virol. 18*, 260-267.

Davies, J. W. (1976). The multipartite genome of brome mosaic virus: Aspects of *in vitro* translation and RNA structure. *Ann. Microbiol. (Inst. Pasteur.) 127A*, 131-142.
Davies, J. W., and Kaesberg, P. (1973). Translation of virus mRNA: Synthesis of bacteriophage Qβ proteins in a cell-free extract from wheat embryo. *J. Virol. 12*, 1434-1441.
Davies, J. W., and Kaesberg, P. (1974). Translation of virus mRNA: Protein synthesis directed by several virus RNAs in a cell-free extract from wheat germ. *J. Gen. Virol. 25*, 11-20.
Davies, J. W., and Samuel, C. E. (1975). Translation of virus mRNA: Comparison of reovirus and brome mosaic virus single-stranded RNAs in a wheat germ cell-free system. *Biochem. Biophys. Res. Commun. 65*, 788-796.
Duda, C. T., Zaitlin, M., and Siegel, A. (1973). *In vitro* synthesis of double-stranded RNA by an enzyme system isolated from tobacco leaves. *Biochem. Biophys. Acta 319*, 62-71.
Dunn, D. B., and Hitchborn, J. H. (1965). The use of bentonite in the purification of plant viruses. *Virology 25*, 171-192.
Efron, D., and Marcus, A. (1973). Translation of TMV-RNA in a cell-free wheat embryo system. *Virology 53*, 343-348.
El Manna, M. M., and Bruening, G. (1973). Polyadenylate sequences in the ribonucleic acids of cowpea mosaic virus. *Virology 56*, 198-206.
Fenner, F. (1976). The classification and nomenclature of viruses. Summary of results of meetings of the international committee on taxonomy of viruses in Madrid, September 1975. *Virology 71*, 371-378.
Fraenkel-Conrat, H. (1976). RNA polymerase from tobacco necrosis virus infected and uninfected tobacco. Purification of the membrane-associated enzyme. *Virology 72*, 23-32.
Fraenkel-Conrat, H., Singer, B., and Williams, R. C. (1957). Infectivity of viral nucleic acid. *Biochem. Biophys. Acta 25*, 87-96.
Francki, R. I. B., and Randles, J. W. (1973). Some properties of lettuce necrotic yellows virus RNA and its *in vitro* transcription by virion-associated transcriptase. *Virology 54*, 359-368.
Funatsu, G., and Fraenkel-Conrat, H. (1964). Location of amino acid exchanges in chemically evoked mutants of tobacco mosaic virus. *Biochemistry 3*, 1356-1362.
Garfin, D. E., and Mandeles, S. (1975). Sequences of oligonucleotides prepared from tobacco mosaic virus ribonucleic acid. *Virology 64*, 388-399.

Geelen, J. L. M. C., Weathers, L. G., and Semancik, J. S. (1976). Properties of RNA polymerases of healthy and citrus exocortis viroid-infected Gynura aurantiaca DC. *Virology 69*, 539-546.

Gierer, A., and Schramm, G. (1956). Infectivity of ribonucleic acid from tobacco mosaic virus. *Nature 177*, 702-703.

Guilley, H., Jonard, G., and Hirth, L. (1975a). Sequence of 71 nucleotides at the 3'-end of tobacco mosaic virus RNA. *Proc. Nat. Acad. Sci. U.S. 72*, 864-868.

Guilley, H., Jonard, G., Richards, K. E., and Hirth, L. (1975b). Sequence of a specifically encapsidated RNA fragment originating from the tobacco-mosaic-virus coat-protein cistron. *Eur. J. Biochem. 54*, 135-144.

Harrison, B. D., Finch, J. T., Gibbs, A. J., Hollings, M., Shepherd, R. J., Valenta, V., and Wetter, C. (1971). Sixteen groups of plant viruses. *Virology 45*, 356-363.

Hershko, A., and Fry, M. (1975). Post-translational cleavage of polypeptide chains: Role in assembly. *Ann. Rev. Biochem. 44*, 775-797.

Higgins, T. J. V., Goodwin, P. B., and Whitfeld, P. R. (1976). Occurrence of short particles in beans infected with the cowpea strain of TMV. II. Evidence that short particles contain the cistron for coat-protein. *Virology 71*, 486-497.

Hunter, T. R., Hunt, T., Knowland, J., and Zimmern, D. (1976). Messenger RNA for the coat protein of tobacco mosaic virus. *Nature 260*, 759-764.

Ikegami, M., and Francki, R. I. B. (1976). RNA-dependent RNA polymerase associated with subviral particles of Fiji disease virus. *Virology 70*, 292-300.

Jackson, A. O., Mitchell, D. M., and Siegel, A. (1971). Replication of tobacco mosaic virus. I. Isolation and characterization of double-stranded forms of ribonucleic acid. *Virology 45*, 182-191.

Jackson, A. O., Zaitlin, M., Siegel, A., and Francki, R. I. B. (1972). Replication of tobacco mosaic virus. III. Viral RNA metabolism in separated leaf cells. *Virology 48*, 655-665.

Kado, C. I., and Knight, C. A. (1968). The coat protein gene of tobacco mosaic virus. 1. Location of the gene by mixed infection. *J. Mol. Biol. 36*, 15-23.

Kaesberg, P. (1975). The RNAs of brome mosaic virus and their translation in cell-free extracts derived from wheat embryo. *INSERM 47*, 205-210.

Kaper, J. M. (1975). The chemical basis of virus structure, dissociation and reassembly. In *Frontiers of Biology, 39*, (Neuberger, A., and Tatum, E. L., eds.), North Holland Publishing Co., New York.

Kassanis, B., and McCarthy, D. (1967). The quality of virus as affected by the ambient temperature. *J. Gen. Virol. 1*, 425-440.

Kassanis, B., and Varma, A. (1975). Sunn-hemp mosaic virus. Descriptions of Plant Viruses. *Commonwealth Mycological Institute, 153*, Kew, Surrey, England.

Keith, J., and Fraenkel-Conrat, H. (1975). Tobacco mosaic virus RNA carries 5'-terminal triphosphorylated guanosine blocked by 5'-linked 7-methyl-guanosine. *FEBS Letters 57*, 31-33.

Knowland, J. (1974). Protein synthesis directed by the RNA from a plant virus in a normal animal cell. *Genetics 78*, 383-394.

Knowland, J., Hunter, T., Hunt, T., and Zimmern, D. (1975). Translation of tobacco mosaic virus RNA and isolation of messenger for TMV coat protein. *INSERM 47*, 211-216.

Kohl, R. J., and Hall, T. C. (1974). Aminoacylation of RNA from several viruses: Amino acid specificity and differential activity of plant, yeast and bacterial synthetases. *J. Gen. Virol. 25*, 257-261.

Lane, L. C. (1974). The bromoviruses. *Adv. Virus Res. 19*, 151-220.

Lane, L. C., and Kaesberg, P. (1971). Multiple genetic components in bromegrass mosaic virus. *Nature New Biol. 232*, 40-43.

Lasky, R. A., and Mills, A. D. (1975). Quantitative film detection of ^{3}H and ^{14}C in polyacrylamide gels by fluorography. *Eur. J. Biochem. 56*, 335-341.

Lister, R. M., and Thresh, J. M. (1955). A mosaic disease of leguminous plants caused by a strain of tobacco mosaic virus. *Nature 175*, 1047-1048.

Morris, T. J. (1974). Two nucleoprotein components associated with the cowpea strain of TMV. *Proc. Am. Phytopathol. Soc. 1*, 83.

Nicolaieff, A., Lebeurier, G., Morel, M.-C., and Hirth, L. (1975). The uncoating of native and reconstituted TMV by dimethylsulphoxide: The polarity of stripping. *J. Gen. Virol. 26*, 295-306.

Nilsson-Tillgren, T., and Kielland-Brandt, M. C. (1976). On the structure of replicating TMV RNA. *Ann. Microbiol. (Inst. Pasteur) 127A*, 23.

Öberg, B., and Philipson, L. (1972). Binding of histidine to tobacco mosaic virus RNA. *Biochem. Biophys. Res. Commun. 48*, 927-932.

Ohno, T., and Okada, Y. (1976). Polarity of stripping of tobacco mosaic virus by alkali and sodium dodecyl sulfate. *Virology 76*, 429-432.

Ohno, T., Nozu, and Okada, Y. (1971). Polar reconstitution of tobacco mosaic virus (TMV). *Virology 44*, 510-516.

Oxefelt, P. (1976). Biological and physiochemical characteristics of three strains of red clover mottle virus. *Virology 74*, 73-80.

Paterson, R., and Knight, C. A. (1975). Protein synthesis in tobacco protoplasts infected with tobacco mosaic virus. *Virology 64*, 10-22.

Perham, R. N., and Wilson, T. M. A. (1976). The polarity of stripping of coat protein subunits from the RNA in tobacco mosaic virus under alkaline conditions. *FEBS Letters 62*, 11-15.

Perry, R. P. (1976). Processing of RNA. *Ann. Rev. Biochem. 45*, 605-629.

Phillips, G., Gigot, C., and Hirth, L. (1974). Replicative forms and viral RNA synthesis in leaves infected with brome mosaic virus. *Virology 60*, 370-379.

Pinck, L. (1976). Blocked 5' termini in alfalfa mosaic virus RNA. *Ann. Microbiol. (Inst. Pasteur) 127A*, 175-181.

Rees, M. W., and Short, M. N. (1975). The amino acid sequence of the cowpea strain of tobacco mosaic virus protein. *Biochem. Biophys. Acta 393*, 15-23.

Richards, K. E., Morel, M. C., Micolaieff, A., Lebeurier, G., and Hirth, L. (1975). Location of the cistron of the tobacco virus coat protein. *Biochimie 57*, 749-755.

Roberts, B. E., Paterson, B. M., and Sperling, R. (1974). The cell-free synthesis and assembly of viral specific polypeptides into TMV particles. *Virology 59*, 307-313.

Sakai, R., and Takebe, I. (1974). Protein synthesis in tobacco mesophyll protoplasts induced by tobacco mosaic virus infection. *Virology 62*, 426-433.

Salomon, R., Sela, I., Soreq, H., Giveon, D., and Littauer, U. Z. (1976). Enzymatic acylation of histidine to tobacco mosaic virus RNA. *Virology 71*, 74-84.

Scalla, R., Boudon, E., and Rigaud, J. (1976). SDS-Polyacrylamide gel electrophoretic detection of two high molecular weight proteins associated with TMV infected tobacco. *Virology 69*, 339-345.

Schwinghamer, M. W., and Symons, R. H. (1975). Fractionation of cucumber mosaic virus RNA and its translation in a wheat embryo cell-free system. *Virology 63*, 252-262.

Semancik, J. S. (1974). Detection of polyadenylic acid sequences in plant pathogenic RNAs. *Virology 62*, 288-291.

Shatkin, A. J., and Both, G. W. (1976). Reovirus mRNA: Transcription and translation. *Cell 7*, 305-313.

Shih, D. S., and Kaesberg, P. (1973). Translation of brome mosaic viral ribonucleic acid in a cell-free system derived from wheat embryo. *Proc. Nat. Acad. Sci. U.S. 70*, 1799-1803.
Shih, D. S., and Kaesberg, P. (1976). Translation of the RNAs of brome mosaic virus: The monocistronic nature of RNA 1 and RNA 2. *J. Mol. Biol. 103*, 77-88.
Shih, D. S., Lane, L. C., and Kaesberg, P. (1972). Origin of the small component of brome mosaic virus RNA. *J. Mol. Biol. 64*, 353-362.
Shipp, W., and Haselkorn, R. (1964). Double-stranded RNA from tobacco leaves infected with TMV. *Proc. Nat. Acad. Sci. U.S. 52*, 401-408.
Siegel, A., Hari, V., Montgomery, I., and Kolacz, K. (1976). A messenger RNA for capsid protein isolated from tobacco mosaic virus-infected tissue. *Virology 73*, 363-371.
Siegel, A., Zaitlin, M., and Duda, C. T. (1973). Replication of tobacco mosaic virus. IV. Further characterization of viral-related RNAs. *Virology 53*, 75-83.
Simmons, D. T., and Strauss, J. H. (1974). Translation of Sindbis virus 26S RNA and 49S RNA in lysates of rabbit reticulocytes. *J. Mol. Biol. 86*, 397-409.
Symons, R. H. (1976). Studies on the replication of cucumber mosaic virus. *Ann. Microbiol. (Inst. Pasteur) 127A*, 161.
Thouvenel, J.-C., Guilley, H., Stussi, C., and Hirth, L. (1971). Evidence for polar reconstitution of TMV. *FEBS Letters 16*, 204-206.
Van Vloten-Doting, L. (1976). Similarities and differences between viruses with a tripartite genome. *Ann. Microbiol. (Inst. Pasteur) 127A*, 119-129.
Van Vloten-Doting, L., and Jaspars, E. M. J. (1977). Plant covirus systems. 3 component systems. In *Comprehensive Virology* (Fraenkel-Conrat, H., and Wagner, R. R., eds.), Plenum Press, New York, Vol. *II*, pp. 1-53.
Weissman, C. (1974). Making of a phage. *FEBS Letters 43S*, S10-S18.
Wengler, G., and Wengler, G. (1976). Localization of the 26-S RNA sequence on the viral genome type 42-S RNA isolated from SFV-infected cells. *Virology 73*, 190-199.
Whitfeld, P. R., and Higgins, T. J. V. (1976). Occurrence of short particles in beans infected with cowpea strain of TMV. I. Purification and characterization of short particles. *Virology 71*, 471-485.
Whittmann, H. G., and Wittmann-Liebold, B. (1966). Protein chemical studies of two RNA viruses and their mutants. *Cold Spring Harbor Symp. Quant. Biol. 31*, 163-172.

Zaitlin, M., Beachy, R. N., Bruening, G., Romaine, C. P., and Scalla, R. (1976). Translation of tobacco mosaic virus RNA. In *Animal Virology* (Baltimore, D., Huang, A., and Fox, C. F., eds.), Academic Press, Inc., New York, pp. 567-582

Zaitlin, M., and Hariharasubramanian, V. (1972). A gel electrophoretic analysis of proteins from plants infected with tobacco mosaic and potato spindle tuber viruses. *Virology* *47*, 296-305.

Zimmern, D. (1975). The 5' end group of tobacco mosaic virus RNA is m^7G5' ppp 5'Gp. *Nucleic Acids Res.* *2*, 1189-1201.

VIROIDS

T.O. Diener

Plant Virology Laboratory
Plant Protection Institute, ARS
U.S. Department of Agriculture
Beltsville, Maryland 20705

The term "viroid" has been introduced to denote a recently recognized class of subviral plant pathogens (Diener, 1971c). Presently known viroids consist solely of a short strand of RNA with a molecular weight of about 100,000 to 140,000. Introduction of this low-molecular-weight RNA into susceptible hosts leads to replication of the RNA and, in some hosts, to disease.

The first viroid came to light during efforts to purify and characterize the agent of the potato spindle tuber disease (PSTV), a disease which, for many years, had been assumed to be of viral etiology (Diener and Raymer, 1971). Diener and Raymer (1967) reported that the infectious agent of this disease was a free RNA and that virus particles, apparently, were not present in infected tissue. Later, sedimentation and gel electrophoretic analyses conclusively demonstrated that the infectious RNA had a very low molecular weight (Diener, 1971c) and that the agent, therefore, differed basically from conventional viruses.

Four additional plant diseases, citrus exocortis (Semancik and Weathers, 1972a), chrysanthemum stunt (Diener and Lawson, 1973), chrysanthemum chlorotic mottle (Romaine and Horst, 1975)

ISBN 012-601950-9

and cucumber pale fruit (Van Dorst and Peters, 1974, and Peters, unpublished results), also have now been shown to be caused by low-molecular-weight RNAs; *i.e.*, by viroids.

The recognition of viroids as a newly identified class of pathogens raises several questions that have potentially important implications for microbiology, molecular biology, and plant pathology, as well as for veterinary and human medicine.

PROPERTIES OF VIROIDS

Sedimentation Properties and Nuclease Sensitivity

Diener and Raymer (1967) showed that most of the infectious material in crude extracts prepared from potato or tomato leaves affected with the potato spindle tuber disease sedimented in sucrose gradients at a very low rate (about 10S). Treatment of crude extracts with phenol affected neither infectivity nor the sedimentation properties of the agent. Incubation of extracts with nucleases revealed that the agent was sensitive to ribonuclease, but not to deoxyribonuclease. In view of these findings, the authors proposed the agent to be a free nucleic acid.

Somewhat similar results were later reported by Singh and Bagnall (1968) who also worked with the potato spindle tuber disease, by Semancik and Weathers (1968) with citrus exocortis disease, and by Lawson (1968) with chrysanthemum stunt disease.

Absence of Virions

Although there was little doubt in each case that the slowly sedimenting infectious material was free RNA, the question arose as to whether this RNA existed as such *in situ* or whether it was released from conventional virus particles during extraction. A systematic study of this question led to results that were incompatible with the concept that conventional viral nucleoprotein particles exist in PSTV-infected tissue (Diener, 1971b).

Furthermore, comparisons of proteins isolated from PSTV-infected tissue with those isolated from healthy tissue gave no evidence for the synthesis in infected leaves of proteins that could be construed as viral coat proteins under conditions such that coat protein of defective strains of tobacco mosaic virus could readily be demonstrated (Zaitlin and Hariharasubramanian, 1972).

Subcellular Location

Isolation of subcellular particles from PSTV-infected tissue revealed that appreciable infectivity was present only in the original tissue debris and in the fraction containing nuclei. Chloroplasts, mitochondria, ribosomes, and the "soluble" fraction contained no more than traces of infectivity. Furthermore, when chromatin was isolated from infected tissue, most infectivity was associated with it and could be extracted with phosphate buffer as free RNA (Diener, 1971b). These and other experiments suggest that, *in situ*, PSTV is associated with the nuclei and particularly with the chromatin of infected cells.

Recognition of the Low Molecular Weight of Viroids

The low sedimentation rate of PSTV would have been consistent with a viral genome of conventional size ($\geq 10^6$ daltons) only if the RNA had been double- or multi-stranded. Early experiments indicated that the RNA might be double-stranded (Diener and Raymer, 1969), but later results showed that its chromatographic properties are not compatible with this hypothesis (Diener, 1971a).

Determination of the molecular weights of viroids was of great importance in elucidating their structure. This determination was difficult, because the agents occur in infected tissue in very small amounts and were, therefore, difficult to separate from host RNA and to purify in amounts sufficient for conventional biophysical analyses.

A concept elaborated by Loening (1967) made it feasible to determine the molecular weight of PSTV, using infectivity as the sole parameter. Combined sedimentation and gel electrophoretic analyses conclusively showed that the infectious RNA had a very low molecular weight (Diener, 1971c). A value of 5×10^4 daltons was compatible with the experimental results. This conclusion was confirmed by the ability of PSTV to penetrate into polyacrylamide gels of high concentration (small pore size), from which high molecular weight RNA molecules were excluded (Diener and Smith, 1971).

On the basis of its electrophoretic mobility in 5% polyacrylamide gels, Semancik and Weathers (1972a) came to the conclusion that the agent causing citrus exocortis disease was also a low-molecular-weight RNA. They estimated that the RNA had a molecular weight of 1.25×10^5. Sänger (1972), on the other hand, estimated that the exocortis disease agent had a molecular weight of 5 to 6×10^4.

Diener and Lawson (1973) showed by a combination of isokinetic density-gradient centrifugation and electrophoresis in 20% polyacrylamide gels that the agent of chrysanthemum stunt disease was a low-molecular-weight RNA similar to, but distinct from, PSTV.

Purification of PSTV

In all experiments so far described, PSTV was identified solely by its biological activity; and no clearly and consistently recognizable ultraviolet light-absorbing component was correlated with infectivity distribution in sucrose gradients or in polyacrylamide gels. Detailed characterization of PSTV required its isolation in amounts sufficient for conventional biophysical and biochemical analyses. Large-scale isolation of PSTV, together with improvements in separation techniques, have made this goal attainable (Diener, 1972a).

As shown in Fig. 1a, electrophoresis in 20% polyacrylamide gels of highly concentrated, low-molecular-weight RNA preparations from healthy plants revealed the presence of 5S RNA and at least three other RNA species of low molecular weight (I, III, and IV). Electrophoresis of identically prepared samples from PSTV-infected plants revealed the same components; namely 5S RNA and components I, III, and IV; but, in addition, another prominent ultraviolet-light-absorbing component, II, was evident (Fig. 1b). Bioassays of individual gel slices demonstrated that infectivity coincided with component II (Fig. 1b). This coincidence, the high level of infectivity, and the fact that component II was not recognizable in preparations from healthy leaves constituted strong evidence that component II was PSTV.

Pure PSTV has been prepared using electrophoresis in 20% polyacrylamide gels as the last step in the purification scheme, followed by elution from gel slices and reconcentration of the RNA by chromatography on hydroxyapatite (Diener, 1973b). Quantities so far produced have been sufficient to determine several properties of the RNA by conventional means.

Thermal Denaturation Properties

The total hyperchromicity shift of PSTV in 0.01 × SSC (0.15 M sodium chloride-0.015 M in sodium citrate, pH 7.0) was about 24% and the T_m about 50°C (Diener, 1972a). The thermal denaturation curve indicated that PSTV was not a regularly base-paired structure, such as double-stranded RNA, since in this case, denaturation would be expected to occur over a narrower temperature range and at higher temperatures (Miura *et al.*, 1966). PSTV could, however, be an irregularly base-paired, single-stranded RNA molecule, similar to transfer RNA, in which single-stranded regions alternate with base-paired regions.

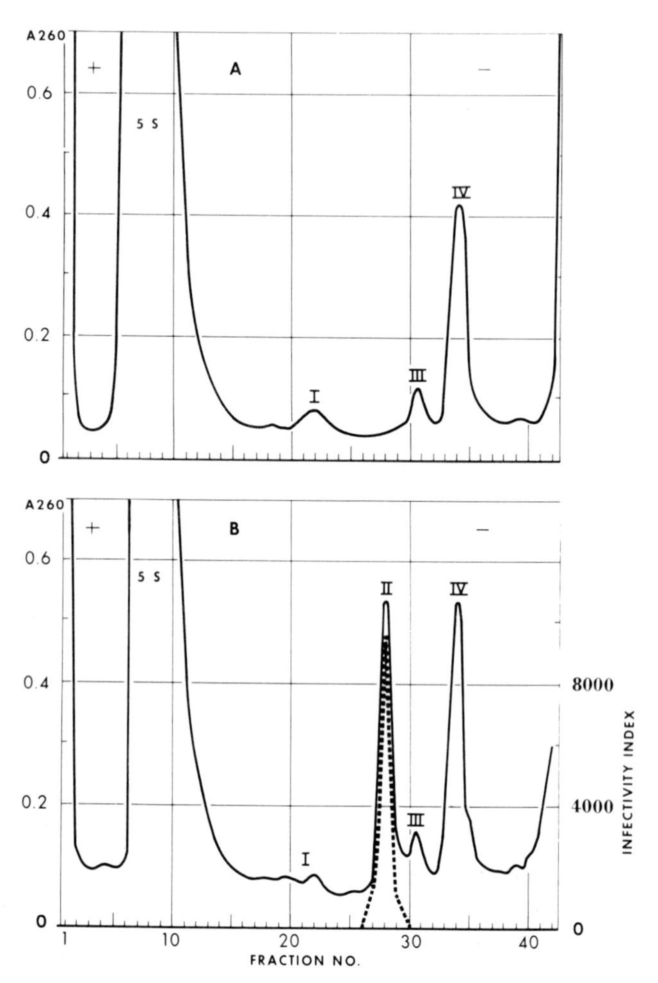

Fig. 1. a. Ultraviolet light-absorption profile of an RNA preparation from healthy tomato leaves after electrophoresis in a 20% polyacrylamide gel for 7.5 hrs at 4°C (5 mA per tube, constant current) (Diener, 1972a). b. Ultraviolet light-absorption (———) and infectivity distribution (-----) of an RNA preparation from PSTV-infected tomato leaves after electrophoresis in a 20% polyacrylamide gel (same conditions as in Fig. 1a). 5S = 5S ribosomal RNA; I, III, IV = unidentified minor components of cellular RNA; II = PSTV; A_{260} = Absorbance at 260 nm. Electrophoretic movement from right to left (Diener, 1972a).

Molecular Weight of PSTV

With the availability of purified PSTV, a redetermination of its molecular weight based, not on its biological activity, but on its physical properties, became possible. For this purpose, a method described by Boedtker (1971) appeared particularly promising, as it permitted the determination of the molecular weight of an RNA independent of its conformation. Application of this method to PSTV led to a molecular weight estimate of 7.5 to 8.5×10^4 (Diener and Smith, 1973).

Visualization of PSTV

In view of the purity of available PSTV preparations, it appeared feasible to visualize PSTV by electron microscopy and to determine its molecular weight by direct length measurements of the RNA in electron micrographs. This, indeed, proved feasible (Sogo *et al.*, 1973).

Figure 2 shows an electron micrograph of a mixture of a double-stranded DNA; namely, coliphage T7-DNA, and PSTV. Length measurements indicated that T7-DNA is about 280 times longer than PSTV. It is also apparent that the width of PSTV is similar to that of T7-DNA. If one assumes a molecular weight of 25×10^6 for T7-DNA and assumes that PSTV in urea is formed by two more or less base-paired strands (either as a hairpin or a double helix), then one obtains a molecular weight estimate for PSTV of 8.9×10^4.

In other experiments, mixtures of PSTV and a single-stranded viral RNA were examined and measured. PSTV appeared thicker than this single-stranded viral RNA, and from length comparisons, a molecular weight of 7.9×10^4 was obtained (Sogo *et al.*, 1973).

The molecular weight estimates obtained by electron microscopy are, therefore, in excellent agreement with the values obtained from analysis of heat-denatured, formylated PSTV in polyacrylamide gels.

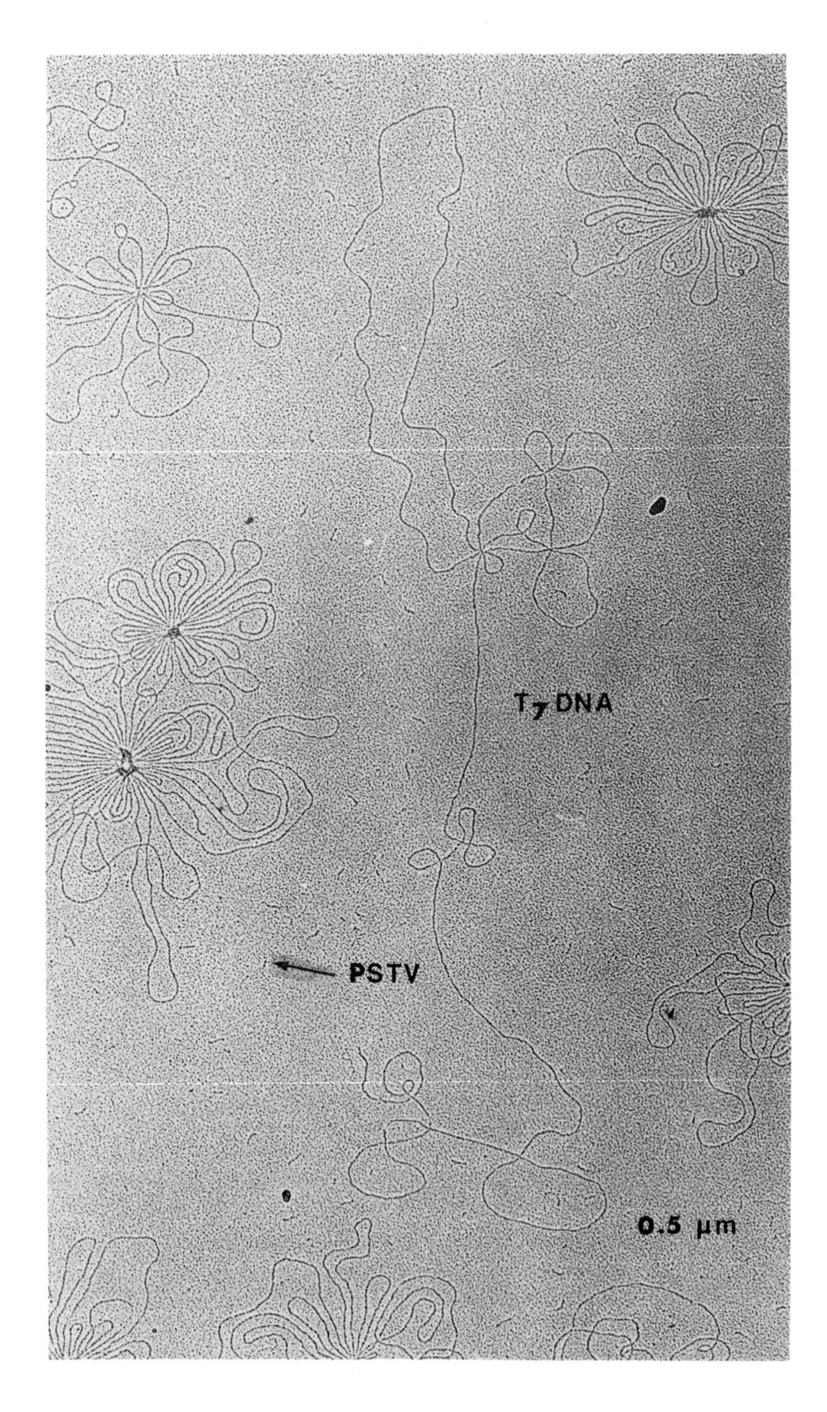
T_7 DNA
PSTV
0.5 μm

Inactivation by Ultraviolet Light

In view of the low molecular weight of PSTV, it was of interest to determine its sensitivity to irradiation with ultraviolet light. Although one might expect that a small molecule, such as PSTV, would be considerably more resistant to ultraviolet irradiation than a conventionally sized viral RNA or DNA, the effect of size on ultraviolet sensitivity of nucleic acids is not well understood (Adams, 1970).

Exposure of purified PSTV, of tobacco ringspot virus (TRSV), and of its satellite (SAT) (Schneider, 1969) to ultraviolet radiation of 254 nm showed that the inactivation dose for PSTV and SAT was 70 to 90 times as large as that for TRSV (Diener *et al.*, 1974). Although other explanations are possible, this marked difference in sensitivity to ultraviolet radiation is probably a consequence of the smaller size (smaller target volume) of PSTV and SAT-RNA, as compared with TRSV-RNA.

Composition

In view of these results, little doubt existed that the infectious RNA had a very low molecular weight. It was conceivable, however, that PSTV was not a single molecular species, but rather a population of several RNA molecules of similar length, with different nucleotide sequences, which together may constitute a viral genome of more or less conventional size. As discussed previously (Diener, 1972b; 1973a), existing knowledge does not support this model, and the high resistance of PSTV to ultraviolet radiation (Diener *et al.*, 1974) may be interpreted as

Fig. 2. Electron micrograph of PSTV mixed with a double-stranded DNA, coliphage T7-DNA. Native T7-DNA (0.8 μg/ml) was mixed with purified PSTV (0.4 μg/ml) previously heated for 10 min at 63°C in the presence of 8 M urea, followed by quenching in ice water. Note that double-stranded T7-DNA and PSTV have similar widths. (Courtesy of T. Koller and J.M. Sogo, Swiss Federal Institute of Technology, Zurich.)

further evidence for the smallness of the PSTV genome.

More definitive results have recently been obtained by two-dimensional fingerprinting of highly purified PSTV and citrus exocortis viroid (CEV) labeled *in vitro* with ^{125}I (Dickson *et al.*, 1975). These analyses demonstrated that each of these RNAs had a complexity compatible with the size estimate of 250 to 350 nucleotides, thus supporting the concept that each viroid consisted of a single RNA species of defined sequence. These studies also demonstrated that PSTV and CEV do not have the same primary sequence. This result contradicted conclusions drawn by some workers from biological experiments which had suggested that the two viroids were independent isolates of the same pathogen (Semancik and Weathers, 1972b; Semancik *et al.*, 1973; Singh and Clark, 1973).

Messenger RNA and Aminoacylation Properties

In vitro experiments with several cell-free protein synthesizing systems indicated that neither purified PSTV (Davies *et al.*, 1974) nor CEV (Hall *et al.*, 1974) act as messenger RNAs in these systems. With CEV, it has been shown that the RNA would not serve as an amino acid acceptor (Hall *et al.*, 1974).

REPLICATION OF VIROIDS

By what mechanisms RNAs of such a small size are replicated in susceptible host cells is, at present, unknown. In view of the small amount of genetic information that PSTV introduces into its host cells, one might plausibly assume PSTV to be analogous to a satellite RNA that requires a helper virus for its own replication. However, efforts to demonstrate the presence of such a helper virus in uninoculated tomato plants gave negative results (Diener, 1971c). In view of these and other results (Diener *et al.*, 1972), it appeared most unlikely that a conventional helper virus was necessary for the replication of PSTV. Hence,

PSTV, in spite of its small size, appeared to replicate autonomously in susceptible cells. Because the RNA could code for only 80 to 120 amino acids and because it was not translated in cell-free protein-synthesizing systems, it appeared that PSTV was replicated by preexisting host enzymes.

Two schemes of PSTV replication are most readily compatible with present views on cellular and viral RNA synthesis. The first scheme postulates that PSTV is replicated on a DNA template, which either is already present in uninfected hosts in repressed form, or is synthesized as a consequence of infection with PSTV. The second scheme postulates that PSTV is replicated independently of DNA; *i.e.*, that its replication is analogous to that of many viral RNAs, proceeding via a complementary RNA transcribed from PSTV.

A distinction between these two schemes should be made possible by the use of drugs that specifically inhibit RNA transcription from DNA templates. These drugs should inhibit PSTV replication if the first scheme is correct. We used actinomycin D because this antibiotic inhibits cellular RNA synthesis in plant cells but does not seriously interfere with the replication of several plant viral RNAs.

Leaf strips from healthy or PSTV-infected tomato plants were incubated in solutions containing ^{3}H-uracil or ^{3}H-UTP. Extraction of nucleic acids and analysis by polyacrylamide gel electrophoresis revealed tritium label incorporation into a component with electrophoretic mobility identical with PSTV in extracts from infected, but not in extracts from healthy leaves. No tritium incorporation into PSTV could be detected when leaf strips were pretreated with actinomycin D under conditions which reduced uptake of ^{3}H-uracil only 10 to 12% but inhibited cellular RNA transcription 87 to 99%. These results suggested that, in infected tomato leaves, PSTV replication required the continued synthesis of one or more cellular RNA species or that PSTV replication proceeded via a DNA intermediate (Diener and Smith, 1975).

Similar results were obtained with an *in vitro* RNA synthesizing system, using purified cell nuclei from healthy or PSTV-infected tomato leaves as an enzyme source. Isolation of low-molecular-weight RNAs from the *in vitro* reaction mixtures and analysis by a gel electrophoresis revealed that PSTV replication was sensitive to incubation with actinomycin D (Takahashi and Diener, 1975). These results also demonstrated that PSTV was replicated in the nucleus of infected cells. The observed sensitivity of PSTV replication to actinomycin D, in both *in vitro* and *in vivo* systems suggested that PSTV may be transcribed from a DNA template.

Recent experiments (Hadidi *et al.*, 1976), in which purified, ^{125}I-labeled PSTV was used as a hybridization probe to detect nucleotide base sequences in PSTV which were complementary to DNA isolated from healthy and PSTV-infected host plants, gave additional support to the hypothesis that PSTV may be transcribed from a DNA template. These experiments showed that infrequent DNA sequences complementary to PSTV exist in both uninfected and infected tomato cells. DNA titration experiments revealed that at least 60% of the PSTV molecule was represented by complementary sequences in the DNA of several normal solanaceous host species and that phylogenetically diverse plants contain DNA base sequences with less homology to PSTV. PSTV-infected tomato plants did not possess new PSTV sequences at detectable levels (Hadidi *et al.*, 1976). These results support the hypothesis that synthesis of PSTV, and presumably of other viroids, is a normal capacity of the host genome which, however, is completely repressed in normal plants. If so, viroids act as derepressors of this latent capacity and could, therefore, be considered as abnormal regulatory molecules.

CONCLUSIONS

With the knowledge now at hand, it is evident that viroids constitute a novel class of pathogens which are clearly distinct in several respects from conventional viruses. Aside from the lack of a protein coat and the smallness of their genomes, viroids differ from viruses in that infection with viroids apparently does not lead to the synthesis of novel, pathogen-specific proteins. Also, PSTV is the first known RNA pathogen of plants which possesses complementarity to its host DNA. Having these properties, viroids lend themselves admirably to studies, on the molecular level, of the mechanisms of pathogenesis and control of gene regulation in eucaryotic cells.

From a practical standpoint it can be predicted that many infectious diseases of plants and possibly of animals, the etiology of which is now obscure, will be found to be caused by agents similar to presently known viroids. Some of these plant diseases may be of considerable economic importance. A case in point is cadang-cadang disease of coconut palms, which poses a serious threat to the coconut industry, particularly in the Philippines. Although information is incomplete, recent results demonstrated the association of a viroid-like, low-molecular-weight RNA with extracts from diseased tissue and its absence in extracts from healthy tissue, suggesting that cadang-cadang disease may be of viroid etiology (Randles, 1975).

Methods developed for the study and purification of viroids have already been adapted as a valuable diagnostic aid for indexing elite or basic potato seed stocks in certification programs (Morris and Wright, 1975).

SUMMARY

The spindle tuber disease of potato, which was previously

believed to be of viral etiology, has been shown to be caused by a free infectious RNA of low molecular weight (about 8×10^4). This agent is the first representative of a newly recognized class of pathogens, the viroids, which have been characterized by the absence of a dormant phase (virions) and by genomes that are much smaller than those of known viruses. In spite of the small amount of genetic information which viroids introduce into their host cells, they are able to replicate and, in some hosts, to produce disease.

The potato spindle tuber viroid (PSTV) has been purified, and some of its physical and chemical properties have been determined. Thermal denaturation properties and nuclease sensitivity indicate that the RNA is a single-stranded molecule with extensive regions of base-pairing. Incubation of ^{125}I-labeled PSTV with T-1 or pancreatic ribonucleases, followed by RNA-fingerprinting demonstrated that purified preparations of PSTV contain a single molecular species. Molecular weight estimates obtained by fingerprinting are compatible with those derived from gel electrophoresis of denatured PSTV.

In leaf strips and in purified nuclei isolated from PSTV-infected tomato, incorporation of ^{3}H-uracil or ^{3}H-UMP into PSTV was shown to be inhibited by pretreatment of the leaf strips and nuclei with actinomycin D. These results demonstrated that PSTV was synthesized in the cell nucleus and that replication may have occurred from a DNA template.

Four additional plant diseases, chrysanthemum stunt, citrus exocortis, cucumber pale fruit, and chrysanthemum chlorotic mottle also are now known to be caused by low-molecular-weight RNAs; *i.e.*, by viroids. RNA fingerprint patterns of PSTV and citrus exocortis viroid differed from one another, indicating that the two RNA's did not have the same primary sequence.

REFERENCES

Adams, D.H. (1970). The nature of the scrapie agent. A review of recent progress. *Pathologie-Biologie 18*, 559-577.

Boedtker, H. (1971). Conformation independent molecular weight determination of RNA by gel electrophoresis. *Biochim. Biophys. Acta 240*, 448-453.

Davies, J.W., Kaesberg, P., and Diener, T.O. (1974). Potato spindle tuber viroid. XII. An investigation of viroid RNA as a messenger for protein synthesis. *Virology 61*, 281-286.

Dickson, E., Prensky, W., and Robertson, H.D. (1975). Comparative studies of two viroids: Analysis of potato spindle tuber and citrus exocortis viroids by RNA fingerprinting and polyacrylamide-gel electrophoresis. *Virology 68*, 309-316.

Diener, T.O. (1971a). A plant virus with properties of a free ribonucleic acid: Potato spindle tuber virus. In *Comparative Virology*, (Maramosch, K., and Kurstak, E., eds.), pp. 433-478. Academic Press, New York.

Diener, T.P. (1971b). Potato spindle tuber virus: A plant virus with properties of a free nucleic acid. III. Subcellular location of PSTV-RNA and the question of whether virions exist in extracts or *in situ*. *Virology 43*, 75-89.

Diener, T.O. (1971c). Potato spindle tuber "virus." IV. A replicating, low molecular weight RNA. *Virology 45*, 411-428.

Diener, T.O. (1972a). Potato spindle tuber viroid. VIII. Correlation of infectivity with a ultraviolet light-absorbing component and thermal denaturation properties of the RNA. *Virology 50*, 606-609.

Diener, T.O. (1972b). Viroids. *Advan. Virus Res. 17*, 295-313.

Diener, T.O. (1973a). Potato spindle tuber viroid: A novel type of pathogen. *Perspect. Virol. 8*, 7-30.

Diener, T.O. (1973b). A method for the purification and reconcentration of nucleic acids eluted or extracted from polyacrylamide gels. *Anal. Biochem. 55*, 317-320.

Diener, T.O., and Lawson, R.H. (1973). Chrysanthemum stunt: A viroid disease. *Virology 51*, 94-101.

Diener, T.O., and Raymer, W.B. (1967). Potato spindle tuber virus: A plant virus with properties of a free nucleic acid. *Science 158*, 378-381.

Diener, T.O., and Raymer, W.B. (1969). Potato spindle tuber virus: A plant virus with properties of a free nucleic acid. II. Characterization and partial purification. *Virology 37*, 351-366.

Diener, T.O., and Raymer, W.B. (1971). Potato spindle tuber "virus." *Descriptions of Plant Viruses. No. 66*. Commonweal. Mycol. Inst. Assoc. Applied Biol., Kew, Surrey, England.

Diener, T.O., and Smith, D.R. (1971). Potato spindle tuber viroid. VI. Monodisperse distribution after electrophoresis in 20 percent polyacrylamide gels. *Virology 46*, 498-499.

Diener, T.O., and Smith, D.R. (1973). Potato spindle tuber viroid. IX. Molecular weight determination by gel electrophoresis of formylated RNA. *Virology 53*, 359-365.

Diener, T.O., and Smith, D.R. (1975). Potato spindle tuber viroid. XIII. Inhibition of replication by actinomycin D. *Virology 63*, 421-427.

Diener, T.O., Schneider, I.R., and Smith, D.R. (1974). Potato spindle tuber viroid. XI. A comparison of the ultraviolet light sensitivities of PSTV, tobacco ringspot virus, and its satellite. *Virology 57*, 577-581.

Diener, T.O., Smith, D.R., and O'Brien, M.J. (1972). Potato spindle tuber viroid. VII. Susceptibility of several solanaceous plant species to infection with low molecular weight RNA. *Virology 48*, 844-846.

Hadidi, A., Jones, D.M., Gillespie, D.H., Wong-Staal, F., and Diener, T.O. (1976). Hybridization of potato spindle tuber viroid to cellular DNA of normal plants. *Proc. Nat. Acad. Sci. U.S. 73*, 2453-2457.

Hall, T.C., Wepprich, R.K., Davies, J.W., Weathers, L.G., and Semancik, J.S. (1974). Functional distinctions between the ribonucleic acids from citrus exocortis viroid and plant viruses: Cell-free translation and aminoacylation reactions. *Virology 61*, 486-492.

Lawson, R.H. (1968). Some properties of chrysanthemum stunt virus. *Phytopathology 58*, 885.

Loening, U.E. (1967). The fractionation of high-molecular-weight ribonucleic acid by polyacrylamide-gel electrophoresis. *Biochem. J. 102*, 251-257.

Miura, K.I., Kimura, I., and Suzuki, N. (1966). Double-stranded ribonucleic acid from rice dwarf virus. *Virology 28*, 571-579.

Morris, T.J., and Wright, N.S. (1975). Detection on polyacrylamide gel of a diagnostic nucleic acid from tissue infected with potato spindle tuber viroid. *Amer. Potato J. 52*, 57-63.

Randles, J.W. (1975). Association of two ribonucleic acid species with cadang-cadang disease of coconut palm. *Phytopathology 65*, 163-167.

Romaine, C.P., and Horst, R.K. (1975). Suggested viroid etiology for chrysanthemum chlorotic mottle disease. *Virology 64*, 86-95.

Sänger, H.L. (1972). An infectious and replicating RNA of low molecular weight: The agent of the exocortis disease of citrus. *Advan. Biosci. 8*, 103-116.

Schneider, I.R. (1969). Satellite-like particle of tobacco ringspot virus that resembles tobacco ringspot virus. *Science 166*, 1627-1629.

Semancik, J.S., Magnuson, D.S., and Weathers, L.G. (1973). Potato spindle tuber disease produced by pathogenic RNA from citrus exocortis disease: Evidence for the identity of the causal agents. *Virology 52*, 292-294.

Semancik, J.S., and Weathers, L.G. (1968). Ecocortis virus of citrus: Association of infectivity with nucleic acid preparations. *Virology 36*, 326-328.

Semancik, J.S., and Weathers, L.G. (1972a). Exocortis disease: Evidence for a new species of "infectious" low molecular weight RNA in plants. *Nature, New Biology 237*, 242-244.

Semancik, J.S., and Weathers, L.G. (1972b). Pathogenic 10S RNA from exocortis disease recovered from tomato bunchy-top plants similar to potato spindle tuber virus infection. *Virology 49*, 622-625.

Singh, R.P., and Bagnall, R.H. (1968). Infectious nucleic acid from host tissues infected with potato spindle tuber virus. *Phytopathology 58*, 696-699.

Singh, R.P., and Clark, M.C. (1973). Similarity of host response to both potato spindle tuber and citrus exocortis viruses. *FAO Plant Protection Bull. 21*, 121-125.

Sogo, J.M., Koller, T.H., and Diener, T.O. (1973). Potato spindle tuber viroid. X. Visualization and size determination by electron microscopy. *Virology 55*, 70-80.

Takahashi, T., and Diener, T.O. (1975). Potato spindle tuber viroid. XIV. Replication in nuclei isolated from infected leaves. *Virology 64*, 106-114.

Van Dorst, H.J.M., and Peters, D. (1974). Some biological observations on pale fruit, a viroid-incited disease of cucumber. *Neth. J. Plant Path. 80*, 85-96.

Zaitlin, M., and Hariharasubramanian, V. (1972). Gel electrophoretic analysis of proteins from plants infected with tobacco mosaic and potato spindle tuber viruses. *Virology 47*, 296-305.

THE CROWN GALL PROBLEM

Milton P. Gordon

Department of Biochemistry
University of Washington
Seattle, Washington

Stephen K. Farrand
Daniela Sciaky
Alice Montoya
Mary-Dell Chilton
Don Merlo
Eugene W. Nester

Department of Microbiology
University of Washington
Seattle, Washington

This paper surveys what we currently know about crown gall tumors. Because the research in this area is making rapid progress, there are areas of controversy. It is gratifying to note that general agreement now prevails on a number of aspects of the problem, and we indicate in this survey the areas in which major questions remain unresolved.

BACKGROUND

Tumors in plants can be induced by genetic imbalance, by viruses, or by bacteria. In all three cases, tumors arise at wound sites, and tumor formation may result from aberration of the normal wound healing response of the plant.

Genetic tumors have been studied most extensively in certain interspecies hybrids of the genus *Nicotiana* (Smith, 1972). Tumors occur not only at sites of injury, but also in locations of mechanical stress. These tumors can also be induced by such di-

ISBN 012-601950-9

verse causes as X-radiation, ultraviolet radiation or even crowding of seedlings. There is no understanding of this process at a molecular level.

Tumors in plants are induced by at least 12 viruses (Black, 1972). The most thoroughly studied is the wound tumor virus, which contains double-stranded RNA. It is not known whether there is a DNA copy of the viral RNA inserted into the host cell genome as is the case for RNA tumor viruses of animals.

Crown gall tumors are incited by a bacterium. Smith and Townsend (1907) isolated the causative agent, which is now termed *Agrobacterium tumefaciens*. Two related diseases, cane gall and hairy root disease, are caused by other *Agrobacterium* species (Lippincott and Lippincott, 1975). Crown gall afflicts gymnosperms and dicotyledonous angiosperms. The causative agent, *Agrobacterium tumefaciens*, is a gram negative soil bacterium, rod-shaped, with one to six flagella. Requirements for tumor induction are a fresh wound site and living bacteria. A number of reports have appeared over the years claiming that various subcellular components of virulent bacteria can induce tumors, *e.g.* bacterial DNA, bacterial RNA, and even an *Agrobacterium* phage called PS8. These claims are not widely accepted because it has proven difficult to reproduce the results.

After the inducing bacteria have been in direct contact with the plant cells at the wound site for several days, viable bacteria are no longer necessary for tumor formation. The plant cells are transformed and escape from normal control mechanisms. The arrangement of the transformed cells becomes highly irregular. It has been postulated by Braun that during the process of tumor induction, some sort of material, a tumor inducing principle (TIP), passes from the bacteria to the wounded plant tissue (Braun, 1947). This idea of a TIP has been central to many investigations of crown gall, and the question of its identity remains unsolved.

The gross morphology of the crown gall tumor is dependent upon the inducing strain of *A. tumefaciens*. Figure 1 shows a completely unorganized tumor produced by *A. tumefaciens* B6-806 in *Nicotiana tabacum* L. cv. Xanthi-nc. The tissue appears to be completely disorganized on a macroscopic basis. Microscopically, the tissue shows some vestiges of vascular elements. In contrast, the teratomatous tissue induced by *A. tumefaciens* T37 consists of highly irregular stem and leaf-life structures. Teratomata are said to be only "partially transformed," while unorganized tumors like that in Fig. 1 are "completely transformed." An explanation of this phenomenon at the molecular level remains a challenging task for the future.

In the second decade of this century, Jensen addressed the question of whether crown gall can be transmitted from axenic transformed plant tissue to normal tissue (Jensen, 1910; 1918). The results were negative and ultimately focussed attention on crown gall as consisting of a cellular disorder analogous to neoplastic transformation in animals; it was not just one of many

Figure 1. Left: A teratoma produced on N. tabacum *L. cv. Xanthi-nc by* A. tumefaciens *T37. Right: A completely disorganized crown gall tumor produced on the same host by* A. tumefaciens *B6-806.*

infectious or parasitic processes that give rise to tissue proliferation in plants. Recently, however, there have been reports that crown gall can be transferred from bacterial-free tumor tissue to healthy tissue under certain circumstances; hence, even today this question is not completely settled (Meins, 1973).

CHARACTERISTICS OF TRANSFORMED TISSUES

Growth Requirements

The growth requirements of crown gall tissues on a defined medium are very simple: several vitamins, inorganic salts, and sucrose or some other carbon source. The most important characteristic that distinguishes tumor cells from normal cells is the lack of need for exogenous sources of either cytokinin or an auxin. It is important to point out that there exists a series of normal plant tissues that exhibit a graduated series of growth requirements for cytokinins and auxins. By growing normal tobacco tissue on media containing high levels of indoleacetic acid, Syono and

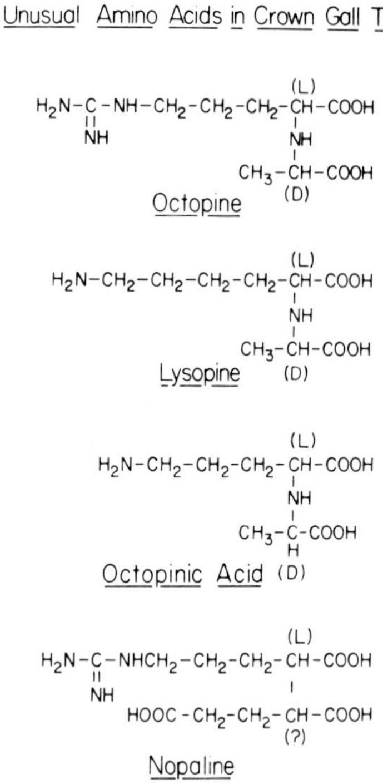

Unusual Amino Acids in Crown Gall Tumors

```
                              (L)
H2N-C-NH-CH2-CH2-CH2-CH-COOH
    ||                         |
    NH                         NH
                               |
                        CH3-CH-COOH
              Octopine       (D)

                              (L)
  H2N-CH2-CH2-CH2-CH2-CH-COOH
                               |
                               NH
                               |
                        CH3-CH-COOH
              Lysopine       (D)

                              (L)
      H2N-CH2-CH2-CH2-CH-COOH
                               |
                               NH
                               |
                        CH3-C-COOH
                               H
         Octopinic Acid      (D)

                              (L)
H2N-C-NHCH2-CH2-CH2-CH-COOH
    ||                         |
    NH
         HOOC-CH2-CH2-CH-COOH
                              (?)
              Nopaline
```

Figure 2. Unusual amino acids found in crown gall tumors.

Furuya (1974) were able to obtain tissues that no longer required auxins. Linsmaier-Bednar found that treatment of normal tissues in culture with potent animal carcinogens, substituted aminofluorenes, resulted in the formation of tissues that no longer required a cytokinin (Bednar and Linsmaier-Bednar, 1971). When normal tissue is kept in culture for a number of passages, it is sometimes possible to isolate from it tissue that will grow in the absence of auxin and cytokinin. Such tissue is said to be "habituated," and there are reports that grafts of habituated tissues also produce overgrowths upon grafting into normal plants (Braun and Morel, 1950). Furthermore, one should not lose sight of the fact that intact plants make quantities of auxins and cytokinins that are sufficient for their growth. Thus, although independence from exogenous auxins and cytokinins is a characteristic of crown gall tumors, it is not an exclusive property.

	Production by Tumor	Not Present in Tumor
Utilized by the Bacteria	Most Strains	EU6 AT 181 (Many tissue culture lines lose Octopine trait)
Not Utilized by the Bacteria	A203 NOP⁻	542 AT 1 AT4

Figure 3. Correlation between utilization of octopine and nopaline by various strains of A. tumefaciens *and production of these compounds by the resulting tumors. Most strains induce tumors which produce the same abnormal amino acids as are utilized by the inducing organism. Strains EU6 and AT181 utilize nopaline but the tumors do not contain this amino acid. Strain A203 nop⁻ cannot utilize nopaline, but induces a nopaline producing tumor.*

Unusual Components

A second characteristic of crown gall tumors is the production of a number of unusual amino acids not found in normal tissues (Fig. 2). A crown gall tissue that contains unusual amino acids will contain either octopine or nopaline, but not both. These unusual amino acids are derived from basic amino acids by condensation with the appropriate α-keto acid presumably derived from products of the tricarboxylic acid cycle. Again, tumor formation and the response of the plant to wounding are interwoven. We have shown that the levels of free arginine, lysine, and histidine, common basic amino acids, are elevated about 1000 fold upon wounding leaves of *Kalanchoe daigremontiana*. Apparently, in some crown gall tissues, a new metabolic pathway is set in operation whereby the basic amino acids are modified by condensation with α-keto acids. There is now general agreement that these unusual amino acids are characteristic only of some transformed cells and do not occur in normal tissues. (The previous observation from this laboratory of the presence of octopine in normal tissues cannot be repeated (R. Johnson *et al.*, 1974).)

There is an interesting correlation between the ability of a virulent strain of *A. tumefaciens* to utilize octopine or nopaline and the production of these amino acids by the tumor. This correlation holds true for a large number of bacterial strains; there are only a few exceptions. Morel first noted this correlation, and he suggested that the production of octopine or nopaline by the tumor tissue was due to the transfer of these traits by bacterial DNA (Morel, 1970). The exceptions to the correlation (Fig. 3) are difficult to reconcile with this model. *A. tumefaciens* strains EU6 and AT181 utilize nopaline, but the tumors incited by these strains do not produce this amino acid. Strain A 203 nop is a mutant that cannot utilize nopaline; nevertheless, this strain induces a tumor that produces nopaline.

A detailed investigation of the enzymes involved in octopine and nopaline metabolism is warranted. At the present time our knowledge of enzymology of these processes is rudimentary. The bacterial and tumor enzymes appear to be grossly different. The bacterial enzyme reactions are irreversible and degrade octopine and nopaline. The enzymes are membrane bound and linked to a cytochrome-like component. The plant tumor enzyme is readily reversible, utilizes NADH, and, at least *in vivo*, seems to operate in a synthetic capacity (Lippincott and Lippincott, 1975; Bomhoff, 1974). It is certainly not unusual for anabolic and catabolic pathways to be different.

Presence of Foreign DNA

It has been suggested that the fundamental characteristic that distinguishes crown gall tissue from normal tissue is the presence of foreign DNA in the transformed cells. Several years ago, a number of papers appeared in which bacterial DNA and DNA from the *Agrobacterium* phage, PS8, were reported in crown gall tissues (Lippincott and Lippincott, 1975). The presence of foreign DNA in plant tumors is a most attractive idea. The persistence of the various tumorous traits could be explained by the presence of replicating macromolecules of bacterial origin. We have tested for the presence of foreign DNA in tumor tissue by means of DNA reassociation kinetic measurements. This procedure measures the rate of renaturation in solution of a low concentration of a highly labeled denatured "probe" bacterial DNA in the presence of "driver" (tumor or normal) plant DNA in high concentration. The presence of homologous bacterial DNA sequences in the plant tumor DNA would increase the rate of renaturation of the probe bacterial DNA. A limitation of this technique is that if the inserted DNA was a single copy of a small fraction of the bacterial genome then it would be impossible to detect its presence experimentally. We have looked for bacterial and phage PS8 DNA in tumors induced by a number of different strains of *A. tumefaciens* in *Nicotiana tabacum* L. cv. Xanthi-nc. In all of these experiments we did not detect any foreign DNA. Reconstitution experiments, wherein known amounts of phage or bacterial DNA were deliberately added to reaction mixtures, showed that if there had been one complete bacterial or PS8 genome per three tumor cells, we would have found it. Thus, there is little if any bacterial or phage DNA present in the DNA from crown gall cells. There appears to be general agreement on this point (Chilton *et al.*, 1974, 1975; Eden *et al.*, 1974; Farrand *et al.*, 1975; Drlica and Kado, 1975; Dons, 1975; Merlo and Kemp, 1976). It should be emphasized that bacterial plasmid DNA constitutes too small a fraction of total bacterial DNA to have been detected in these experiments.

Reversibility of Plant Tissue Transformation

A most dramatic description of the reversibility of a crown gall tumor was recently published by Braun and Wood (1976). Cloned tissues of a tobacco teratoma were grafted onto cut shoots of a different variety of tobacco. In many instances, the morphology and function of the grafts were normal, although when returned to tissue culture, the normal appearing tissue reverted to teratomatous morphology and grew without hormones. The tissues, although normal in appearance, can be thought of as organized tumors. It is most interesting that the seeds produced by normal appearing tissue were normal and gave rise to normal plants. Thus, the transformation induced by the teratoma-evoking strain of *A. tumefaciens* T37, is reversible. The mechanism of this reversal remains a problem for the future.

BACTERIAL PLASMIDS

Background

Recently one of the landmark advances in the area of the induction of crown gall tumors was made in the laboratory of Dr. J. Schell. He and his coworkers found that the presence of a large plasmid in *A. tumefaciens* was associated with virulence (Zaenen *et al.*, 1974; Van Larebeke *et al.*, 1974). Eleven virulent strains of *A. tumefaciens* contained large plasmids whereas eight avirulent strains did not. A bacterial plasmid is a covalently closed circular piece of bacterial DNA which replicates independently of the bacterial chromosomal DNA. In bacteria, plasmids often carry non-essential traits such as resistance to various antibiotics, resistance to heavy metals, and metabolism of unusual compounds such as hydrocarbons, camphor, and other properties. Some plasmids are readily transferred between strains of the same species, and in some cases even between unrelated species. The virulence associated plasmid in *A. tumefaciens* has been found to range in size from 98 to 156×10^6 daltons. In our laboratories and in the laboratories of Drs. Schell and Schilperoort, studies have been made of an interesting strain of *A. tumefaciens*, C58, that had

TABLE I
Plasmid homologies

Strain	Degree of homology (percentage of hybridization) C58	A6
A. tumefaciens		
C58	100	7
27	84	12
223	58	13
0362	94	10
IIBV7	28	27
A_6	8	100
15955	9	88
CG1C	4	80
B6-806	6	83
B_2A	12	100
A. rhizogenes		
TR7	16	7
11325	63	11
A. rubi		
13335	84	6
A. radiobacter		
84	28	5
6467	17	18

been found by Hamilton and Fall (1971) to lose virulence upon incubation at 37°. It was found that the loss of virulence of C58 is associated with the loss of the covalently closed circular form of the plasmid (Van Larebeke *et al.*, 1974; Watson *et al.*, 1975). Hybridization reactions showed that the plasmid DNA base sequences are absent in the resulting non-virulent organisms; hence the plasmid sequences have not been integrated into the bacterial chromosome (Watson *et al.*, 1975).

Relatedness of Plasmids

If large plasmids are associated with virulence, one might expect that large plasmids present in various virulent strains of *A. tumefaciens* would be closely related. Plasmids from a number of strains of *A. tumefaciens*, *A. rubi*, *A. Rhizogenes*, and *A. radiobacter* were isolated for measurement of the relatedness of virulence plasmids. These strains were obtained from laboratories situated on different continents. The plasmids from *A. tumefaciens* strains A6 and C58 were made highly radioactive by a process called "nick translation" (Maniatis *et al.*, 1975). These labeled plasmids were then used as probes in DNA-driven reactions with high concentrations of unlabeled plasmids from a large number of strains to determine the degree of relatedness of the sequences (Currier and Nester, 1976).

Two major conclusions can be drawn from these data (Table I). First, it is rather surprising that a number of the virulence associated plasmids possess only a small fraction of common sequences. The two labeled probes, C58 and A6, for example, have a maximum of about 8% of their sequences in common. Second, most, but not all, of the *A. tumefaciens* plasmids fall into two classes, a class showing a high degree of relatedness to A6 (80-100%) and a class showing a high degree of relatedness to C58 (58-94%). The first group of plasmids carries genetic information for utilization of octopine (Chilton *et al.*, 1976), while the second group carries information for the utilization of nopaline (Van Larebeke *et al.*, 1974; Watson *et al.*, 1975). Plasmids of the second group are also related to large plasmids present in various strains of *Agrobacterium radiobacter* (D. Merlo, unpublished data). These latter organisms are avirulent isolates which are closely related to *A. tumefaciens* and often found in tumors in association with virulent organisms. Strains that are related are not necessarily found in the same geographic area. The data in Table I also indicate that a third group of plasmids exists that are not related by nucleotide sequence homology to the two main classes or to each other.

The close relationship among octopine-type plasmids is borne out by analysis of restriction endonuclease digest fingerprints (Fig. 4). Nopaline-type plasmids, in contrast, exhibit diverse fingerprints (Fig. 5).

A Possible "Virulence Gene"

The small amount of DNA that the virulence plasmids from A6 and C58 have in common is of interest, for this material could be closely associated with the ability of these strains to induce tumors. Accordingly, DNA enriched for sequences common to A6 and C58 plasmids was prepared by annealing labeled, sheared A6 plasmid to nonlabeled C58 plasmid which had been fixed to nitrocellulose filters. Only A6 sequences that formed high melting duplexes with the C58 plasmid DNA were retained by the filters. These sequences were subsequently recovered by denaturation. The labeled A6 plasmid DNA enriched for common sequences was then tested for homology with plasmid DNA of other strains. In each case these enriched sequences showed at least 2 to 5 times greater homology with plasmid DNA from other strains (Table 2), than did the original labeled A6 plasmid DNA. This experiment shows that sequences which are common to the dis-

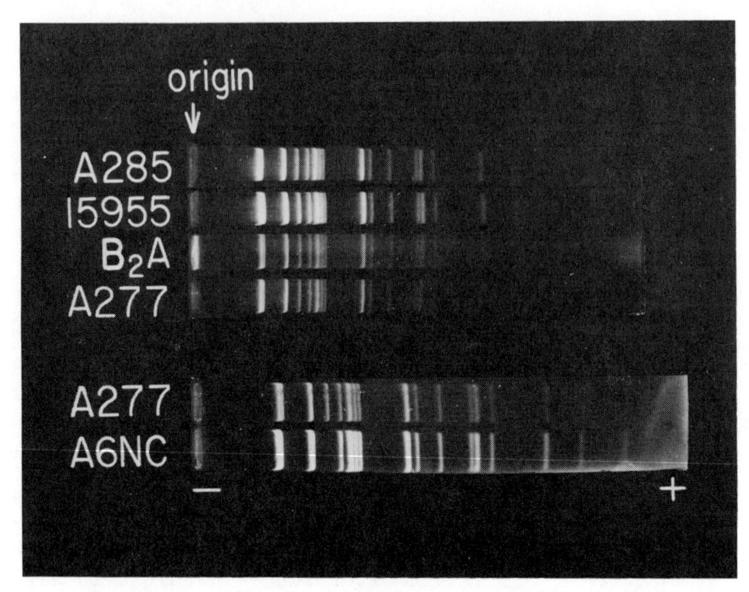

Figure 4. DNA fingerprints of agarose gels of SMI I digests of plasmids isolated from octopine-utilizing A. tumefaciens *strains. By a genetic cross* in planta *(Watson* et al., *1975), the virulence plasmids of donor strains B6-806 and B6-V87 (L. Moore), were transferred to plasmidless recipient strain A136 to yield virulent octopine-utilizing exconjugant strains A277 and A285, respectively. Plasmid DNA was isolated as described elsewhere (Currier and Nester, 1976b), from strains A6NC, A277, B2A, 15955 and A285 (Currier and Nester, 1976a, give the origins of the strains).*

Plasmid DNA (2 μg) in 50 μl of 20 mM Tris, pH 9.0, 15 mM KCl, 6 mM $MgCl_2$ and enough Sma I endonuclease to give a complete digest was incubated at 30°C for 30 min., heated to 67°C for 10 min. and chilled. After addition of 5 μl 0.07% bromphenol blue, 7% sodium dodecyl sulfate, 33% glycerol, the sample was heated to 67°C and 20 μl of 0.7% agarose in electrophoresis buffer (0.08 M Tris, 0.04 M sodium acetate, 4 mM EDTA, pH 8.0), was added. Digests were loaded into wells (1.2 mm × 1 mm × 7 mm) of a 0.7% agarose horizontal slab gel (19.3 cm × 21.6 cm × 0.7 cm) and electrophoresed at 60 volts for 15 minutes, then at 30 volts for 18-22 hrs. The gel was soaked in 1 liter of 10 μg/ml ethidium bromide for 30 min., and the bands were visualized by a Blak-Ray Transilluminator (Ultraviolet Products, Inc.). The photograph was made with Kodak Panatonic X film through orange and UV filters.

The four digests in the upper part of the figure are from a single gel, while those in the lower part are from a second gel. Note that the A277 digest is repeated as a standard. The only difference found in these patterns is that two bands are missing in A6NC. The resolution of the highest molecular weight doublet is also diminished in A6NC compared to the other strains.

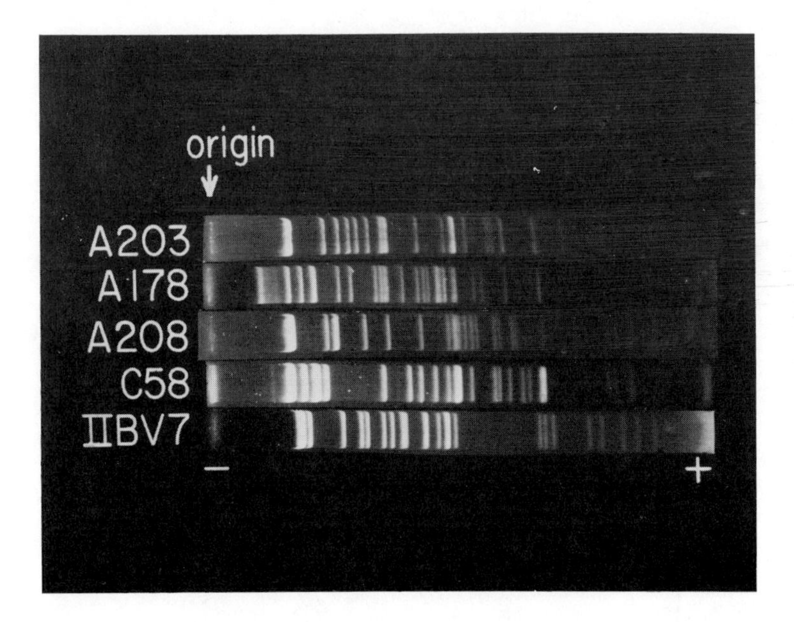

Figure 5. DNA fingerprints of agarose gels of Sma I digests of plasmids isolated from nopaline-utilizing A. tumefaciens *strains. Plasmid DNA from* A. tumefaciens *strains C58, A208, A178, A203, and 11BV7 were digested, electrophoresed, and visualized as in Fig. 4. The faint bands seen closest to the origin in C58 and A178 digests have been shown by subsequent experiments to be partial digests. Strains A208, A178, and A203 were derived from crosses* in planta *in which the donor strain was T37, K27, and 223, respectively. Strain T37 was obtained from J. A. Lippincott. Origins of all other strains are reported elsewhere (Currier and Nester, 1976).*

tantly related plasmids of strains C58 and A6 have a high degree of homology with virulence associated plasmids from a number of other strains.

There are many possibilities opened up by these experiments. If these sequences contain the information necessary for virulence, they are logical sequences for the "tumor inducing principle" and it is possible that at least part of these sequences may be present in the tumor. In addition, when coupled with information necessary for self-replication, the sequences common to C58 and A6 plasmids may produce a small virulence inducing plasmid.

TABLE 2

Strain	Plasmid homology with (percentage hybridization) A6 plasmid	Percent reaction with labeled A6 plasmid detected with C58 plasmid
A. tumefaciens		
A6	100	100
C58	13	54
181	12	54
AT1	19	36
27	18	32
0362	15	31
223	12	29
11325 Fhiz	10	29
TR7 Rhiz	6	28
11BV7	28	87
A. rubi		
13335	6	54
A. rhizogenes		
TR7	10	29
11325	6	28

Further Considerations

A number of broad questions are raised by these studies with plasmids. First, why is plasmid borne information necessary for tumor induction? Can this information be carried and expressed when present on the bacterial chromosome? Second, why are tumor inducing plasmids so large, even though virulence plasmids from different strains have only about 10% of their sequences in common? Virulence plasmids could code for as many as 100 to 150 proteins. Are all of these genes required for virulence? Will small tumor-associated plasmids be found? What is the contribution of the chromosomal DNA to virulence? Will organisms other than *A. tumefaciens* which carry these plasmids be able to induce crown gall tumors?

Transfer of Plasmids

In view of the close association of large plasmids with virulence, it is of interest to develop techniques that will determine what information is carried by the plasmid DNA. The simplest way to ascertain what traits are carried by a bacterial plasmid is to cure the strain of the plasmid. This can readily be done with the thermosensitive plasmid in C58 (Van Larebeke, 1974; Watson *et al.*, 1975). Plasmids cannot be eliminated readily from most other strains, hence other procedures must be used to determine plasmid borne information. One very useful way to determine plasmid borne information is to move the plasmid into another cell and ascertain what new properties are expressed by the recipient cell. Actually, Kerr had developed a method for the transfer of virulence before the association of virulence with large plasmids was known (Kerr, 1969). His procedure consists of incubating a virulent and a nonvirulent strain of *A. tumefaciens* together in a plant tumor for several weeks, after which time it is possible to isolate from the tumor colonies of the avirulent strain which have now acquired virulence. The details of the conjugation that occurs in the tumor are not known; however, the recipient bacteria acquire both virulence and a virulence plasmid. This procedure is useful, but it has some inherent difficulties: (a) several weeks of incubation time are required; (b) during this long period of time, it is difficult to ensure that no contamination has occurred; (c) confinement of potentially harmful, newly created strains is difficult.

In our laboratory a procedure has been developed whereby the virulence plasmid can be transferred *in vitro* in a short period of time (Chilton *et al.*, 1976). The basis of this procedure is the observation that RP4, a *Pseudomonas* plasmid which carries several drug resistance traits, will often facilitate conjugational transfer of DNA (Stanisich and Holloway, 1971). In the present experiments, RP4 was transferred from *E. coli* into a

TABLE 3
Correlation of Octopine or Nopaline Utilization and Production of Compounds by Tumors

Strain	Utilizes	Incites tumor which produces
A. tumefaciens		
C58	Nopaline	Nopaline
C58 exconjugant*	Nopaline	Nopaline
15955	Octopine	Octopine
15955 exconjugant*	Octopine	Octopine
542	Neither	Neither
542 exconjugant*	Neither	Neither
K27 and 223	Octopine + Nopaline	Nopaline only
K27 and 223 conjugants*	Nopaline	Nopaline
EU6	Nopaline	Neither
EU6 exconjugant*	Nopaline	Neither

*The recipient in all of these crosses was A136, a rifamycin resistant, nalidixic acid derivative of heat-cured C58.

strain of *A. tumefaciens* which contained a virulence plasmid. The virulence plasmid in the latter was mobilized by the presence of RP4, and upon conjugation with an appropriately marked strain of *A. tumefaciens* both the RP4 and the virulence plasmid were transferred.

Information Carried by Plasmids

In spite of their size, the large plasmids associated with virulence have thus far been shown to transfer only a limited number of traits. Whenever the donor strain was virulent, the exconjugant was found to have acquired virulence. Sometimes the virulence appears to be modified and is either decreased in severity on certain hosts or greatly increased in severity as compared to the virulence of the donor.

The sensitivity to agrocin 84, a bacteriocin produced by strain K84, is carried by the thermosensitive plasmid in strain C58. When C58 is cured of its plasmid, the resulting nonvirulent organisms are no longer killed by agrocin 84. When these organisms re-acquire the C58 plasmid, the sensitivity to agrocin 84 reappears (Watson *et al.*, 1975; Van Larebeke *et al.*, 1975). Certain virulence plasmids also confer upon their host the ability to exclude the *Agrobacterium* phage AP1 (Van Larebeke *et al.*, 1975).

The virulence plasmid carries the genes for nopaline utilization in all cases, and for octopine utilization in biotype 1 strains (Table 3). In addition, it determines whether the bacterial strain will incite an octopine tumor, a nopaline tumor, or a tumor which produces neither compound. Evidence that the nopaline production and nopaline utilization traits are distinct is provided by strain EU6. This strain, a nopaline utilizer, incites tumors which produce no nopaline. The nopaline utilization trait is clearly plasmid borne, since it is transferred to the exconjugant. The lack of nopaline production by tumors incited by EU6 and its exconjugant shows that production is genetically distinct from utilization.

Is the Plasmid DNA or Part of It in the Transformed Plant Tissues?

In view of the fact that one characteristic of plasmids is their ready transfer from one bacterium to another, the question can be raised as to whether the entire plasmid or at least part of the plasmid is transferred to the recipient plant cell in the course of tumor induction. The production of octopine or nopaline by the tumor could thus be due to plasmid-mediated transfer of information (Table 3). In other words, is the postulated "tumor inducing principle" the plasmid DNA or part of the plasmid DNA? This question, of course, is a more modern query of Morel's hypothesis. The previous hybridization reactions which did not detect bacterial or phage DNA in tumors would not have detected

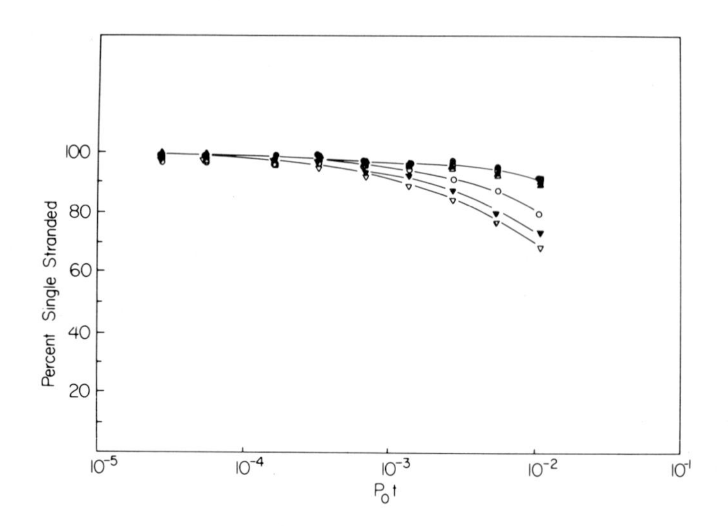

Figure 6. Kinetics of reassociation of ^3H labeled plasmid DNA in the presence of plant tumor or control DNAs. Plasmid DNA, isolated as described elsewhere (Currier and Nester, 1976b), from A. tumefaciens *strain B6-806, was labeled by nick translation to a specific activity of 2×10^7 cpm/μg, using ^3H-deoxythymidine triphosphate (New England Nuclear, 40-60 Ci/mM), as the labeled precursor. This labeled probe DNA (10 ng/ml) was allowed to reassociate in 20 mM $NaClO_4$, 0.015 mM NaH_2PO_4 in the presence of sheared, unlabeled driver DNAs as follows: ● Salmon DNA (3.14 mg/ml), Normal tobacco Callus DNA (3.23 mg/ml, ▲ B6-806 Tumor Clone E8 DNA (2.87 mg/ml), △ B6-806 Tumor Clone E9 DNA (3.23 mg/ml) ▼ Mixture of 2.55 mg/ml E8 Tumor DNA + 42 ng/ml B6-806 plasmid DNA, ▽ Mixture of 3.55 mg/ml E9 Tumor DNA + 58 ng/ml B6-806 plasmid DNA, ○ Mixture of 3.14 mg/ml salmon DNA + 19 ng/ml B6-806 plasmid DNA.*

The three mixtures of authentic plasmid DNA with tumor or salmon DNA are constructed to simulate the model of 1 copy of the plasmid per diploid tumor cell (tumor mixtures), or 1 copy per three diploid tumor cells (salmon mixture). The data rule out the presence of at least one copy of the B6-806 virulence plasmid per diploid tumor cell. (See text).

the presence of plasmid DNA in tumors, because the plasmid DNA constitutes such a small proportion, 3%, of the total labeled bacterial DNA used as probe.

In studies of this type two considerations are of primary importance. First, it is important to use cloned tissues. It has been suggested that fewer than 10 habituated or tumor cells could provide 100 normal cells with sufficient auxin and cytokinin for growth (Dr. F. Meins, Jr., personal communication). If a similar situation prevailed in uncloned tumor tissue, the result would be that any foreign DNA would be greatly diluted out. We used cloned tissue of tumors induced by *A. tumefaciens* B6-806. All of the tissues were octopine producers. A second consideration which is of importance is the use of a highly labeled probe DNA which would permit the detection of sufficiently small quantities of foreign plasmid DNA in tumor.

The kinetics of reassociation of bacterial plasmid DNA driven by plant tumor or control DNAs are shown in Fig. 6. Labeled bacterial plasmid probe DNA reassociates at the same rate in the presence of "driver" concentrations of plant tumor DNA and control DNAs. If a *complete* complement of plasmid of *A. tumefaciens* B6-806 had been present at the level of one copy per 2 or 3 diploid tumor cells, the labeled bacterial "probe" DNA would have reassociated faster, as demonstrated by the reconstruction experiments. As an additional control, ^{32}P-labeled PS8 bacteriophage DNA, a "probe" not expected to be in tumor cells, was included in the reassociation mixtures; these experiments confirmed that plant tumor DNA had no nonspecific influence on DNA reassociation rates (data not shown).

Subsequent to the performance of these experiments, agarose gel patterns of restriction endonuclease digests of the plasmid DNA of B6-806 and an exconjugant derived from it indicated the presence of two plasmids in B6-806, only one of which was necessary for virulence (Fig. 7). This means that the probe DNA used in the hybridization experiment in Fig. 6 was a mixture of two plasmids, each approximately 125×10^6 daltons; accordingly, the sensitivity of the experiment for detection of the *virulence* plasmid is diminished slightly. However, one complete copy of a plasmid DNA per diploid cloned tumor cell should have been detected.

There are several further approaches to the detection of plasmid DNA in plant tissue utilizing hybridization techniques. First, since the virulence associated plasmids are so large, about 1×10^8 daltons, even a small fraction of the virulence associated plasmid could code for a number of proteins. In our laboratory, we are currently treating the virulence associated plasmid DNA derived from B6-806 with a restriction endonuclease, separating the DNA segments by molecular weight on agarose gels, and looking for the presence of each individual segment in the tumor DNA. Cleaving the plasmid DNA into a number of DNA segments should increase the sensitivity of detection of foreign DNA ten to twenty fold. This increased sensitivity should be sufficient to detect an amount of DNA that could code for one protein.

A second approach that we are currently employing is the attempt to detect in cloned tumors the DNA sequences that are common to the distantly related plasmids present in *A. tumefaciens* strains A6 and C58 (Table 2). A further experimental possibility would be to clone the tumor producing portions of the DNA of the virulence plasmid and look for the sequences in DNA from transformed plant cells.

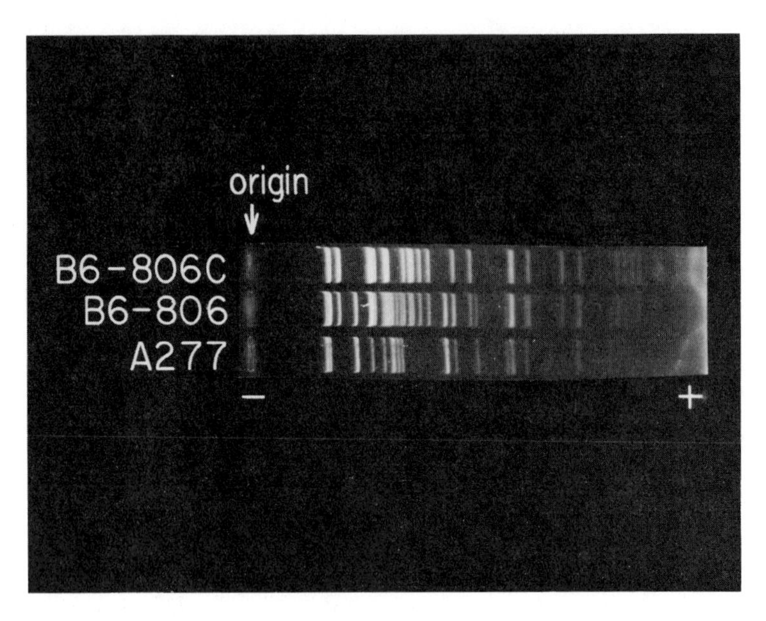

Figure 7. DNA fingerprints of agarose gels of Sma digests of plasmids from A. tumefaciens *strains B6-806, B6-806C, and A277. Virulent, octopine-utilizing strain B6-806 was mutagenized with N-methyl-N'-nitro-N-nitrosoguanidine and, in subsequent screening for mutants unable to utilize octopine, B6-806C, an avirulent derivative was isolated. The origin of strain A277, a virulent exconjugant derived from B6-806, is described in the legend of Fig. 4. Plasmids were isolated and digest prepared, and electrophoresed as in Fig. 4.*

Note that the digest of B6-806 contains the sum of the bands found in the two derivative strains, A277 and B6-806C, whereas the latter two have strikingly different fingerprints. Previous studies found plasmid molecules of only one contour length in strain B6-806 (Currier and Nester, 1976a), consistent with molecular weight 125 × 10^6. These results indicate that B6-806 contains two large plasmids of similar size, one determining octopine utilization and virulence, and the other cryptic.

POSSIBLE NATURE OF CROWN GALL

Alteration of Plant Cells by Incorporation of Foreign DNA

The disease may be associated with insertion of foreign DNA. In spite of the reversibility of crown gall mentioned above, there is still the possibility that foreign DNA is present. There are examples of abortively transformed animal cells which revert to normal phenotype but which still contain SV40 viral DNA (Smith *et al.*, 1972). It must be emphasized that although the presence of foreign DNA in a tumor tissue opens up many approaches, the question still remains as to why the tissue is transformed and becomes neoplastic. The foreign gene hypothesis is most attractive because of its analogy to virally induced tumors in animals and because it affords so many experimental approaches. One would hope that nature has not conspired to design a process whereby the foreign DNA is only temporarily present but is subsequently lost when the transformed tissue is placed in axenic culture. The loss of octopine production by some tumors in our laboratory could be interpreted as the loss of some putative exogenous DNA, but this entire phenomenon is deserving of a more systematic study.

Undetected Infectious Agent

In spite of intensive work that has failed to find it, there could still be some sort of infectious entity such as an RNA virus or a viroid involved in crown gall. The recent report of the transmissibility of crown gall to normal tissue is consistent with this idea (Meins, 1973). The experiments of Braun and Wood which demonstrated the reversal of the crown gall transformation are reminiscent of the classical procedures for freeing plant tissues of viruses.

Loss of Transcriptional or Translational Control

A third possible explanation for the transformation observed in crown gall is the loss of transcriptional or translational control due to the loss of some control elements or replication of viroid-like material that damages control mechanisms. This is an example of an epigenetic change and could also be involved even if crown gall is associated with a genetic change.

SUMMARY

1. The formation of crown gall, as well as other known plant tumors, is associated with the response of a plant to wounding.
2. The ability of a strain of *A. tumefaciens* to induce tumors at a wound site is associated with the presence of a large bacterial plasmid in the strain.
3. The tumor inducing activity, octopine/nopaline utilization by the bacterium, octopine/nopaline production by the tumor, and several other traits of a strain of *A. tumefaciens* are coded for by the large plasmid the strain carries. The octopine/nopaline utilization and octopine/nopaline production by the tumor plant cells are genetically distinct traits.
4. The virulence associated plasmid is not present in transformed tissue at a level of one whole copy per diploid tumor cell. Whether there is less than one whole plasmid per tumor cell or a specific small fragment of the plasmid present per tumor cell remains to be seen.
5. Plasmids in various strains of *A. tumefaciens* are often distinctly related as determined by DNA homology, but many of them can be grouped into two large families of related members.
6. Studies of the DNA fingerprint agarose gel patterns of restriction endonuclease digests of plasmids from B6-806, an avirulent derivative B6-806C, and a virulent exconjugant A277, demonstrate that B6-806 harbors two large plasmids of similar size, only one of which is essential to virulence.
7. Large and small plasmids commonly occur in avirulent *Agrobacterium* isolates. Some of these carry the nopaline utilization trait and exhibit homology with the virulence plasmid of strain C58.

ACKNOWLEDGMENTS

This work was supported by funds from the National Cancer Institute Grant No. CA 13015-05, and funds from the American Cancer Society Grant No. NP 194.

Don Merlo was supported by Fellowship No. 1-F32-CA0 5222-01 from the National Cancer Institute.

Our thanks to Dr. George Melchers and Dr. M. D. Sacristan-Alaily of the Max-Planck-Institut für Biologie, Tübingen, West Germany, for clones of B6-806 tumors.

REFERENCES

Bednar, T. W., and Linsmaier-Bednar, E. M. (1971). Induction of cytokinin-independent tobacco tissues by substituted fluorenes. *Proc. Nat. Acad. Sci. U.S. 68*, 1178-1179.

Black, L. M. (1972). Plant tumors of viral origin. In *Progress in Experimental Tumor Research, 15*, (Braun, A. C., and Karger, S., eds.), pp. 110-137. Lippincott, Philadelphia.
Bomhoff, G. H. (1974). Thesis, University of Leyden, Netherlands.
Braun, A. C. (1947). Thermal studies on the factors responsible for tumor initiation in crown gall. *Amer. J. Bot. 34*, 234-240.
Braun, A. C., and Morel, G. (1950). A comparison of normal, habituated, and crown gall tumor tissue implants in the european grape. *Amer. J. Bot. 37*, 499-506.
Braun, A. C., and Wood, H. N. (1976). Suppression of the neoplastic state with the acquisition of specialized functions in cells, tissues, and organs of crown gall teratomas of tobacco. *Proc. Nat. Acad. Sci. U.S. 73*, 496-500.
Chilton, M. D., Currier, T. C., Farrand, S. K., Bendich, A. J., Gordon, M. P., and Nester, E. W. (1974). *Agrobacterium tumefaciens* DNA and PS8 bacteriophage DNA not detected in crown gall tumors. *Proc. Nat. Acad. Sci. U.S. 71*, 3672-3676.
Chilton, M. D., Farrand, S. K., Eden, F., Currier, T., Bendich, A. I., Gordon, M. P., and Nester, E. W. (1975). In *Modifications of the Information Content of Plant Cells*. (Markham, R. M., ed.) p. 297, North-Holland Publishing Co., Amsterdam.
Chilton, M. D., Farrand, S. K., Levin, R., and Nester, E. W. (1976). RP4 promotion of transfer of a large *Agrobacterium* plasmid which confers virulence. *Genetics, 83*, 609-618.
Currier, T. C., and Nester, E. W. (1976a). Evidence for diverse types of large plasmids in tumor-inducing strains of *Agrobacterium*. *J. Bacteriol. 126*, 157-165.
Currier, T. C., and Nester, E. W. (1976b). Isolation of covalently closed circular DNA of high molecular weight from bacteria. *Anal. Biochem. 76*, 431-441.
Dons, J. J. M. (1975). Crown gall, a plant tumor. Thesis, University of Leiden, Netherlands.
Drlica, K. A., and Kado, C. I. (1974). Quantitative estimation of *Agrobacterium tumefaciens* DNA in crown gall tumor cells. *Proc. Nat. Acad. Sci. U.S. 71*, 3677-3681.
Eden, F. C., Farrand, S. K., Powell, J. S., Bendich, A. J., Chilton, M. D., Nester, E. W., and Gordon, M. P. (1974). Attempts to detect deoxyribonucleic acid from *Agrobacterium tumefaciens* and bacteriophage PS8 in crown gall tumors by complementary ribonucleic acid/deoxyribonucleic acid-filter hybridization. *J. Bacteriol. 119*, 547-553.
Farrand, S. K., Eden, F. C., and Chilton, M. D. (1975). Attempts to detect *Agrobacterium tumefaciens* and bacteriophage PS8 DNA in crown gall tumors by DNA·DNA filter hybridization. *Biochem. Biophys. Acta. 390*, 264-275.
Hamilton, R. H., and Fall, M. Z. (1971). The loss of tumor-initiating ability in *Agrobacterium tumefaciens* by incubation at high temperature. *Experentia 27*, 229-230.

Jensen (1910). Von echten Geschwülsten bei Pflanzen. Quoted in *Progress in Experimental Tumor Research, 15,* (Braun, A. C., and Karger, S., eds.), p. 67. Lippincott, Philadelphia, 1972.

Jensen (1918). Undersøgelser verdrørende nogle svulstlignede dannelser hos planter. Quoted In *Progress in Experimental Tumor Research, 15,* (Braun, A. C., and Karger, S., eds.), p. 67. Lippincott, Philadelphia, 1972.

Johnson, R., Guderian, R. H., Eden, F., Chilton, M. D., Gordon, M. P., and Nester, E. W. (1974). Detection and quantitation of octopine in normal plant tissue and in crown gall tumors. *Proc. Nat. Acad. Sci. U.S. 71*, 536-538.

Kerr, A. (1969). Transfer of virulence between isolates of *Agrobacterium. Nature 223*, 1175-1176.

Lippincott, J. A., and Lippincott, B. B. (1975). The genus agrobacterium and plant tumorigenesis. *Ann. Rev. Microbiol. 29*, 377-380.

Mariatis, R., Jefferey, A., and Kleid, D. G. (1975). Nucleotide sequence of the rightward operator of phage λ. *Proc. Nat. Acad. Sci. U.S. 72*, 1184-1188.

Meins, F., Jr. (1973). Evidence for the presence of a readily transmissible oncogenic principle in crown gall teratoma cells of tobacco. *Differentiation 1*, 21-25.

Merlo, D. J., and Kemp, J. D. (1976). Attempts to detect *Agrobacterium tumefaciens* DNA in crown gall tumor tissue. *Plant Physiol. 58*, 100-106.

Morel, G. (1970). Le problème de la transformation tumerale chez les végétaux. *Physiol. Veg. 8*, 189.

Smith, E. F., and Townsend, C. O. (1907). Plant tumors of bacterial growth. *Science 25*, 671-673.

Smith, H. H. (1972). Plant genetic tumors. In *Progress in Experimental Tumor Research, 15*, (Braun, A. C., and Karger, S., eds.), pp. 138-164. Lippincott, Philadelphia.

Smith, H. S., Gelb, L. D., and Martin, M. A. (1972). Detection and quantitation of simian virus 40 genetic material in abortively transformed BALB/3T3 clones. *Proc. Nat. Acad. Sci. U.S. 69*, 152-156.

Stanisich, V. A., and Holloway, B. W. (1971). Chromosome transfer in *Pseudomonas aeruginosa* mediated by R factors. *Genet. Res. 17*, 169-172.

Syono, K., and Furuya, T. (1974). *Plant Cell Physiol. 15*, 7.

Van Larebeke, N., Engler, G., Holsters, M., Van den Elsacker, S., Zaenen, I., Schilperoort, R. A., and Schell, J. (1974). Large plasmid in *Agrobacterium tumefaciens* essential for crown gall-inducing ability. *Nature 252*, 169-170.

Van Larebeke, N., Genetello, Ch., Schell, J., Schilperoort, R. A., Hermans, A. K., Hernalsteens, J. P., and Van Montagu, M. (1975). Acquisition of tumor-inducing ability by non-octogenic agrobacteria as a result of plasmid transfer. *Nature 255*, 742-743.

Watson, B., Currier, T. C., Gordon, M. P., Chilton, M. D., and Nester, E. W. (1975). Plasmid required for virulence of *Agrobacterium tumefaciens*. *J. Bacteriol. 123*, 255-264.
Zaenen, I., Van Larebeke, N., Teuchy, H., Van Montagu, M., and Schell, J. (1974). Supercoiled circular DNA in crown-gall inducing *Agrobacterium* strains. *J. Mol. Biol. 86*, 109-127.

CROWN-GALL: BACTERIAL PLASMICS AS ONCOGENIC ELEMENTS FOR EUCARYOTIC CELLS

J. Schell
M. Van Montagu
A. Depicker
D. De Waele
G. Engler
C. Genetello
J. P. Hernalsteens
M. Holsters
E. Messens
B. Silva
S. Van den Elsacker
N. Van Larebeke
I. Zaenen

Laboratorium voor Genetica
Riuks Universiteit Gent
B9000 Gent, Belgium

Crown-gall is a plant disease caused by a gram-negative bacterium, *Agrobacterium tumefaciens*. The disease is the result of the stable transformation of normal plant cells into typical tumor cells. Since the agent causing the neoplasmic transformation is in this case a bacterium, the possibility existed to use this system as a model for the fundamental study of oncogenicity. Particularly since it could be expected that many of the questions could be answered by a genetical and biochemical analysis that could be carried out to a large extent in bacteria, such as:

--What is the nature of the tumor-inducing principle (TIP)?

--Are specific genes and gene products involved and, if so, how many genes and how do they function?

The advantages in this respect of working with bacteria as compared to most eucaryotic systems are obvious. Furthermore, the system could be expected to yield several other advantages, both fundamental and applied.

ISBN 012-601950-9

The possibility existed that crown-gall would turn out to be a first example of a natural system in which genetic elements from bacterial origin would play a measurable role in the regulation of gene expression in eucaryotic cells.

Also, it could be expected that one would be able to uncover genetic mechanisms and structures that have the capacity to cross the barriers that restrict the transfer and expression of procaryotic DNA in eucaryotic cells. By elucidating these mechanisms one could hope to find effective and controllable ways to achieve genetic engineering involving the introduction of specified genes from bacteria into plants.

Such were the reasons that prompted us to study *Agrobacterium tumefaciens* and, as should be obvious from the data presented at this symposium, there are now good reasons to believe that most of these expectations were justified.

THE SEARCH FOR AN IDENTIFIABLE GENETIC ELEMENT RESPONSIBLE FOR THE ONCOGENIC PROPERTIES OF *A. TUMEFACIENS*: THE TI-PLASMIDS

Crown-gall is in most cases a very stable, genetically inherited change in the regulatory mechanisms that control cellular growth, division and differentiation. It is relatively easy to envision such a stable change either by mutation or by the introduction of a specific piece of new genetic information into the transformed plant cells.

Since differentiation itself is also a fairly stable condition and, if one accepts the notion that differentiation can be achieved by gene-regulation alone (no mutations, no addition or withdrawal of genetic material), it is conceivable that the crown-gall bacterium could produce a signal that causes a series of relatively irreversible events that eventually lead to a stably altered regulation pattern. Nevertheless, we choose to think that the tumor-inducing capacity of *Agrobacterium* would be due to a specific genetical element carried by these bacteria; this hypothesis was both relatively simple and could be tested experimentally. If indeed *Agrobacterium* carried such oncogenic elements, it was not unreasonable to think that these would be extra-chromosomal elements. Two types of extrachromosomal elements are known to occur in bacteria: the prophages of temperate bacteriophages and autonomously replicating circular DNA plasmids. In a previous symposium we have already described the evidence that rules out prophages as the oncogenic elements in *Agrobacterium* (Schell, 1975).

Discovery and General Properties of *Agrobacterium* Plasmids

As described previously (Schell, 1975; Zaenen *et al.*, 1974) all oncogenic *Agrobacterium* strains harbor a large closed circular DNA plasmid. The general properties of these plasmids can be summarized as follows: *(1) Various* Agrobacterium *strains contain plasmids with different molecular sizes ranging from 95 to 156 magadaltons (Zaenen* et al., *1974). (2) Many strains contain more than one plasmid.* This was shown to be the case in various ways:

(a) by contour length measurements of open circular molecules with the electron microscope (Zaenen *et al.*, 1974).

(b) by the study of plasmid DNA fragments obtained after digestion of the purified plasmids with restriction endonucleases and separation of the DNA fragments by electrophoresis on agarose gels (Thompson *et al.*, 1974).

(c) by a comparison of the plasmids in donor and acceptor strains after experiments in which plasmids from donor strains were introduced into plasmid-free acceptor strains either by conjugation or by transformation with purified plasmid DNA preparations.

Thus *Agrobacterium* strain 925 was shown by contour length measurements to contain two plasmid populations of 133 and 96 megadaltons.

In *Agrobacterium* strain K14 plasmids were observed with the following molecular sizes: 125, 150, and occasionally 180 megadaltons. That only one of these plasmids was involved in the oncogenic process could be demonstrated by a conjugation experiment involving K14 as donor and the plasmid cured *A. tumefaciens* C58-C1 strain as acceptor (Van Larebeke *et al.*, 1974). As will be described more extensively in the next section, it was possible to show that the oncogenic ex-conjugant acceptor strains resulting from this cross contained only one plasmid with a molecular size of 125 megadaltons. That this plasmid was identical to one of the plasmids in the K14 donor strain could easily be demonstrated by comparing the DNA fragment fingerprints obtained after digestion of these plasmids with the restriction endonuclease EcoRl (derived from the *E. coli* strain RY13 (Sugden *et al.*, 1975).

The results obtained are illustrated in Fig. 1. Strain B6S3 also could be shown to contain two different plasmids; only one was essential for oncogenicity.

Contour length measurements yielded two populations of plasmids with molecular sizes of 111 and 122 megadaltons. The plasmids from this strain were extracted, purified and used in a transformation system using the plasmid-cured strain C58-C1 as an acceptor. The oncogenic transformants obtained were shown to contain only one plasmid with a molecular size of 122

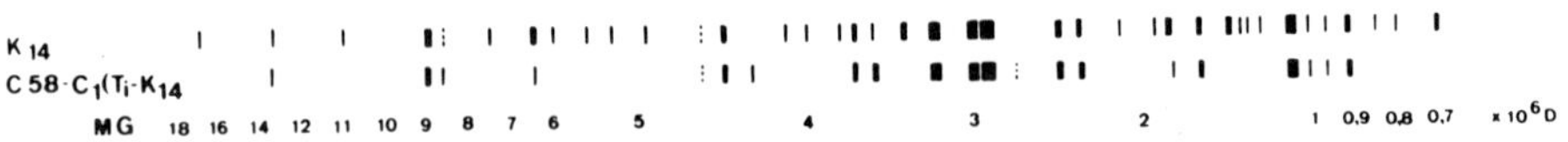

K 14
C 58 · C_1(T_i·K_{14}
MG 18 16 14 12 11 10 9 8 7 6 5 4 3 2 1 0.9 0.8 0.7 $\times 10^6$ D

FIGURE 1. Fingerprints of EcoRl digest of the plasmid DNA extracted from Agrobacterium *strain K14 and from a conjugant C58-Cl(Ti-K14).*

megadaltons. When the DNA fragment patterns after digestion with EcoRl were compared between the donor B6S3 strain and the C58-Cl(Ti-B6S3) transformants, it was obvious, as can be seen in Fig. 2, that the B6S3 donor strain had a band-pattern complexity typical for two different plasmids, whereas the transformants contained only one plasmid.

On the other hand a number of strains such as C58, ACH5, TT111 and A6 turned out to contain only one type of plasmid DNA according to the same criteria.

(3) At least two general types of plasmids exist in Agrobacterium *strains.*

If one compared the fingerprints of EcoRl digests of various plasmid DNAs, one can readily observe that they fall into at least two distinct groups. As can be seen in Fig. 3, the plasmids isolated from four different octopine-utilizing strains give very similar banding patterns (correction has to be made for the presence of two plasmids in strain B6S3). Whereas the Ti-plasmids isolated from three different nopaline-utilizing strains are clearly different from the previous ones.

It is also obvious that there is considerable variation between different plasmids from nopaline-utilizing strains whereas the plasmids from octopine-utilizing strains appear to have a high degree of similarity. The fact that the capacity to catabolize the unusual amino acids octopine [N^2-(D-1-carboxyethyl)-L-arginine] or nopaline [N^2-(1,3-dicarboxypropyl)-L-arginine] is determined by plasmid-borne genes may explain

C 58 · C_1(T_i·B_6S_3)
B_6S_3
MG 18 16 14 12 11 10 9 8 7 6 5 4 3 2 1 0.9 0.8 0.7 $\times 10^6$ D

FIGURE 2. Fingerprint of EcoRl digest of plasmid DNA extracted from strain B6S3 and from the transformant C68-Cl(Ti-B6S3).

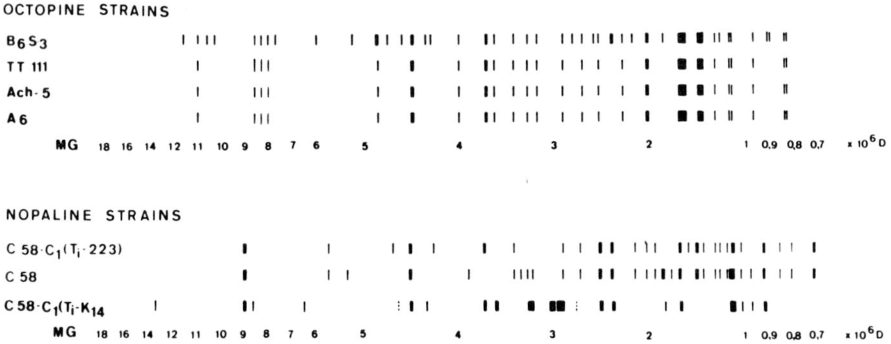

OCTOPINE STRAINS

B6S3

TT 111

Ach-5

A6

MG 18 16 14 12 11 10 9 8 7 6 5 4 3 2 1 0,9 0,8 0,7 $\times 10^6$ D

NOPALINE STRAINS

C 58-C_1(T_i-223)

C 58

C 58-C_1(T_i-K_{14}

MG 18 16 14 12 11 10 9 8 7 6 5 4 3 2 1 0,9 0,8 0,7 $\times 10^6$ D

FIGURE 3. EcoR1 fingerprints of various plasmid DNAs.

the correlation between this phenotype and the type of plasmid carried by these strains.

Demonstration that the Ti-plasmid is Essential For Tumor Formation

To answer the question whether or not the large plasmids present in oncogenic *A. tumefaciens* strains are implicated in the transformation process, two types of experiments were performed:

(1) Curing of the plasmid: Hamilton and Fall (1971) found that a particular *A. tumefaciens* strain, C58, irreversibly lost its capacity to induce crown-gall tumors after growth at an elevated temperature. Making use of this observation, it was possible to demonstrate that this loss of oncogenicity was in fact due to the loss of the Ti-plasmid (Schell, 1975; Van Larebeke *et al.*, 1974; Watson *et al.*, 1975).

(2) Introduction of Ti-plasmids into plasmid-free non-oncogenic strains.

This was achieved in several ways:

a) By conjugation *in planta* using a technique first described by Kerr (1969, 1971). Kerr had observed that oncogenicity could be transferred from oncogenic to non-oncogenic strains by infecting plants with both types of strains. In view of our hypothesis that the oncogenic properties of *A. tumefaciens* would be determined by plasmid-linked genes, we speculated that Kerr's observations could be due to a plasmid transfer. This was shown to be the case in various *in planta* crosses (Watson *et al.*, 1975; Van Larebeke *et al.*, 1975).

b) By direct conjugation between donor strains harboring a Ti-plasmid and receptor strains free of Ti-plasmid. Recently Kerr *et al.* (1977) and Genetello *et al.* (1977) demonstrated that Ti-plasmids have sex-factor activity i.e. they are able to promote their own transfer via a conjugation mechanism.

To test for plasmid transfer we made use of the fact that the capacity of *Agrobacterium* strains to utilize the unusual amino acids octopine and nopaline are coded for by Ti-plasmid linked genes (Bomhoff *et al.*, 1976; Chilton *et al.*, 1976). This was shown in various ways:

i) Strain C58 is a nopaline-utilizing strain, i.e. it can grow on a mineral medium containing nopaline as a sole N- and C-source. It cannot grow on a mineral medium with octopine as the sole N- and C-source. When this strain is cured of its Ti-plasmid (C58-Cl), it is no longer able to utilize nopaline (Watson *et al.*, 1975; Bomhoff *et al.*, 1976).

ii) When the Ti-plasmid from an octopine-utilizing strain (such as B6S3 or ACH5) is transferred to the plasmid-cured C58-Cl strain by *in planta* conjugation (Bomhoff *et al.*, 1976) or by direct conjugation (Kerr *et al.*, 1977; Genetello *et al.*, 1977) or by transformation, this strain now acquires the capacity to utilize octopine (but not nopaline) as a sole N- and C-source.

iii) When the Ti-plasmid from a nopaline utilizing strain (such as K14) was introduced in a naturally non-oncogenic, plasmid-free *Agrobacterium* namely *A. radiobacter* S1005, by *in planta* conjugation (Bomhoff *et al.*, 1976), this strain now acquired the capacity to utilize nopaline (but not octopine) as a sole N- and C-source.

Direct conjugation experiments were performed by mixing donor (Ti-plasmid containing) and receptor (C58-Cl *ery-r, rif-r*) bacteria in a ratio of 10:1. The mixture was filtered on millipore filters and the filters were incubated for 1 to 7 days on solid media containing 2 mg/ml octopine, when using octopine utilizing donors, and 2 mg/ml nopaline when using nopaline utilizing donors. In both cases conjugant colonies were easily detected by resuspending the cells from the filters and plating on selective media (mineral medium plus either octopine or nopaline as a sole N- and C-source plus 100 μg/ml rifampicin and 150 μg/ml erythromycin) that selected against the rifampicin, erythromycin sensitive donor cells and for either octopine or nopaline utilizing receptor cells. It is important to note that apparently the sex-factor activity of the Ti-plasmids is normally repressed and can be activated specifically by either octopine or nopaline. Indeed, if the conjugation is performed in the absence of the specific guanidines, no Ti-plasmid transfer was observed. In each case it was demonstrated that these conjugants had acquired the Ti-

C 58-C1(Ti-B6S3)
B6S3
Ach-5
C 58-C1(Ti-ACH-5)
MG 18 16 14 12 11 10 9 8 7 6 5 4 3 2 1 0,9 0,8 0,7 $\times 10^6$ D

FIGURE 4. EcoR1 fingerprints obtained from the octopine-utilizing donor strains B6S3 and ACH5 and from the conjugants C58-C1 (Ti-B6S3) and C58-C1 (Ti-ACH5).

plasmid of the donor strain. This can be clearly illustrated by the fingerprints obtained after extraction of the plasmid DNA from both the donor and the conjugant strains and digestion with ECOR1 (Fig. 4). The results obtained with nopaline-utilizing donors are not represented but again it could be demonstrated that donor and conjugant plasmids were identical. Also, in all cases, the conjugant strains had become fully capable of inducing crown-gall tumors.

c) By transformation with purified Ti-plasmid DNA. The most direct way to demonstrate that the Ti-plasmids are responsible for the tumor-inducing capacity of *Agrobacterium* consists in making a preparation of purified Ti-plasmid DNA and to introduce this purified DNA in a suitable non-oncogenic, plasmidless receptor strain. Again the cured C58-C1 strain was used as the receptor strain. The Ti-plasmid DNA was extracted and purified from the octopine utilizing strains B6S3 and ACH5 (Ledeboer *et al.*, 1976). Possible transformants were selected as octopine utilizing colonies capable of growing on a mineral medium with octopine as their sole N-source. Most of the transformation techniques described in the literature for gram negative bacteria did not work for *Agrobacterium*. Finally, the following technique reproducibly yielded transformants with a frequency of about 5×10^{-8} (no. of transformants/no. of receptor cells). Concentrated receptor bacteria (10^{10} cells/ml) in 10^{-2}M Tris-HCl buffer (pH 7.4) were mixed with 20 to 30 μg/ml of purified Ti-plasmid DNA. This mixture was immediately frozen for 5 minutes at -80°C (dry ice and ethanol) and subsequently thawed for 25 minutes at 37°C. The cells were then diluted and plated on mineral medium with octopine as their sole N-source (100 μg octopine/ml). All the transformants thus obtained contained a Ti-plasmid identical to the donor plasmid except when plasmid DNA from strain B6S3 was used. As previously discussed, this strain contains two different plasmids only one of which was found in the transformants. EcoR1 fingerprints of the plasmids from the transformants C58-C1 (Ti-B6S3) and C58-C1 (Ti-ACH5) and from the donor DNA extracted from strains B6S3 and ACH5 were as shown in

Fig. 4 and were thus identical to the similar strains obtained after direct conjugation. All of the transformants tested turned out to be fully oncogenic.

These and other experiments by Gordon *et al.* (this volume) have clearly demonstrated that the Ti-plasmid not only is essential for oncogenicity but also that it can confer tumor-inducing capacity to either cured or naturally plasmid-free *Agrobacteria*.

WHAT IS THE PRECISE ROLE OF THE TI-PLASMID IN TUMOR-FORMATION?

Two general models can be thought of to explain why the Ti-plasmids are essential to oncogenicity:

1) A direct role: the Ti-plasmid DNA could carry some specific gene(s) (oncogenes ?) that are transferred, upon infection of the plants, from the vector bacteria to the plant genome. The integration, replication and expression of these gene(s) would than be responsible for the stable neoplasmic transformation. Alternatively some Ti-plasmid genes could code for a signal that is released from the infecting bacteria and sets in motion a series of dedifferentiation steps that ultimately lead to the neoplasmic condition of the transformed cells.

2) An indirect role: it is conceivable that all the Ti-plasmid DNA does is to code for some bacterial functions that allow the proper transfer from bacteria to plant cells of the tumor inducing principle.

There are several ways to distinguish between these different models. The first is to test whether or not it is possible to cause crown-galls with purified Ti-plasmid DNA. This, to the best of our knowledge, has not been extensively tried. There are many reasons why such an experiment might fail, such as lack of uptake, nonspecific degradation of the DNA, lack of the proper mechanisms for DNA transfer and integration in the plant genome, and so on.

Another obvious way is to see whether or not one can find Ti-plasmid DNA sequences in the DNA of transformed plant cells by DNA/DNA or DNA/cRNA hybridization techniques. Many such experiments have been described and are still in progress. Since these experiments have been reviewed and discussed in the paper by Gordon *et al.* (this volume) we shall not describe them here. All that we want to point out is that up to now there is no good evidence either for or against the hypothesis that a specific segment of the Ti-plasmid is present in crown-gall cells. More accurate techniques should and are being used now to resolve this question. What appears to be sure, however, is that there is not a complete Ti-plasmid present in transformed plant cells.

A third way is a genetic approach. If indeed some plasmid linked genes are actually transferred, replicated and expressed in the transformed plant cells, one might hope to find a Ti-plasmid coded phenotype both in the vector bacterium and in the transformed plant cells. The unusual amino acids, octopine and nopaline, do indeed represent such an instance. The history of the discovery of these amino acids starts in the literature with the publication in 1956 of two short communications, one by Morel (1956) describing the presence of some unknown guanidines in crown-gall tissues and mentioning the isolation of gamma guanido butyric acid. It is probably this compound, that is also found in normal plant tissues, that was mistaken for octopine 15 years later. In the same volume Lioret described the existence of a new amino-acid, later called lysopine (Lioret, 1956). In 1960 Lioret (Biemann *et al.*, 1960) described the chemical structure of lysopine and in 1966 the same author published an extensive study of the physiological significance of the presence of lysopine in crown-gall tissues (Lioret, 1966) and concluded that lysopine appears to be a normal metabolite of plants also present in normal tissues. In the meantime Morel and his collaborators continued to study the products of arginine metabolism in crown-gall tissues. Their work led in 1964 to the identification of octopine in crown-gall tissues (Ménagé and Morel, 1964). These authors also had discovered another guanidine compound in crown-gall tissues of *Opuntia vulgaris* which they later called nopaline. The main step in the discovery of the importance of these compounds for the understanding of the mechanism of tumoral transformation is the publication by Goldmann *et al.* (1968) in which it is reported that not only are octopine and nopaline specific for crown-gall tissues but that it is the infecting bacteria and not the plant species that determines the kind of guanidine (octopine or nopaline) that will be formed by the tumor. These guanidines could not be found in the oncogenic bacteria, however a striking correlation does exist (Petit *et al.*, 1970). It was observed that *Agrobacterium* strains that induce octopine tumors could specifically catabolize this compound but could not degrade nopaline and *vice versa* that nopaline inducing strains could degrade nopaline but not octopine. On the basis of these observations the authors proposed the hypothesis of a transfer of genetic material from the oncogenic bacteria to the transformed plant cells. These observations have later been confirmed and extended (Lippincott *et al.*, 1973; Bomhoff, 1974). Although there have been a number of reports that octopine also can be found in normal tissues, this can not be repeated and there is now general agreement that the original observations of Morel's group were correct and that both octopine and nopaline can only be found in crown-gall tissues (Holderback and Beiderbeck, 1976). If indeed both the specificity of degradation by the bacteria and the specificity of synthesis of these guanidines by the transformed plant cells are due to the same genes, we speculated that two predictions could be made:

1) In view of the involvement of the Ti-plasmid in the oncogenic transformation, we expected that the bacterial genes that would be transferred to the plant cells should be plasmid linked. Therefore, both the genes specifying octopine-nopaline degradation in the bacteria and the genes specifying synthesis in the transformed plant cells should be located on the Ti-plasmid.

2) If oncogenicity and specificity of guanidine synthesis are both due to transferred Ti-plasmid genes one might expect them to be part of the same plasmid segment.

Both these predictions have been tested and turn out to be correct. Indeed it was relatively easy to demonstrate that:

a) Plasmid cured strains are unable to catabolize octopine or nopaline (Bomhoff *et al.*, 1976; Gordon *et al.*, this volume).

b) The introduction of a Ti-plasmid derived from an "octopine" strain (i.e. a strain that catabolizes octopine and induces tumors that contain octopine) into a plasmidless acceptor strain results in the acquisition by the recipient strain of both the capacity to degrade octopine and the capacity to induce tumors that specifically synthesize octopine. In the same way, the introduction of a Ti-plasmid derived from a "nopaline" strain into the same acceptor strain results in the acquisition by the recipient strain of the capacity to catabolize nopaline and to form tumors that contain nopaline (Bomhoff *et al.*, 1976). Furthermore, we have shown that the genes essential for oncogenicity and the genes specifying guanidine metabolism are relatively closely linked on the Ti-plasmid. This was done by a study of plasmid deletion mutants. We made use of our previous observation (Schell, 1975; Engler *et al.*, 1975) that the sensitivity of a number of *A. tumefaciens* strains towards an agrocin produced by the *A. radiobacter* strain K84 (Kerr and Htay, 1974) is determined by the Ti-plasmid. By selecting for agrocin K84 resistance, two types of mutants were found: 1) plasmid-cured strains and 2) plasmid deletion mutants.

One of these plasmid-deletion mutants, called K14 R1 (derived from the "nopaline" strain K14) was extensively studied. Heteroduplex DNA molecules (Fiandt *et al.*, 1971) produced by reannealing denatured plasmid DNA derived from the wild type strain with plasmid DNA from the K14 R1 deletion mutant showed in electron micrographs a single stranded deletion loop with a length of about 10 μm (corresponding to a segment of about 20 $\times 10^6$ daltons). Furthermore, the bacterial strain harboring this deleted plasmid had irreversibly lost the capacity to induce tumors and the capacity to catabolize nopaline. It was obvious, therefore, that the genes coding for both of these phenotypes were located on the same plasmid segment. An extensive phenotypic study of a number of independent agrocine K84 resistant mutants further corroborated this conclusion. Table 1 summarizes the phenotypic properties of different mutants. The main conclusions are that:

(1) there is linkage between the genes coding for these different properties since they are often co-deleted;

(2) a tentative gene order can be proposed: oncogenicity--agrocin sensitivity--nopaline metabolism.

(3) mutants such as R3 (not oncogenic but able to catabolize nopaline) and R4 (unable to catabolize nopaline and inducing tumors without nopaline) demonstrate that there is no essential correlation between oncogenicity and guanidine metabolism. The existing correlation can adequately be explained by the cotransfer to the transformed plant cells of these linked plasmid genes.

A number of observations have been made that seem to contradict this gene transfer model:

1) strains such as EU6 and AT181 can utilize nopaline but the tumors induced by these strains do not produce nopaline (Gordon *et al.*, this volume; Schilperoort *et al.*, 1975).

2) several mutants have been found that cannot utilize octopine but induce tumors that contain octopine (J. Tempé, P. Klapwijk, J. Kemp, personal communication). Also the strain A203 nop (Gordon *et al.*, this volume) has an analogous phenotype.

If one assumes, however, that more than one gene is involved in guanidine metabolism and that the products of these genes interact with host cell gene products, it is possible to reconcile these observations with the proposed model. In summary, the described observations strongly favor the hypothesis that the Ti-plasmid is essential for oncogenicity because it contains genes that can be transferred to--and replicated and expressed in--the transformed plant cells.

TABLE 1

Phenotypic Properties of Various Types of Agrocin K84 Resistant Mutants of Strain K14

	Growth on M.M. + nopaline	Presence of nopaline in tumor	Agrocin K84 sensitivity	Oncogenicity	Ti-plasmid	
					Presence	Size in megadalton
K14 WT	+	+	S	+	+	125
R1	–	–	R	–	+	100
R2	+	+	R	+	+	125
R3	+		R	–	+	n.m.
R4	–	–	R	+	+	n.m.
R5	–	–	R	–	+	100

Growth on minderal medium (M.M.) plus nopaline was tested by looking for full colony growth on a mineral medium with 100 μg/ml of nopaline as sole N-source.

The presence or absence of nopaline in the tumors was tested as described in Bomhoff et al. *(1976).*

All the other properties were tested as described in Van Larebeke et al. *(1977).*

n.m. = not measured

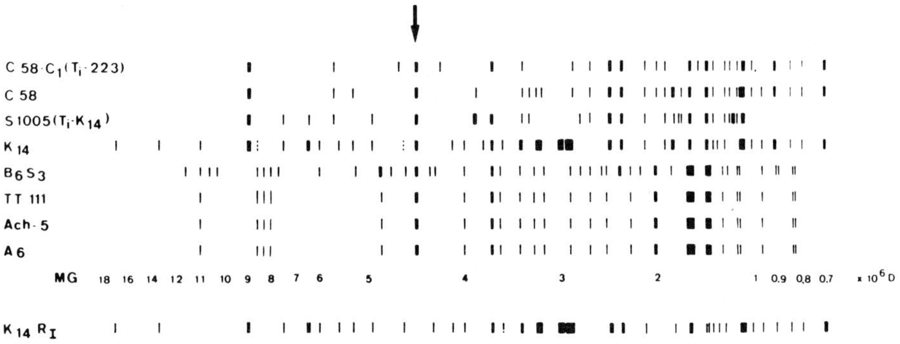

FIGURE 5. EcoR1 fingerprints of various Ti-plasmids. The arrow indicates the 4.5 × 10^6 dalton fragment missing from $K_{14}R_1$.

WHAT IS THE SIZE OF THE TI-SEGMENT THAT IS PRESUMABLY TRANSFERRED?

The work of Gordon *et al.* (this volume) implies that only a relatively small segment of the Ti-plasmids can be expected to be transferred to the plant genome. The main reasons for this are 1) Ti-plasmids from various oncogenic strains only have a small degree of homology and 2) a complete Ti-plasmid per transformed plant cell genome should have been detected had it been present. The question therefore arises: what segment of the Ti-plasmid can be expected to be transferred? Obviously the definitive answer to this question will also have to come from DNA/DNA hybridization experiments involving various Ti-plasmid segments (e.g. fragments obtained after digestion with restriction endonucleases). As a first attempt to identify such a segment we wanted to know whether we could find a DNA fragment that would be present in all oncogenic Ti-plasmids but absent in the K14R1 deletion mutant.

One such fragment was detected in the fingerprints of Eco-R1 digests of various Ti-plasmids as can be seen in Fig. 5. It is a fragment of about 4.5 × 10^6 daltons.

CAN THE TRANSFERRED SEGMENT OF THE TI-PLASMID BE CONSIDERED AS A "TRANSPOSON"?

Many bacteria contain a special class of genetic elements, called IS-elements (Insertion Sequence). These IS-elements are discrete DNA sequences with a defined length of between 800 and 1400 base pairs (Starlinger and Saedler, 1975). Among other properties, these IS elements are known to promote integration of episomes, such as the F-factor, in the bacterial chromosome (Davidson *et al.*, 1975) and to form so-called transposons. The known transposons consist of one or more antibiotic resistance genes, flanked on each side by DNA sequences that are homologous but inverted with respect to each other. As a result of this structure, DNA fragments that contain a transposon will form (after denaturation and self-annealing of single strands) typical mushroom-like structures with double stranded stems (the inverted repeats) and single stranded loops (the genes in between the inverted repeats). These units have been called transposons because they are able to undergo transposition, as a unit, from one DNA molecule (e.g. a plasmid) to an other DNA molecule (e.g. the chromosome, another plasmid or bacteriophage). Transposition of these units is independent of the normal bacterial recombination system and does not require detectable sequence homology with the recipient DNA.

It is now postulated that the inverted repeats bordering the transposons, are in fact IS-elements (Cohen and Kopecko, 1976). One possible model, therefore, that we thought could explain the special properties of the Ti-plasmid would be that the DNA segment containing the genes controlling octopine or nopaling metabolism and the genes responsible for oncogenicity, would be part of a transposon. This hypothetical transposon would have the particular property that it could undergo transposition from a bacterial plasmid to the plant genome.

Preliminary evidence in favor of this hypothesis was obtained in two ways:

1) By looking for inverted repeats in Ti-plasmids.

Such inverted repeats were indeed observed after denaturation and self annealing of some Ti-plasmids and looking for the typical "mushroom" structures with the electron microscope. Table 2 summarizes these observations. The following comments should be made: a) in the DNA of each of the plasmids of bacterial strain K14 a characteristic inverted repeat was observed; b) the inverted repeat in the 125×10^6 dalton plasmid of stain K14 could not be found in an avirulent mutant of this strain. These mutants arise spontaneously at a very high rate in strain K14: in some populations grown for several genera-

TABLE 2
Inverted Repetitions Observed in a Number of Ti-plasmids Isolated From Strain K14 and From a Conjugant S1005(Ti-K14) Harboring the Ti-plsmid From K14[a]

Strain	Mean contour length (μm) ± SD	Mean length of inverted repetition(μm)	Mean length single stranded loops(μm)
K14	Plasmid Type I (Ti-plasmid) 63.0 ± 2.8	0.32	3.55
	Plasmid Type II 75.7 ± 3.2	0.36	2.2
K14R1	Plasmid Type I (deleted Ti-plasmid) 50.9 ± 1.6	No inverted repetitions observed	-
	Plasmid Type II 74.6 ± 1.8	0.33	2.3
K14 avirulent	Plasmid Type I (Ti-plasmid) 62.8 ± 2.9	No inverted repetitions observed	-
	Plasmid Type II 75.55 ± 3.1	0.35	2.3
S1005(TI K14) (conjugant)	63.7 ± 3.2	0.30	3.8

[a]*The inverted repeats were observed under the E.M. as described in Hu* et al. *(1975). The inverted repetition typical for the Ti-plasmid of K14 was absent both in the Ti-deletion mutant K14R1 which has been shown to have lost both the oncogenic and guanidine determinants (see text) and in the K14 avirulent, an avirulent mutant of strain K14.*

tions from a single colony isolate, up to 20 percent of the cells turned out to be avirulent mutants; c) the inverted repeats could not be observed in the Ti-plasmids of such oncogenic strains as B6S3 and C58. It is possible, however, that they were not observed because the segment in between the inverted repeats was too long and resulted in unstable "mushroom" structures.

2) One of the well-documented properties of IS-elements is that they interact with other IS-elements and thus promote integration of different DNA molecules into one continuous DNA structure.

Many bacterial RTF (Restistance Transfer Factors) factors are known to contain IS-elements (Starlinger and Saedler, 1976). If the Ti-plasmid also contains similar IS-elements it is conceivable that they might interact and thus form a co-integrated plasmid. We tested this possibility by looking for a co-integrate between the Ti-plasmid and the P-like wide host range, RTF factor RP4 (Datton *et al.*, 1971). It had been observed independently in several laboratories (Chilton *et al.*, 1976; Hooykaas *et al.*, 1976; Van Larebeke *et al.*, 1977) that this RP4 plasmid can readily be introduced into *Agrobacterium* and that it can "mobilize" the Ti-plasmid. Indeed in a conjugation between a donor harboring both the RP4 and the Ti-plasmid and a plasmid-free receptor, one finds that amongst the receptor cells that have received the RP4 plasmid about 1/1000 have also received a complete Ti-plasmid. It was our purpose to see whether spontaneous co-integrated recombinants could occur between RP4 and Ti-plasmids. This was done by a conjugation between RP4 and Ti-containing donors and the plasmid-free C58-C1 receptor. First selection was made for receptors that had received the RP4 factor (this factor codes for resistance to kanamycin, tetracycline and carbenicillin). Amongst these, a screening was made for receptors that had received the Ti-plasmid as well as the RP4 factor (either by looking for agrocin K84 sensitivity or for octopine or nopaline utilization). Several of these C58-C1(RP4)(Ti) receptors were cloned, cultured, and used as donors in a second cross. As receptors we used a C58-C1 either resistant to erythromycin and rifampicin (first cross), or resistant to spectynomycin and streptomycin (second cross). In this second cross about 10 percent of the donors tested had jointly transmitted the RP4 and Ti-markers.

That these donors indeed contain a co-integrated Ti::RP4 plasmid was demonstrated in several ways:

a) There is 100% linkage between the RP4 markers and the Ti-plasmid markers [oncogenicity, utilization of octopine and synthesis of octopine in tumors (Ti-B6S3::RP4) or utilization of nopaline and synthesis of nopaline in tumors (Ti-C58::RP4), sensitivity to agrocin K84 (Ti-C58::RP4), exclusion of phage AP1]. This was shown because in a series of crosses using receptors of a previous cross as donors for a subsequent cross, there was always 100 percent co-transfer of RP4 and Ti-markers independent of whether the selection for the conjugants was made for RP4 markers or for Ti-plasmid markers.

b) Both these donors and conjugants obtained with these

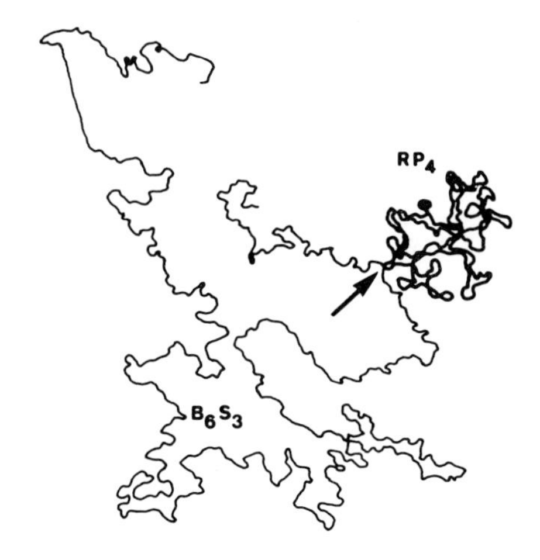

FIGURE 6. Heteroduplex molecule formed between the co-integrated plasmid DNA and purified RP4 DNA. Arrow indicates a double stranded circle of the size of RP4.

donors, contain a very large plasmid that has the molecular size of a complete co-integrate between the Ti-plasmid and RP4.

c) By making heteroduplex molecules after denaturation and reannealing between the co-integrated plasmid DNA and purified RP4 DNA, one observes, under the electron microscope, single stranded molecules of the co-integrated plasmid that show a double stranded circle of precisely the size of RP4 (Fig. 6).

Both the fertility and the host range (i.e., the different bacterial species to which an RTF factor can be transmitted by conjugation) of the co-integrated Ti::RP4 plasmids are apparently identical to those of the RP4 factor itself. This means that the Ti-plasmid can readily be transmitted in a natural way to most gram-negative bacteria!

It was also observed that a small fraction of the conjugants had only received the RP4 markers and none of the Ti-plasmid markers. Since all of the known RP4 markers were present it is assumed that in some of the donor cells segregation occurred releasing from the co-integrated plasmid the original RP4 factor. If confirmed, this will be an argument in favor of the hypothesis that the co-integrated plasmid originated by recombination involving IS-elements since it is known that such recombinations are reversible. These observations therefore:

1) lend some support to the hypothesis that the Ti-plasmids

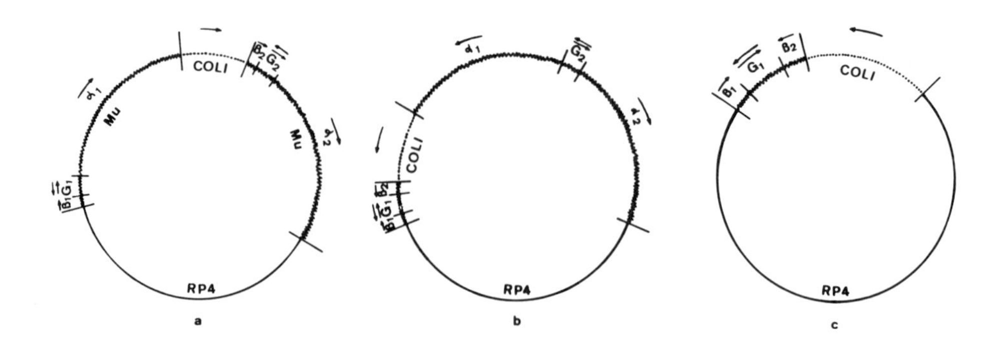

FIGURE 7. Molecular structures of hybrid RP4 plasmids that contain a strong amber suppressor. (a) A RP4 hybrid plasmid containing an E. coli *DNA segment jointed to two Mu DNA segments. (b) A RP4 hybrid plasmid with a rearrangement of the Mu DNA segments. (c) A thermostable RP4 hybrid plasmid derivative of the hybrid shown in 7b. The bulk of the Mu DNA has been deleted; the strong amber suppresor remains.*

may contain some IS_like sequences and will allow experimental confirmation of this idea.

(2) Demonstrate that fairly stable recombinant plasmids between the Ti-plasmid and wide host range fertility factors can occur in nature. It can, therefore, be assumed that Ti-plasmids ought not to be confined to *Agrobacterium*.

(3) Open the way to a number of important experiments in the field of plant improvement by genetic engineering techniques.

PROSPECTS FOR FUTURE RESEARCH AND APPLICATIONS

The Introduction of a Nonsense Suppressor Gene Into *Agrobacterium*

To perform a detailed genetic study of the genes on the Ti-plasmid that are involved in the oncogenic process, it is very important to be able to work with conditional mutants such as suppressible nonsense mutations. Therefore it is essential to have nonsense suppressing (Su^+) derivatives of *Agrobacterium*. Since we were unable to find Su^+ mutants of *Agrobacterium* it was decided to introduce a well-characterized suppressor gene from *E. coli* into *Agrobacterium*. It has been shown by Faelen and Toussaint (1976) that upon induction of phage Mu, in an

E. coli strain harboring a plasmid, a rather large, random fragment of the bacterial chromosome can be transposed onto a plasmid. With this technique they constructed a set of F' lac factors. Genetic evidence indicated that the transposed segment was surrounded by two complete Mu genomes, integrated with the same 5'-3' orientation.

With the same technique we transposed a strong amber suppressor from ϕ 80su$^+$ and from *E. coli* onto an RP4 plasmid. Hybridization with respectively Mu, ϕ 80su$^+$ and RP4 DNA, proved that the transposed DNA segment is indeed surrounded by two Mu phages as indicated in Fig. 7a.

Electron microscopic homoduplex analysis, however, showed that in this plasmid one not only observes a G loop inversion (Hsu and Davidson, 1974) but that also the whole DNA segment located between the two G segments can invert, yielding a structure represented in Fig. 7b. This plasmid was constructed with a thermoinducible Mu. We then selected for bacteria surviving at 42°C, still having all the RP4 markers and the transposed *E. coli* sequence (in our case the strong amber suppressor). Those bacteria contained a plasmid (Fig. 7c) from which both α segments and the inner G were excised. As such a plasmid no longer harbors any Mu functions "lethal" for the host cell, one can transfer this plasmid by conjugation to most gram-negative bacteria, including *Agrobacterium*.

Furthermore, it will probably be possible to obtain Mu mediated gene transpositions in different bacterial species, closely related to *E. coli*, but which are not a natural host for Mu as they lack the Mu absorption sites. Indeed by inserting a thermoinducible phage Mu in RP4 it is possible to conjugate this RP4::Mu into many gram-negative bacteria. We noticed that in strains such as *Serratia*, *Proteus*, and *Klebsiella* infective Mu phage was produced upon thermal induction. It can be expected that all Mu based techniques for making insertion and deletion mutants as well as gene transpositions can be used with these bacteria. If so, it must be possible to integrate into RP4 any fragments of the chromosome of these enterobacteriaceae. The transposed set of genes could then, after conjugation, be transferred to *Agrobacterium*.

On the Possibilities of the Ti-plasmid As a Natural Vector For the Genetic Engineering of Plants

The main problem in the genetic engineering of higher organisms is to find a way to introduce specific genetic information into eucaryotic cells in such a way that this new information is stably integrated, replicated, and expressed. It would seem that the Ti-plasmid contains a segment that is transferred, replicated, and expressed in transformed plant cells. It is therefore conceivable that if specific genes were introduced in this transferable segment (the hypothetical "transposon") they would likewise be maintained and expressed in the plant cells.

We want to test this possibility with the nitrogen-fixation (*Nif*) genes of *Klebsiella* and *Rhizobium*. The *Nif* genes of *Klebsiella* have recently been translocated onto the RP4 factor (Dixon *et al.*, 1976). In view of the possibility of integrating RP4 into the Ti-plasmid, one can now integrate the RP41 factor (containing the *Nif* genes) into the Ti-plasmid and test for whether or not crown-galls made with such a Ti::RP41 plasmid can fix nitrogen.

Furthermore, the Ti::RP4 co-integrate that we have isolated can readily be introduced into various *Rhizobium* strains. We want to test whether one can thus obtain *Rhizobium* strains that will induce tumors capable of fixing nitrogen.

ACKNOWLEDGMENTS

The authors wish to thank Dr. R. Schilperoort, Dr. A. Kerr and Dr. J. Tempé for many helpful discussions about this work. Dr. N. Datta and Dr. S. Cohen contributed excellent suggestions about the use of the RP4 plasmid. Dr. M. Gordon kindly provided the manuscript of his paper for this symposium.

This work was supported by grants from the "Kankerfonds van de A.S.L.K." and from the "Fonds voor Kollektief Fundamenteel Onderzoek" (n°10316) to J.S. and M.V.M. M.H. is a "aangesteld Navorser" of the "Nationaal Fonds voor Wetenschappelijk Onderzoek". A.D. and D.D.W. are indebted to the Belgian I.W.O.N.L. for a fellowship.

REFERENCES

Biemann, K., Lioret, C., Asselineau, J., Lederer, E., and Polonsky, J. (1960). Sur la structure chimique de la lysopine. Nouvel acide aminé isolé de tissu de crown-gall. *Bull. Soc. Chim. Biol. 42*, 979-991.

Bomhoff, G. H. (1974). Studies on crown gall--a plant tumor. Investigations on protein composition and on the use of guanidine compounds as a marker for transformed cells. Thesis, University of Leiden, The Netherlands.

Bomhoff, G. H., Klapwijk, P. M., Kester, H. C. M., Schilperoort, R. A., Hernalsteens, J. P., and Schell, J. (1976). Octopine and nopalaine synthesis and breakdown genetically controlled by a plasmid of *Agrobacterium tumefaciens*. *Mol. Gen. Genet. 245*, 177-181.

Chilton, M.-D., Farrand, S. K., Levin, R., and Nester, E. W. (1976). RP4 promotion of transfer of a large *Agrobacterium* plasmid which confers virulence. *Genetics 83*, 609-618.

Cohen, S. N., and Kopecko, D. J. (1976). Structural evolution of bacterial plasmids: role of translocating genetic elements and DNA sequence insertions. *Fed. Proc. 35*, 2031-2036.

Datta, N., Hedges, R. W., Shaw, E. J., Sykes, R. B., and Richmond, M. H. (1971). Properties of an R factor from *Pseudomonas aeruginosa*. *J. Bact. 108*, 1244-1249.

Davidson, N., Deonier, R. C., Hu, S., and Ohtsubo, E. (1976). Electron microscope heteroduplex studies of sequence relations among plasmids of *Escherichia coli*. *Microbiol. 1*, 56.

Dixon, R., Cannon, F., and Kondorosi, A. (1976). Construction of a P plasmid carrying nitrogen fixation genes from *Klebsiella pneumoniae*. *Nature 260*, 268-271.

Engler, G., Hernalsteens, J. P., Holsters, M., Van Montagu, M., Schilperoort, R. A., and Schell, J. (1975). Agrocin 84 sensitivity: A plasmid determined property in *Agrobacterium tumefaciens*. *Mol. Gen. Genet. 138*, 345-349.

Faelen, M., and Toussaint, A. (1976). Bacteriophage Mu-1: a tool to transpose and to localize bacterial genes. *J. Mol. Biol. 104*, 525-539.

Fiandt, M., Hradecna, Z., Lozeron, H. A., and Szybalski, W. (1971). Electron micrographic mapping of deletions, insertions, inversions, and homologies in the DNAs of coliphages lambda and phi 80. In *The Bacteriophage Lambda* (Hershey, A. D., ed.), pp. 329-354. Cold Spring Harbor Laboratory, Cold Spring Harbor, N. Y.

Genetello, C., Van Larebeke, N., Holsters, M., Depicker, A., Van Montagu, M., and Schell, J. (1977). Ti plasmids of *Agrobacterium* as conjugative plasmids. *Nature 265*, 561-563.

Goldmann, A., Tempé, J., and Morel, G. (1968). Quelques particularitiés de diverses souches d'*Agrobacterium tumefaciens*. *C. R. Soc. Biol. 162*, 630-631.

Hamilton, R. H., and Fall, M. Z. (1971). The loss of tumor-initiating ability in *Agrobacterium tumefaciens* by incubation at high temperature. *Experientia 27*, 229-230.

Holderbach, E., and Beiderbeck, R. (1976). Octopingehalt in normale und tumorgeweben einiger höherer pflanzen. *Phytochemistry 15*, 955-956.

Hooykaas, P. J., Klapwijk, P. M., Nuti, M. P., Schilperoort, R. A., and Rörsch, A. (1976). Isolation and characterization of *Agrobacterium tumefaciens* mutants affected in the utilization of octopine, octopine acid, and lysopine. *J. Gen. Microbiol. 96*, 155-164.

Hsu, M. T., and Davidson, N. (1974). Electron microscope heteroduplex study of the heterogeneity of mu phage and prophage DNA. *Virology 58*, 229-239.

Hu, S., Ohtsubo, E., Davidson, N., and Saedler, H. (1975). Electron microscope heteroduplex studies of sequence relations among bacterial plasmids: identification and mapping of the insertion sequences IS1 and IS2 in F and R plasmids. *J. Bacteriol. 122*, 764-775.

Kerr, A. (1969). Transfer of virulence between isolates of *Agrobacterium*. *Nature 223*, 1175-1176.

Kerr, A. (1971). Acquisition of virulence by non-pathogenic isolates of *Agrobacterium radiobacteria*. *Physiol. Plant Pathol. 1*, 241-246.

Kerr, A., and Htay, K. (1974). Biological control of crown gall through bacteriocin production. *Physiol. Plant Pathol. 4*, 37-44.

Kerr, A., Manigault, P., and Tempé, J. (1977). Transfer of virulence *in vivo* and *in vitro* in *Agrobacterium*. *Nature 265*, 560-561.

Ledeboer, A. M., Krol, A. J. M., Dons, J. J. M., Spier, F., Schilperoort, R. A., Zaenen, I., Van Larebeke, N., and Schell, J. (1976). On the isolation of Ti-plasmid from *Agrobacterium tumefaciens*. *Nucleic Acids Res. 3*, 449-463.

Lioret, C. (1956). Sur la mise en évidence d'un acide aminé non identifié particulier aux tissus de "crown-gall." *Bull. Soc. Fr. Physiol. Végêtale 2*, 76.

Lioret, C. (1966). Physiological significance of the presence of lysopine in *Scorzonera* crown-gall tissues cultivated *in vitro*. *Physiol. Végêtale 4*, 89-103.

Lippincott, J. A., Beiderbeck, R., and Lippincott, B. B. (1973). Utilization of octopine and nopaline by *Agrobacterium*. *J. Bacteriol. 116*, 378-383.

Ménagé, A., and Morel, G. (1964). Sur la présence d'octopine dans les tissues de Crown-gall. *C.R. Acad. Sci.*, Paris *259*, 4795-4796.

Morel, G. (1956). *Bull. Soc. Fr. Physiol. Végêtale 2*, 75.

Petit, A., Delhaye, S., Tempé, J., and Morel, G. (1970). Recherches sur les guanidines des tissus de Crown Gall. Mise en évidence d'une relation biochimique specifique entre les souches d'*Agrobacterium tumefaciens* et les tumeurs qu'elles induisent. *Physiol. Végêtale 8*, 205-213.

Schell, J. (1975). The role of plasmids in crown-gall formation by *A. tumefaciens*. In *Genetic Manipulations with Plant Material* (Ledoux, L., ed.), pp. 163-181. Plenum Press, N.Y.

Schilperoort, R., Kester, H., Klapwijk, P., Rörsch, A., and Schell, J. (1975). In *Proceedings Study Week Agriculture and Plant Health*, 8-12 September 1975, Gembloux.

Starlinger, P., and Saedler, H. (1976). IS-elements in microorganisms. *Current Topics in Microbiology and Immunology 75*, 111-152.

Sugden, B., De Troy, B., Roberts, R. J., and Sambrook, J. (1975). Agarose slab-gel electrophoresis equipment. *Anal. Biochem. 68*, 36-46.

Thompson, R., Hughes, S. G., Broda, P. (1974). Plasmid identification using specific endonucleases. *Mol. Gen. Genet. 133*, 141-149.

Van Larebeke, N., Engler, G., Holsters, M., Van den Elsacker, S., Zaenen, I., Schilperoort, R. A., and Schell, J. (1974). Large plasmid in *Agrobacterium tumefaciens* essential for crown gall-inducing ability. *Nature 252*, 169-170.

Van Larebeke, N., Genetello, C., Schell, J., Schilperoort, R. A., Hermans, A. K., Hernalsteens, J. P., and Van Montagu, M. (1975). Acquisition of tumor-inducing ability by nononcogenic agrobacteria as a result of plasmid transfer. *Nature 255*, 742-743.

Van Larebeke, N., Genetello, C., Hernalsteens, J. P., Depicker, A., Zaenen, I., Messens, E., Van Montagu, M., and Schell, J. (1977). Transfer of Ti-plasmids between *Agrobacterium* strains by mobilization with the conjugative plasmid RP4. *Mol. Gen. Genet. 152*, 119-124.

Watson, B., Currier, T. C., Gordon, M. P., Chilton, M.-D., and Nester, E. W. (1975). Plasmid required for virulence of *Agrobacterium tumefaciens*. *J. Bacteriol. 123*, 255-264.

Zaenen, I., Van Larebeke, N., Teuchy, H., Van Montagu, M., and Schell, J. (1974). Supercoiled circular DNA in crown-gall inducing *Agrobacterium* strains. *J. Mol. Biol. 86*, 109-127.

INDEX

S

T

U, V, W

X, Y, Z